PROCESS TECHNOLOGY

PLANT OPERATIONS

USED TEXT

DEC 1 3 2010

LMC BOOKSTORE

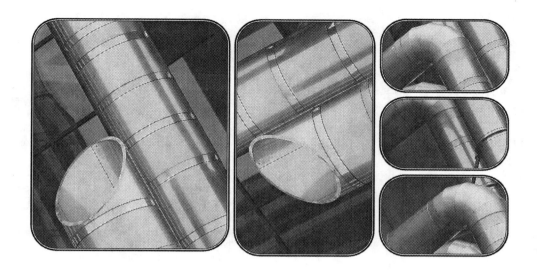

PROCESS TECHNOLOGY
PLANT OPERATIONS

Michael Speegle

DELMAR
CENGAGE Learning

Australia • Brazil • Japan • Korea • Mexico • Singapore • Spain • United Kingdom • United States

Process Technology Plant Operations
Michael Speegle

Vice President, Technology and Trades ABU: David Garza

Director of Learning Solutions: Sandy Clark

Senior Acquisitions Editor: David Boelio

Product Manager: Sharon Chambliss

Marketing Director: Deborah S. Yarnell

Marketing Coordinator: Mark Pierro

Director of Production: Patty Stephan

Senior Production Manager: Larry Main

Content Project Manager: Jennifer Hanley

Technology Project Specialist: Linda Verde

Editorial Assistant: Andrea Domkowski

© 2007 Delmar, Cengage Learning

ALL RIGHTS RESERVED. No part of this work covered by the copyright herein may be reproduced, transmitted, stored or used in any form or by any means graphic, electronic, or mechanical, including but not limited to photocopying, recording, scanning, digitizing, taping, Web distribution, information networks, or information storage and retrieval systems, except as permitted under Section 107 or 108 of the 1976 United States Copyright Act, without the prior written permission of the publisher.

For product information and technology assistance, contact us at **Cengage Learning Customer & Sales Support, 1-800-354-9706**

For permission to use material from this text or product, submit all requests online at **www.cengage.com/permissions**
Further permissions questions can be emailed to **permissionrequest@cengage.com**

Library of Congress Control Number: 2005031954

ISBN-13: 978-1-4180-2863-3

ISBN-10: 1-4180-2863-0

Delmar
Executive Woods
5 Maxwell Drive
Clifton Park, NY 12065
USA

Cengage Learning is a leading provider of customized learning solutions with office locations around the globe, including Singapore, the United Kingdom, Australia, Mexico, Brazil, and Japan. Locate your local office at **international.cengage.com/region**

Cengage Learning products are represented in Canada by Nelson Education, Ltd.

For your lifelong learning solutions, visit **www.cengage.com/delmar**

Visit our corporate website at **www.cengage.com**

Notice to the Reader

Publisher does not warrant or guarantee any of the products described herein or perform any independent analysis in connection with any of the product information contained herein. Publisher does not assume, and expressly disclaims, any obligation to obtain and include information other than that provided to it by the manufacturer. The reader is expressly warned to consider and adopt all safety precautions that might be indicated by the activities described herein and to avoid all potential hazards. By following the instructions contained herein, the reader willingly assumes all risks in connection with such instructions. The publisher makes no representations or warranties of any kind, including but not limited to, the warranties of fitness for particular purpose or merchantability, nor are any such representations implied with respect to the material set forth herein, and the publisher takes no responsibility with respect to such material. The publisher shall not be liable for any special, consequential, or exemplary damages resulting, in whole or part, from the readers' use of, or reliance upon, this material.

Printed in the United States of America
2 3 4 5 6   14 13 12 11 10

ED178

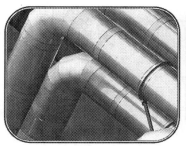

Table of Contents

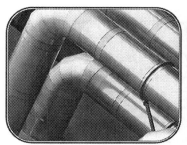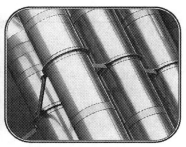

Preface

Operators in the processing industry have been taken for granted for many years, perhaps because at one time all that was required was a high school degree. That opinion and requirement has changed in the last 10 years. Being an operator in the processing industry is not an easy job today because process technology is high tech and the technology keeps changing and the learning curve never ends. I worked in the process industry for 18 years, almost half of that time as a technician and half as a training coordinator and supervisor. Often, when I went out to the processing units to speak with operators, I was amazed at what they did and didn't know. Most operators were conscientious individuals with intentions of doing a good job. Some of them pushed buttons and did things because that was what they were told to do when they were trained. Often *why* they pushed that button and *what if* and *what happens* was not explained. That did not necessarily make them bad operators. The operators were simply a product of the culture of their plant.

Today, management's expectation of operators, now called *technicians,* is technical competence. This expectation is verified by the battery of examinations that test for technical knowledge and interpersonal skills that most processing industries subject their potential employees to. San Jacinto College Central is very fortunate to have a tremendous amount of industry around it. My process technology advisory committee has 22 local industry representatives on it. They are very pleased with the knowledge and skills of the graduates they are hiring with two-year degrees in process technology. In fact, one major integrated oil company presented evidence that a well-trained student with a two-year degree in process technology had the knowledge and skills level of an operator with three years operating experience and no formal schooling in the process technology curriculum. Plus, the company estimated about $15,000 in reduced training expenses per new hire with an Associate in Applied Science (AAS) degree. This means the colleges teaching this curriculum must be doing a good job.

The first two times I taught Process Technology III: Operations, I was frustrated because I did not have a good textbook. Since I have authored several other textbooks, I decided to write this one because it would make my job as an instructor easier. This book has been written specifically for the process industries—industries such as refining, petrochemicals, electric power generation, food processing and canning, and pulp and paper. Millions of dollars have been invested and continue to be invested in the processing industries. The investment is made with a belief and commitment to a return on that investment, in other words, a profit. If a company is producing the right product at the right time with little competition, profits come easily. However, when a company has a lot of competition, as many refineries and chemical plants do, plus strict health, safety, and environmental constraints, profits do not come easily. Companies strive to protect their investments in people and equipment. Hence, the need for good training and training materials. I hope this textbook fulfills the function of a good training resource.

Process Technology III: Operations is the capstone course for the process technology curriculum. This course brings together all the concepts and knowledge the students are supposed to have acquired from approximately 10 other process technology courses. The capstone course integrates concepts from each of those courses and weaves them all together into a mosaic that says,

> Young man/woman, you are a technician responsible for expensive equipment, proper operating conditions, quality control and improvement, profits and losses, problem solving, community involvement, the environment, and many other things. The company that hires you will not survive without your contributions in all these areas. You endanger your job and the survival of the company if you opt to have no technical expertise and just be a pair of hands that pushes buttons or opens and closes valves.

This textbook, likewise, has sought to integrate important concepts from those courses. I have been tempted to incorporate everything, including the kitchen sink, but then realized that this book would become too unwieldy and expensive. Instead, I have tried to focus on the many daily routines and tasks a technician is required to do, and to give examples of why integrity, attention to detail, attention to quality, and being productive are important attributes of a good technician.

As usual, I struggled with my arrangement of the chapters and after several rearrangements, quit and left them as they now appear. The first six chapters could apply to just about any profession. The technical skills and knowledge, safety knowledge, economic savvy, quality consciousness, and effective communications required of workers today are necessities for a vital workforce. These chapters come first to remind the student that they are not going into a process plant to do grunt work. Instead, they are being hired as workers with a good overview of the safety, technical, economic, and interpersonal skills required of the job. Chapters 7 through 9 combine process chemistry and physics with safety and health to reveal how they are interrelated. One can directly affect the other. Chapters 10 through12 discuss the duties of technicians, from making rounds to equipment maintenance responsibilities. Quite a few pages are spent on equipment lubrication because of its importance. Chapters 13 through 15 discuss the material handling of bulk materials, its importance to public and plant safety, and the significant amount of company assets tied up in bulk material storage. Chapters 16 through 18 describe the process, rationale, and duties during unit shutdown, turnaround, and startup, with a good emphasis on safety and economics. The last two chapters fit nicely together: abnormal situations and troubleshooting.

I believe this book meets almost all of the course objectives as determined by the Gulf Coast Process Technology Alliance for Process Technology III: Operations, especially when labs that require the student operate a process unit and use and monitor instruments, is included. As I mentioned earlier, I could have added more chapters, ones on instrumentation and control, or reactor and distillation systems, but I did not because it is assumed that when a student takes this course they have had the other 10 courses in the process technology curriculum (safety and health, systems, physics, chemistry, two instrumentation courses, etc.). As I stated earlier, the intent of this book is to bring all of the previously acquired knowledge together in a general overview to complete the education of the student. This overview should be a final preparation before finding employment and applying the knowledge and skills obtained in this program and needed to make them a valuable asset to industry.

My writing style is deliberate. I try to communicate through writing the same way I do through speaking in the classroom. It is not my intention to impress people with a didactic style, but to convey information with words. And finally, I would like to thank several people for their helpfulness in furnishing their wisdom and some materials to help make this book possible. So, thanks Mark Demark, Lou Caserta, G. C. Shah, Glen Johnson, Gaylene Webb, Richard Westerlage, and Michael Sobbotik. Also, I want to credit the workers and management of Air Products in Deer Park, Huish Detergents in LaPorte, Odfjell Terminal in Seabrook, Penreco in Dickinson, and John Payne of British Petroleum for allowing me to tour their plants and get a better understanding of some process equipment and operating methods.

Mike Speegle

CHAPTER 1

Process Technology Today

Learning Objectives

After completing this chapter, you should be able to

- *Discuss the core values needed by process technicians to meet management's goals and objectives.*

- *Discuss technician responsibilities for environmental compliance.*

- *List the roles of today's process technician.*

- *List and discuss a variety of skills and knowledge categories required by technicians today.*

- *Describe the difference between a fixed and rotating schedule.*

INTRODUCTION

The role of the process technician in the processing industry today has dramatically changed from what it was just 20 years ago. Their role will continue to evolve due to changes in technology and competitive pressures from competition that is both national and international in scope.

The role of a process technician is to assist the operations division in a process plant to assess, adapt, and coordinate process manufacturing and maintenance activities to meet the business production schedules and product specifications. In the past, management assumed all those responsibilities and technicians were not asked to *think*, but to simply *do as they were told*. Management, usually the first-line supervisor, was responsible for many tasks now assigned to process technicians. However, unlike in the past, it is currently important for all operations personnel to proactively identify opportunities to improve process operations, detect and remove threats to steady-state operations, seek continuous operating improvements, and compensate and correct abnormal operations. Because many process units are highly automated, when they are lined out they literally run themselves. The primary function of today's process technician is to detect and correct abnormal situations. Today's process technician has become or is in the process of becoming a highly skilled jack-of-all trades. However, skills and knowledge are not sufficient alone. They must be accompanied by strong personal values and integrity.

CORE VALUES AND COMPETENCIES OF TODAY'S WORKERS

Core Values

Core values are essential values. They are not nice-to-have values; *they are essential.* Companies cannot survive in today's fiercely competitive environment unless their workforce has certain core values. Individuals without these core values are often termed "high maintenance," which is management's way of saying you have to keep looking over their shoulders to ensure the individuals are working and doing their work correctly. They always have to be told what to do and how to do it. In today's world, management does not have the time or resources to do that.

Certain core values of a processing unit's technicians, such as the ones listed here, are necessary for the unit to meet its goals and obligations:

- Integrity—Technicians perform their duties ethically and responsibly.
- Safety—Individuals are proactive in creating and maintaining a safe, healthy and incident free environment.
- Environmental awareness—Technicians are knowledgeable and vigilant about understanding threats to the environment and surrounding community.
- Diversity—Technicians embrace and nurture an inclusive work environment.
- Responsibility—Technicians accept responsibility for their actions and consider the impact of their actions on team members and the site organization.
- Performance—Individuals are committed to excellence.

Many individuals have some or all of these core values to some degree. What core values or degrees of values an individual lacks will not magically appear when they are hired in the processing industry. These core values do not just happen. Instead, management often screens for these core values through extensive testing and an interviewing process before hiring. They then foster and nurture a work atmosphere that promotes these values through acknowledgement, awards, bonuses, or promotions.

Technician Competencies

A list of general competencies for today's process technician consists of the following:

- Business basics—Understand the economics of the process unit and product, the corporate vision, and operating philosophy.
- Systems thinking (grasping the "Big Picture")—Understand how people, equipment, material, and technology interact to affect their unit operations and downstream and upstream operations.

- Loss prevention—Understand the critical role of preventative maintenance and mechanical integrity to successful and profitable operations.
- Continuous improvement—Realize things can always be done better and continuous improvement is necessary for operational survival.
- Problem solving—Use a systematic process to solve operational problems rather than guessing and jumping to conclusions.
- Process control—Understand how automatic controllers and advanced process controls manipulate process variables.
- Basic mathematics—Understand and use decimals, fractions, percentages, charts and graphs, and basic algebra.

In response to the economic pressures of the changing industrial scene, process technicians have assumed a multifaceted role. Today's technicians no longer confine themselves to just production work; they are active in safety committees, health issues, public relations, quality and environmental concerns, continuous improvement of process units, preventative maintenance, and problem solving. This is a huge difference from 20 or 30 years ago when technicians were rarely delegated responsibilities outside of production.

Note: The author of this textbook attended a Critical Issues and Best Practices conference for process technology in September of 2004. A senior manager of one of the largest chemical companies in the United States told his audience of educational representatives to go back and tell their students that being an operator was no longer a blue-collar job. Process technology is high technology and his company no longer considers this a blue-collar job. His company now requires a two-year associates degree in process technology.

TECHNICIANS AND THE ENVIRONMENT

Unlike operators in the past, process technicians are often required to assist safety, health, and environmental (SH&E) personnel with air, water, and solid waste regulations, plus safety and hygiene regulations. Technicians write incident reports regarding spills and releases, assist in cleaning up the spills, assist in incident investigations, and conduct safety audits. Table 1-1 reveals some of the many ways technicians are involved with environmental compliance due to the Environmental Protection Agency's mandates found in the Clean Air Act, Clean Water Act, and Resource Conservation and Recovery Act.

ROLES OF TODAY'S PROCESS TECHNICIANS

Major changes have occurred in the processing industry over the last 20 to 30 years. The Gulf Coast Process Technology Alliance (GCPTA), an alliance in the refining and petrochemical industry created in 1997 of industry representatives, trade associations, and educational institutions, has identified some of the changes (see Figure 1-1). Process technicians today must

- Accept a much more diverse workforce.
- Self-train themselves using computer-based training materials.
- Understand computerized controls and automation.
- Understand and comply with safety and environmental regulations.
- Function as a team member with good interpersonal skills.
- Support and contribute to process quality and a quality improvement process.
- Become involved with process hazard analysis.
- Apply analytical skills to process troubleshooting.
- Assist in site security.

Table 1-1   Technician Responsibilities for Environmental Compliance

Fugitive emissions monitoring
Incident reporting
Sampling and testing
Assist when there is an inspection by a regulatory agency
Conduct routine inspections of wastewater systems
Report upsets in the wastewater system
Keep the solid waste storage areas clean
Ensure all waste containers are properly labeled and leak-free
Conduct inspections of the waste management areas
Minimize releases to the environment
Understand and comply with all federal, state, and company rules and regulations related to the environment

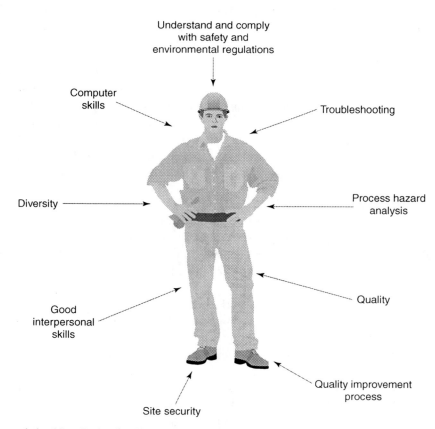

Figure 1-1   New Roles for Process Technicians

The skill and knowledge requirements for a process technician have changed because of the sophistication of new automated control systems and the complexity of the growing list of responsibilities required of workers in the industry for their companies to remain competitive. Process technicians can no longer be people who could come in off the street without a core of critical knowledge or experience and be trained for the job in a few weeks. The fact that operators are now more frequently referred to as process technicians implies a change in the role of the operator.

Technical is defined as "having special or practical knowledge of a mechanical or scientific subject." Process technicians today are required to have *special or practical knowledge and skills of a mechanical or scientific nature.* They are being hired for *knowledge.* Part of the reason for that is because of the role's increased responsibilities involving production, quality, safety, and the environment. The United States is a highly technical society, one of the most high-tech nations in the world. The better jobs in such a society require an educated workforce with the ability to learn and keep learning. The ability to learn is critical because technology keeps evolving.

Physical strength is no longer a major requirement of technicians today because plants are highly automated. Plants now seek employees who can think, analyze, solve problems, and respond in correct ways with minimal supervision. The old days of a supervisor being responsible for everything and telling everyone what to do are gone forever. Companies hire technicians capable of understanding the chemistry used in their process and who have some college math, analytical skills, and communication skills. Technicians should also be familiar with computers and some computer programs. They must understand the economics of their process. They will be required to continually improve their knowledge and skills because changing technology will require them to continue learning and schooling. After receiving training, process technicians are responsible for running their unit economically, safely, and efficiently. It is *their* unit; its proper operation is in *their* hands. Figure 1-2 emphasizes that today process technicians are hired because of attributes they possess from the neck up whereas in the past they were hired because of attributes from the neck down.

Process control systems have increased in complexity as electronic and pneumatic control systems were replaced by computerized **distributed control systems (DCS)**. The computer can also drive regression analysis and neural network systems, which can model processes to predict product and process responses instead of just measuring them. Control rooms jammed with panels, gauges, switches, meters, and charts are long gone and replaced with something that resembles a space shuttle control panel. Processes run faster, more safely, and produce higher quality product using the latest methods of statistical quality control. Improvements in information technology make it possible for the technicians to know almost immediately what each piece of equipment under their control is doing. Modeling programs tells the technicians what the process change they made will do to the product to be produced in the next hour.

More Skills, More Knowledge

Earlier we said that the refining and petrochemical industry was largely responsible for changing the process technician's job from one that requires a manual laborer to one that requires a skilled technician. Historically, batch distillation units that produced only kerosene were replaced by more complex batch units that produced everything from fuel gases to heavy tars. These more complex batch units gave way to continuous processes that

Strength

Process control

Procedures
Regulations
Day orders
Problems
Work orders
Quality
Production

Knowledge

Figure 1-2   Strength Worker Versus Knowledge Worker

were mostly manually controlled units producing the simpler petrochemical derivatives. Because the continuous flow process could operate around the clock and around the calendar, it instituted shift work and rotating shift schedules.

As more complex operations such as catalytic cracking and reforming were introduced, more complex instrumentation and controls were needed to operate the plants safely and economically. Pneumatic units replaced manual controllers as operations became more complex. These were subsequently replaced by electronic controllers, then by digital controllers, and eventually by computerized systems. Each change required the process technician to have more technical training to understand the new controls and how they affected the process operation. An evolutionary process occurred and the operator of yesterday who was originally hired for strength and stamina evolved into a process technician of sophisticated skills and knowledge.

The increasing technical and regulatory environment requires a process technician to possess varied skills. Some of the more important ones are discussed in the following paragraphs.

Technical Expertise. Process technicians today must possess technical expertise. In the past they were not expected to design process improvements, be involved in quality, understand instrumentation and control systems, be aware of environmental issues, or be involved in visits by governmental agencies. They are now. Their value to the company is in terms of their ability to continue to learn new technical knowledge and skills.

Regulatory Knowledge. With new requirements, laws, and regulations being enacted technicians must be aware of these regulations and adhere to them in their daily work. In the past a chemical spill may not have been considered a serious concern. Today, process technicians must document spills, classify them, and report them to the proper agencies. Failure to do so can result in the company or technician being fined, imprisoned, or both.

Communication Skills. Lack of communication and poor communication are constant complaints in all businesses. Process technicians must communicate effectively with fellow team members and other plant personnel. They should be capable of good verbal and written communication skills. Information should be clear, concise, and easily understood so that it can be acted on without error. The written report technicians prepare at the end of their shift should summarize their activities in a way that others can easily understand. The procedures and guidelines that they help write should be written so that misunderstandings and mistakes are eliminated. Many reports and logbooks are legal documents that can be referred to in case of accidents or audits by regulatory agencies.

Computer Literacy. Technicians must also be very familiar with computers and several types of computer programs. They use computers for issuing maintenance work orders, tabulating records and data from the unit and the laboratory, and for maintaining personnel records such as timesheets, payroll, and vacation schedules. Plus, much of a technician's training takes place via computers. And more importantly, technicians must understand the control schemes for their units, many of which will be on computers or local microprocessors. Plants today have much of their operations controlled by computers using sophisticated programs on their distributive control systems (DCS) for this purpose.

Interpersonal Skills. Process technicians must develop the interpersonal skills needed to work as an effective team member. Each crew must function as a team to do its job effectively. The operations and support groups must work together as a team to resolve process or equipment problems. Technicians may also be asked to work on special teams to troubleshoot or upgrade equipment, to review safety or environmental issues, or to write operating or maintenance procedures. The work force is highly diversified and getting along with individuals of various races, creeds, and personality types is a necessary attribute. Teamwork is an important part of the technician's job today. Individuals that have poor interpersonal skills and personality conflicts hamper teamwork.

Problem Solver. Process technicians must be able to troubleshoot problems in operating equipment and process systems and determine if the equipment is running properly or if maintenance is needed. They must become so familiar with their unit that they can quickly recognize when an operating or mechanical problem occurs and adjust unit conditions to correct for quality and yield loss problems. They should be able to recognize hazardous conditions that require corrective action.

Trainer. Process technicians may be asked to train newly hired process technicians in their roles and responsibilities, which includes safety and environmental training. Much of the training involving a specific process or unit is done one-on-one using experienced technicians.

Quality and Continuous Improvement. Quality and continuous improvement are requirements for survival in today's highly competitive markets. In recent years continuous improvement has become a relentless goal for all organizations. Process technicians know every valve, pipe, vessel, and the in-and-outs of their unit better than anyone else. They are the most qualified for defining the large and small pathways that lead to continuous incremental improvements and higher productivity and profitability.

RETHINKING PROCESS PLANT ROLES AND RESPONSIBILITIES

The primary role of production is to safely and effectively achieve daily production plans. Process industries are seeing and will continue to see a fundamental change in the roles, responsibility, and authority of technicians. In the near future all operations personnel will proactively (it is part of their job) collaborate to set targets, initiate work orders, identify opportunities to improve process operations, identify and compensate for process disturbances, and avoid critical process situations or mitigate the consequences, if the situations are unavoidable. Technicians will also be responsible for monitoring the condition of manufacturing assets (equipment) and performing light maintenance to ensure the day-to-day operation of the plant. Ultimately, technicians will proactively manage their plants to maximize safety and minimize environmental impact while driving the process to optimal production (Figure 1-3). There will be a commitment to the concept of collaborative knowledge workers and individual empowerment that makes optimal production possible. Peter Drucker, a famous management scientist and consultant, had it right when he stated, "Managing the productivity of equipment and of knowledge workers will be the main challenge of the twenty-first century."

Process technicians play a critical role in site security. They are on the frontline manning the foxholes against terrorists. Their knowledge, skills, and alertness are necessary for a successful defense of their production site. A large incident such as a major release of a site's hazardous material can injure people, harm the environment, and seriously damage a company by disrupting operations, thus inviting lawsuits, requiring expensive remediation, and injuring the company's image. Since process technicians must assume more responsibility for the security of their process unit and site, they receive training on perimeter and equipment security, access to control rooms, and important process control software.

THE PROCESS TECHNICIAN AS A SHIFT WORKER

Shift work in the process industry has been around for decades. It is nothing new, but it does have a major impact on a technician's life. It is discussed here briefly because future technicians should have a clear understanding of shift work, including the social and physical demands, and how to follow a schedule.

Shift work occurs in industries such as refineries and petrochemical plants because operations maintains maximum efficiency by continuously running 24 hours a day, 365 days a year. Shift work offers advantages not found with straight day schedules, but it can also be the cause of problems and conflicts. Depending on the particular shift schedule, the shift worker can be off for a week of continuous time each month. This permits more flexibility

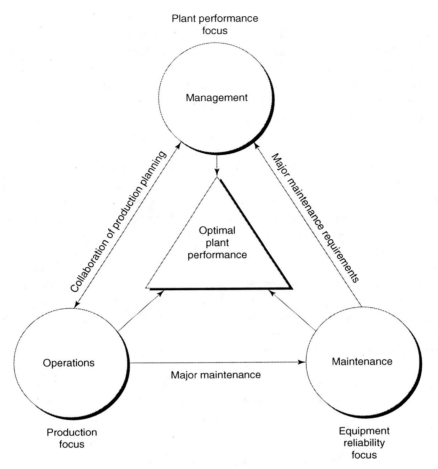

Figure 1-3 Total Plant Performance

in scheduling vacation activities. But work takes precedence when the shift worker is scheduled to work. It may be a holiday, or an important day for their family, but they still have to be at work. This is especially a problem when young children are involved. In most situations when a person is scheduled to work a 12-hour shift it leaves little time at the end of the day for other activities. In addition, the worker's schedule is always at odds with his body's biological clock, and frequent schedule changes can desynchronize these rhythms. Shift work can cause physical and emotional problems if the worker is not successful in adapting to such a schedule.

Fixed and Rotating Schedules

A fixed-shift schedule assigns a worker to one shift indefinitely, while a rotating-shift schedule moves the worker through all shifts on a rotating basis. Each shift schedule has its merits and disadvantages. Manufacturing industries, which can easily shut down their processes, tend toward fixed-shift schedules. The process industries that have large continuous operations that are difficult to shut down favor the rotating shift. The rotating schedule rotates workers through all shifts, days, evenings, and graveyards. Since everyone works all shifts there is no segregation of employees by seniority and the work crews have a better balance of experience.

The 12-hour shift has become the standard for shift workers in the process industries. This means that in a normal workday the operator will work 12 hours and be off 12 hours. Starting and ending times vary with each facility. Work schedules starting at 6:00 a.m. and ending at 6:00 p.m. are common, but some facilities will work from 7:00 a.m. to 7:00 p.m. There are some advantages to such a schedule, such as workers receive more days off, they work fewer weekends, and it requires less commuting time. However, the 12-hour shift raises concerns about worker fatigue affecting quality, safety, and worker health. Plus, such a schedule also leaves little time for family and friends when the technician is working.

A common 12-hour schedule is shown in Figure 1-4. The schedule has the workers working four days, then four off. This schedule uses four crews of workers (A, B, C and D crews), with two crews working and two crews off. One of the crews working is on the night shift, the other on the day shift. Essentially, this schedule has a technician working four days, then off four days, working four nights, then off four days. The cycle is then repeated.

SUMMARY

The process technician's role has changed radically from several decades ago. Today they are responsible for many important roles in the processing industry, including those involving safety, environmental compliance, continuous improvement, product quality, maintenance, and troubleshooting. Because technology is constantly evolving and changing, the technician will be required to constantly learn and upgrade their skills.

The process technician is now required to have *special or practical knowledge and skills of a mechanical or scientific nature.* Today's technician must have skills such as technical

HOUSTON OPERATIONS
12-HOUR SHIFT SCHEDULE 2003

	M	T	W	T	F	S	S	M	T	W	T	F	S	S	M	T	W	T	F	S	S	M	T	W	T	F	S	S
1 Jan		(1)	2	3	4	5	6	7	8	9	10	11	12	13	14	15	16	17	18	19	20	21	22	23	24	25	26	
2 Jan–Feb	27	28	29	30	31	1	2	3	4	5	6	7	8	9	10	11	12	13	14	15	16	17	18	19	20	21	22	23
3 Feb–Mar	24	25	26	27	28	1	2	3	4	5	6	7	8	9	10	11	12	13	14	15	16	17	18	19	20	21	22	23
4 Mar–Apr	24	25	26	27	28	29	30	31	1	2	3	4	5	6	7	8	9	10	11	12	13	14	15	16	17	(18)	19	20
1 Apr–May	21	22	23	24	25	26	27	28	29	30	1	2	3	4	5	6	7	8	9	10	11	12	13	14	15	16	17	18
2 May–Jun	19	20	21	22	23	24	25	(26)	27	28	29	30	31	1	2	3	4	5	6	7	8	9	10	11	12	13	14	15
3 Jun–Jul	16	17	18	19	20	21	22	23	24	25	26	27	28	29	30	1	2	3	(4)	5	6	7	8	9	10	11	12	13
4 Jul–Aug	14	15	16	17	18	19	20	21	22	23	24	25	26	27	28	29	30	31	1	2	3	4	5	6	7	8	9	10
1 Aug–Sep	11	12	13	14	15	16	17	18	19	20	21	22	23	24	25	26	27	28	29	30	31	(1)	2	3	4	5	6	7
2 Sep–Oct	8	9	10	11	12	13	14	15	16	17	18	19	20	21	22	23	24	25	26	27	28	29	30	1	2	3	4	5
3 Oct–Nov	6	7	8	9	10	11	12	13	14	15	16	17	18	19	20	21	22	23	24	25	26	27	28	29	30	31	1	2
4 Nov	3	4	5	6	7	8	9	10	11	12	13	14	15	16	17	18	19	20	21	22	23	24	25	26	(27)	(28)	29	30
1 Dec	1	2	3	4	5	6	7	8	9	10	11	12	13	14	15	16	17	18	19	20	21	22	23	24	(25)	26	27	28
2 Dec	29	30	31																									

1 0630–1830	A	A	A	A	B	B	B	B	C	C	C	C	D	D	D	D	A	A	A	A	B	B	B	B	C	C	C	C	Day:
1830–0630	C	C	C	C	D	D	D	D	A	A	A	A	B	B	B	B	C	C	C	C	D	D	D	D	A	A	A	A	Ngt:
2 0630–1830	D	D	D	D	A	A	A	A	B	B	B	B	C	C	C	C	D	D	D	D	A	A	A	A	B	B	B	B	Day:
1830–0630	B	B	B	B	C	C	C	C	D	D	D	D	A	A	A	A	B	B	B	B	C	C	C	C	D	D	D	D	Ngt:
3 0630–1830	C	C	C	C	D	D	D	D	A	A	A	A	B	B	B	B	C	C	C	C	D	D	D	D	A	A	A	A	Day:
1830–0630	A	A	A	A	B	B	B	B	C	C	C	C	D	D	D	D	A	A	A	A	B	B	B	B	C	C	C	C	Ngt:
4 0630–1830	B	B	B	B	C	C	C	C	D	D	D	D	A	A	A	A	B	B	B	B	C	C	C	C	D	D	D	D	Day:
1830–0630	D	D	D	D	A	A	A	A	B	B	B	B	C	C	C	C	D	D	D	D	A	A	A	A	B	B	B	B	Ngt:

NOTE: CIRCLED DATES ARE SCHEDULED HOLIDAYS

Figure 1-4 Shift Schedule

expertise, communication skills, troubleshooting abilities, and computer literacy. They will use these skills to meet the goals of their process unit, which are safety, production, reliability, and cost.

REVIEW QUESTIONS

1. Describe how a process technician's role has changed in the last 30 years.

2. List three core values desired in a process technician.

3. Discuss the importance of employee safety to the process industries.

4. Explain the importance of environmental compliance to the process industries.

5. Discuss the roles a process technician performs in the area of environmental compliance.

6. List five roles of a process technician today.

7. Explain the importance of those five roles to the process industry.

8. List at least two skills and two knowledge categories desired of process technicians today.

9. Describe how the process technician has a role in site security.

10. What is the primary function of today's process technician?

11. Discuss two ways, social or physical, that shift work affects a process technician.

CHAPTER 2

Safety I:
Process Hazards

Learning Objectives

After completing this chapter, you should be able to

- *List four types of hazards associated with high-pressure systems.*

- *Discuss the hazards of negative pressure.*

- *Discuss the hazards associated with steam.*

- *List some hazards of air on a process unit.*

- *List two hazards of water in a process unit.*

- *Describe several hazards associated with electricity.*

INTRODUCTION

Numerous hazards are involved in the operation of processing plants, especially petro-chemical plants and refineries. Table 2-1 presents recent U.S. statistics about work injuries and death, illustrating how dangerous and costly process operations can be if something were to go wrong. In order not to contribute to these statistics, proper analysis of a problem at an early stage of development will minimize the hazards or difficulties involved in the operation. Accurate procedures and adhering to the procedures used to achieve the desired tasks and products depends upon the discipline and good judgment of the technicians who are completing the tasks. The nature of the processes requires fuel, air, heat, chemicals, and mechanical equipment. Only by understanding how these things represent hazards and by the close control of them can a plant continue to operate without serious injury to workers and damage to equipment. Even though the best and most modern equipment is

Table 2-1   U.S. Statistics of Workplace Injuries and Deaths

3,800,000 injuries per year
5,100 deaths per year
$30,000–40,000 for each disabling injury (includes indirect costs)
$910,000 cost for each work-related death
$250,000 in fines for the average EPA citation
Note: The above statistics are quoted from Hydrocarbon Processing, June 2002, "Is Regulatory Compliance a Burden or an Opportunity?"

used, the equipment will only do what it is made to do at the hands of the technician. Safety of personnel and equipment depend on the critical thinking and judgment of the technician. This chapter will review common process hazards a technician may encounter daily while performing their job.

HAZARDS OF PRESSURE

On April 27, 1865, the side-wheeler steamer *Sultana* steamed up the Mississippi River carrying more than 2,000 Union soldiers, many bound for home after being released from Confederate prison camps. Quick repairs had been made to the vessel's boilers at Memphis. A few miles north of Memphis the boilers blew up and tore the *Sultana* apart, hurling men and parts of the vessels hundreds of feet. An estimated 1,700 soldiers died either from the explosion or from drowning. The pressure at which the Sultana's boiler ruptured so violently would be considered low compared with boiler pressures commonly used today.

It is not necessary to have much pressure to create conditions where serious injuries and damage can occur. It is commonly and mistakenly believed that injury and damage will result only from high pressures, however, there is no agreement on the definition of the term *high pressure* beyond the fact that it is greater than normal atmospheric pressure. For accident prevention purposes, any pressure system must be regarded as hazardous. Hazards lie both in the pressure level and in the total energy involved.

A common definition of ***pressure*** is the force distributed over a surface. Pressure can be expressed in force or weight per unit of area, such as pounds per square inch (psi). Critical injury and damage can occur with relatively little pressure. Workers in processing industries are often surrounded with pipes and vessels under pressure. The hazards most commonly associated with high-pressure systems are:

- Leaks
- Pulsation
- Vibration
- Release of high-pressure gases
- Whiplash from broken high-pressure tubing and hoses

Some strategies for reducing high-pressure hazards include

- Strict adherence to design codes for vessels and piping systems
- Limiting vibration through the use of vibration dampening

- Decreasing the potential for leaks by limiting the number of joints in the system
- Overpressure relief systems
- Engineering controls

When the pressure of a fluid inside a vessel exceeds the vessel's strength, it will fail by rupture. A slow rupture may occur by popping rivets or by opening a crack. Some atmospheric storage tanks are designed to fail along the roof and shell seam, popping the seam and venting the pressure upwards. This maintains the integrity of the shell walls and contains the liquid within. Vessels that experience a rapid rupture literally explode, generating metal fragments and a shockwave with blast effects as damaging as those of exploding bombs. One way to reduce or control pressure in storage tanks is to insulate the tank or paint it with a reflective paint that reflects the sun's radiation.

Pressure vessels do not have to be fired (heated by burners) to be hazardous. The sun can heat outdoor pressure vessels. Portable cylinders, some that contain gases at pressures up to 2,000 pounds per square inch gauge (psig) at room temperature, should be stored only in shaded areas. For example, the vapor pressure of liquid carbon dioxide is 835 psig at 70°F but rises to 2,530 psig at 140°F. Cylinders under pressure inside buildings should not be located near sources of heat, such as radiators or furnaces.

Dynamic Pressure Hazards

Dynamic pressure hazards are hazards caused by a substance in motion. The substance can be a fragment of metal from a ruptured vessel or a whipping hose. A common source of injury is through pressure-gauge failure. Sometimes the thin-walled Bourdon tube or bellows inside the gauge case fails under pressure due to metal fatigue or corrosion. Unless the gauge case is equipped with a blowout back, the face of the gauge will rupture first, hurling out pieces of glass and metal. A person standing in front of the gauge will be injured. A blowout back does not prevent failures but it ensures that no fragments will be propelled forward. Some boiler and furnaces have blowout panels in case there is an explosion due to delayed ignition of unburned fuel gases.

The pressure in a full cylinder of compressed air, oxygen, or nitrogen is over 2,000 psig. A cylinder weighs slightly more than 200 pounds. The force generated by gas flowing through the small opening created when a valve breaks off a cylinder can be 20 to 50 times greater than the cylinder weight. Cylinders with broken valves have taken off like rockets, reaching velocities of 50 feet per second in one-tenth of a second, smashing through buildings and rows of vehicles. A tenth of a second is less time than it takes to blink an eye.

Whipping flexible hoses can also be dynamic pressure hazards. Workers have been killed when the end fitting of a compressed-air line was not properly tightened when the line was connected. The line separated under pressure and lashed about until it hit a worker in the head and crushed his skull. Such types of accidents can occur with high-pressure water hoses (firewater hoses) as well. A whipping line of any kind can break bones and damage equipment. All high-pressure lines and hoses should be restrained from possible whipping. Rigid lines should be preferred to flexible hoses, but if the latter must be used they should be kept as short as possible. If a line or hose gets free, workers should leave the area immediately and shut off flow to the line. They should never attempt to grab and restrain a whipping line or hose.

Other Pressure Hazards

Systems under pressure should not be worked on. Each pressurized vessel or line should be considered hazardous until all pressure has been released. Verify a lack of pressure by checking a gauge directly connected to the vessel or line by opening a test cock. A person working on a nitrogen gas line pressurized to almost 6,000 psig failed to verify the line had been depressurized. He loosened the bolts of a flange, the flange separated slightly, and a thin stream of compressed gas shot out and cut into his leg like a knife. If a pressurized line is suspected of leaking never use fingers or any body part to probe for the leak since high pressures can cut like knife blades. Safe alternatives are to use a piece of cloth on a stick (called a flag) or a soap-and-water solution.

Compressed air can also cause other bodily harm. Dirt, debris, and other particles can be blown by compressed gas into an eye or through the skin. There have also been cases in which compressed air entered the circulatory system through cuts in the skin. Since the skin breathes through pores, compressed air can pass through the skin and into the bloodstream. In a Massachusetts plant, a woodworker covered with sawdust held a compressed air nozzle 12 inches away from the palm of his hand and opened the valve to blow the sawdust off. Within seconds his hand swelled up to the size of a grapefruit. It is important that the Occupational Safety and Health Administration (OSHA) standards be followed if compressed air is to be used for cleaning purposes. OSHA's standards are that (1) pressure is less than 30 psig, (2) effective guarding is used to protect against flying debris, and (3) personal protective equipment is used.

Negative Pressure (Vacuum)

Unintended vacuum or negative pressure can also be dangerous and damaging. Many structures may not be built to withstand reversed stresses. Much of the damage done by high winds during hurricanes and tornadoes is due to negative pressures. Most building are designed to take positive loads but not to resist negative pressure. Negative pressure might be generated on the lee side of a building when winds pass over it. Although the actual difference in pressure is very small, the area over which the total negative pressure will act is very large, so a large force is involved. For example, a roof on a small house may be 1,500 square feet. If the difference in pressure is only 0.05 psi, the force tending to tear off the roof equals $1,500 \times 144 \times 0.05$, or 10,800 pounds (about 5 tons).

Field storage tanks and rail car tanks have collapsed due to unintended vacuum. Condensation of vapors in closed vessels, such as storage tanks, is a source of vacuum pressure that could collapse the vessels. A liquid occupies far less space than does the same weight of its vapor. A vapor that cools and liquefies, such as steam, will decrease the pressure inside the vessel. Unless the vessel is designed to sustain the load imposed by the difference between the outside and inside pressures, or unless a vacuum breaker is provided, the vessel may collapse. Draining an unvented tank may also draw enough vacuum to collapse the tank.

Detection of Pressure Hazards

Finding gas leaks (pressure hazards) can be difficult. After a gas has leaked into the ocean of air around it, obvious symptoms of the leak (odor or a cloud) may disappear. There are several methods for detecting gas leaks:

- Sounds such as a whistling noise may indicate a gas discharge, particularly with highly pressurized gases escaping through small openings. However, these sounds may be hard to hear in a noisy plant. Workers should be careful when searching for gas leaks as high-pressure gases can cut through clothing and flesh.

- Streamers (cloth or plastic banners) may be tied to a stick to help locate leaks.
- Soap solutions smeared over the vessel surface form bubbles when the gas escapes if the leak is small and of low velocity.
- Scents may be added to gases that do not naturally have an odor. This is done with natural gas before it is piped into homes.
- Portable leak detectors may be used.

HAZARDS OF STEAM

Steam is the utility most observed by people passing by refineries and chemical plants. White clouds of escaping steam are seen all over a plant because steam can be used for so many things. Because steam is not combustible nor does it support combustion, is readily available at refineries, and affordable, it is often used as an inert gas.

Steam is used to purge air from vessels and lines prior to startup of a unit. A greater volume of steam may have to be used for purging than if nitrogen gas were used because when steam enters a cold vessel or line, all or part of the steam condenses. Only after the equipment (metal) is thoroughly heated does the steam remain as an inert vapor. With nitrogen, each volume introduced into the equipment forces out an equal volume of a mixture of nitrogen and air. Because of condensation on cold piping and metal, however, steam introduced into a vessel may initially displace very little of the air or gas.

A visible plume of steam at the purge vent is not a reliable sign that a vessel has been thoroughly purged of air. The temperature of a saturated steam-air mixture at any pressure is an indication of its air content. There are tables that can be consulted that indicate the temperature and corresponding air volume percent or other noncondensable in various mixtures at atmospheric pressure.

After vessels and lines that operate at atmospheric pressure or above it have been purged with steam prior to startup, fuel gas or another suitable gas must be backed (pumped) into the vessel when purging is completed. This is done to displace the steam, which if left inside, will condense and form a vacuum. If the vessel is left full of steam with valves closed, condensation can produce a vacuum great enough to collapse the vessel. Also, since valves frequently do not close tightly, the vacuum caused by condensing steam may draw in air. This creates a flammability hazard when hydrocarbons are introduced into the equipment. Fuel gas or other gas lines must not be opened into the purged vessel for backfilling until the steam pressure in the vessel is lower than the gas line pressure. However, as a rule, the steam pressure in the vessel must not fall lower than approximately 5 psig before the backfilling is started.

Steam heating of blocked-in exchangers or steam tracing of pipe or other equipment completely full of liquid can result in dangerously high pressures if a pressure relief valve is not provided. Liquids expand when heated, as Figure 2-1 illustrates. If a full vessel is blocked in and heated something is going to give. For example, in a vessel filled with water and without a vapor space, a 50°F rise above the atmospheric temperature results in a pressure of 2,500 psig, an average of 50 psig for each degree rise in temperature. At higher temperatures, the rate of pressure rise per degree is even greater because the thermal expansion of

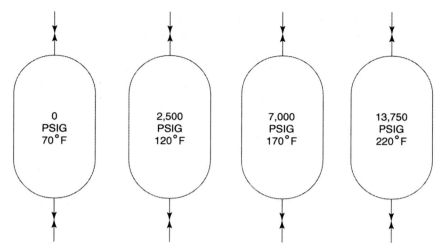

Figure 2-1   Pressure Increase Due to Heated Liquid Expansion

water is greater. The greater the pressure, the greater the potential for an explosion if a relief valve is not provided.

HAZARDS OF AIR

Air is basically composed of approximately 21 percent oxygen and 79 percent nitrogen by volume with a residue of small amounts of other gases. Oxygen (or air) is an important part of our fire triangle and one that can be controlled. Since air is pressing down all around us at approximately 14.7 pounds per square inch, and oxygen is used in some processes, we can expect oxygen (or air) from many sources to be present in chemical processes. Some of these sources are

- Atmospheric air entering open lines and vessels
- Atmospheric air introduced with wash water
- Atmospheric air leakage into vacuum systems
- Atmospheric air leakage through open or defective valves
- Oxygen (or air) in solution in feed stocks or products

Some petroleum products will vaporize enough at room temperature to form combustible mixtures with air. Others will do this at very low temperatures while some must be heated to cause sufficient vaporization. All hydrocarbons are hazardous when their vapors are mixed with air in flammable proportions. The lowest temperature at which enough vapors are given off to form a flammable mixture of vapor and air immediately above the liquid surface is the ***flash point*** of the liquid fuel. Even a small amount of liquid petroleum above its flash point can produce enough vapor to make a large volume of dangerous vapor-air mixture. One gallon of propane vaporized in a 10,000-gallon drum and mixed with air has the energy equivalent of 13 pounds of TNT. Never assume an empty vessel is safe unless it has been purged and gas tested.

HAZARDS OF LIGHT ENDS

Light ends, as discussed here, are either a single hydrocarbon or a hydrocarbon mixture having a Reid Vapor Pressure (RVP) of 18 psi or more. The individual compounds that make

Table 2-2   Flammability Limits of Light Ends

Light End	Volume % of Light End Vapor in Air	
	Lower Limit	**Upper Limit**
Methane	5.3	14.0
Ethane	3.0	12.5
Propane	2.2	9.5
Butane	1.9	8.5
Pentane	1.4	7.8

up the group of hydrocarbons called light ends are listed in Table 2-2. Light ends are very hazardous because they will evaporate rapidly at room temperature and pressure. Special attention is required to remove air from a system before light ends are introduced and to prevent air from entering while light ends are present.

Hydrogen and methane gases are lighter than air. When they leak out of a container they rise into the atmosphere and unless the leak is very large, are diluted in a short time to the point they are not a hazard. However, light ends are often compressed and stored and handled as liquids. This is particularly true of propane, butanes, and pentanes. Liquefied petroleum gas (LPG) is butane or propane, or a mixture of the two. To keep propane a liquid at 100°F, it must be kept at a pressure of at least 189 psi; normal butane, 52 psi; and isopentane, 21 psi. If these pressures are not maintained, the liquid will quickly vaporize. A small amount of liquid leaking from equipment will vaporize into a large vapor cloud that can spread quickly and cause a large explosion and fire. This is why it is so important to keep light ends confined. Heavier hydrocarbons at atmospheric temperature do not have to be handled under pressure and normally do not vaporize enough to be a hazard except in the immediate area of a leak.

The relatively low viscosity of light ends adds to the problem of containing them within pressurized equipment. Their low viscosity makes them prone to leaks. At 100°F, kerosene is about five times as viscous (thick) as propane and six times as viscous as pentane. The low viscosity of light ends allows them to escape through flanged joints and packing much more easily than heavier hydrocarbons. Light ends can flow through openings where even water can not.

The low boiling points at atmospheric pressure of propane (−43.8°F) and butane (31.1°F) create hazards in depressuring equipment, which may contain water or heavy hydrocarbons. Fires and explosions have occurred when vaporizing butane lowered the surrounding temperature below the freezing point of water. In one accident, a reflux drum was being pumped out in preparation for a shutdown. During depressuring, residual light hydrocarbons vaporized, cooling the drum and freezing the water in a drain connection. This prevented a complete purge of the vessel and allowed water and some hydrocarbons to remain in a drum believed to be purged of all water and hydrocarbons. An accident was inevitable.

Another hazard of light ends is frostbite. If liquid butane or propane contacts skin, it can drop the skin temperature below the freezing point of water and cause frostbite.

HAZARDS OF WATER

Water is necessary in processing plants, but if found at the wrong place in a process system it can lead to a disaster. The big danger from water that accidentally enters a refining unit is that it can flash to steam and create pressures inside a vessel that will cause internal damage or a rupture. At 212°F and at atmospheric pressure water expands about 1,600 times in volume when it becomes steam. In other words, one gallon of water would expand to 1,600 gallons of steam. Vaporizing a five-gallon can of water produces enough steam to fill a rail tank car. At pressures lower than atmospheric water boils and turns to steam at temperatures below 212°F and its expansion is much greater.

Water Hazards in Process Units

Water may have been used to purge air from process equipment by flooding. In such a case, water can be a hazard due to its weight. Water is approximately 25 percent heavier than most hydrocarbons. Water flooded (purged) vessels and structures must have sufficient strength to hold the weight of water and their foundations must be able to safely withstand the added weight.

Experienced personnel know the hazard of water is always present during a unit startup. Dangerously high pressures, high enough to rupture almost any system not protected with a pressure relief device, can develop when closed systems containing water heat up. Also, the destruction resulting from a rupture is much greater if the water at the time of the rupture is above its atmospheric boiling point (212°F). After a water purge, careful draining is necessary to prevent (1) trapping water in a vessel or piping or (2) pulling a vacuum when the water drains too fast.

Foam usually forms when water vaporizes below the surface of hot asphalt or heavy oil. The foam volume may be many times that of the oil, and under severe conditions, up to 20 to 30 times the volume of oil. The foamover may overflow not only the tank containing the hot oil but also its firewalls and can spread into adjacent areas and roads. If hot asphalt or heavy oil is pumped into a tank that contains even a small amount of water or an emulsion, violent foaming may result. When hot asphalt is loaded into tank trucks and tank cars at temperatures above 212°F, loading should not be started until an inspection of the tank truck or tank car has been made and the absence of water has been confirmed. Otherwise, a foamover may occur.

Water can be a source of air. Water used for process washing or for flushing can carry air into a hydrocarbon system because water can contain dissolved air which will degas (leave the water) under the right conditions.

HAZARDS OF ELECTRICAL SHOCK

Electricity can be hazardous to the process technician in a variety of forms:

- Sparks and arcs
- Static electricity
- Lightning
- Stray currents
- Energized equipment

Electrical hazards pose dangers because they might ignite mixtures of air and flammable gases or vapors, or cause electrocution. In petroleum refineries and petrochemical plants electric sparks and arcs and electrical shock are the two principal hazards of electricity. Sparks and arcs may ignite mixtures of air and flammable gases or vapors resulting in explosions and fires. Electrical shock can cause fatalities or serious injuries.

Electrical Shock Injury

The amount of current and current path are two important factors affecting the extent of electrical shock injury. The amount of current depends on voltage and body resistance. Body resistance can be high or low, depending on whether the skin is dry (high resistance) or wet (low resistance). Contact area also affects body resistance: A person in a bathtub has both wet skin and a large contact area and is almost certain to be electrocuted if a shock is received. Electrocution may occur when the heart area or the respiratory control center of the brain is in the current's path. The human body can only tolerate a very small amount of current and it is measured in milliamperes (a milliampere is one thousandth of an ampere). A rough gauge of type of injury relative to electric current is revealed in Table 2-3.

The significant factor in fatalities is current flow through the body. A current of less than 1 milliamp (mA) may not even be noticed by a normal man. Above 3 mA, it becomes unpleasant. Above 10 mA the victim is unable to let go. Above 30 mA, asphyxiation will result. Still higher levels lead to heart stoppage and death. These values are for sustained contact. Much higher levels can be tolerated for a fraction of a second. It is important to note that the greatest number of accidental electrocutions occur when people are fatigued.

Electric sparks and arcs occur in the normal operation of certain electrical equipment. They also occur during the breakdown of insulation on electrical equipment. When electricity jumps a gap in air, it is called a spark. We are all familiar with the static spark that jumps from the end of a finger to a metal doorknob after walking across a carpet. The minimum amount of energy which a spark or arc must have to ignite a flammable mixture is extremely small. Most electrical equipment can produce sparks and arcs which have more than enough energy to cause ignition.

Table 2-3   Electric Current and Injury

Current (milliamperes)	Reaction/injury
1	Perception level
5	Shock, disturbing but not painful
8 to 15	Painful shock
15 to 30	Painful shock with control or adjacent muscles lost
50 to 150	Extreme pain, respiratory arrest, severe muscular contractions and difficult breathing
10,000 or more	May be fatal, cardiac arrest, severe burns

Data from National Safety Council; and *Hazard Communication Training Guide*, Office of Environmental Health and Safety (EHS), Princeton University, *http://web.Princeton.edu/sites/ehs/hazard/commguide/8.htm*, 2005.

Working with 220/440 volt switchgear can be dangerous because of the possibility of deadly arc flashes. Arc flashes occur when an arc shorts across components in a system and creates an ultraviolet flash that can permanently blind a person, plus temperatures of 35,000°F can occur, vaporizing metal into a gas and heating the air so that it creates a pressure wave called an "arc blast." Bystanders have been killed standing 8 feet away from the point of the arc flash. Special flash suits are available for operating switchgear capable of an arc flash.

Static Electricity

The principal hazard of static electricity is a spark discharge which can ignite a flammable mixture. Refined flammable liquids, such as gasoline, kerosene, jet fuels, fuel oils, and similar products become charged with static electricity from pumping, flowing through pipes, filtering, splash filling, or by water settling through them. Different liquids generate different amounts of static electricity.

Refined hydrocarbon liquid fuels vary widely in their ability to generate and to conduct static electricity. Generally speaking, the products that are the better conductors also are better generators of static, but because they are better conductors, the static electricity generated is discharged more readily. The discharge process is called *relaxation* and relaxation time is often expressed as the time required for a given charge to decrease to half its original value. If this time is very short, large static potentials in bulk fuel are not created because the relaxation process limits the charge that can build up. The poorer conductors (generally the cleaner products) are also poorer generators, but because their relaxation times are so much longer, large charges can be generated. Very poor conductors, having excessive relaxation times, are also such poor generators that hazardous static potentials might not be created.

Three common industrial situations that generate a static charge are (1) pumping a non-conductive liquid through a pipe, (2) mixing immiscible liquids, and (3) allowing non-conductive liquids to free-fall into a tank or large container. Petroleum products flowing through pipelines become electrostatically charged. Fortunately, most of these static charges do not become dangerously large unless the flow velocity is high. If water is present in the hydrocarbon, the hazard increases greatly because even small amounts of water in flowing oil can cause a dangerous build-up of static charges. Static sparks produced in this manner have caused many accidents. Electrostatic charging of oils also occurs when droplets of water settle out through oil in tankage. Also, electrostatic charging of oils occurs if non-conductive liquids are allowed to free-fall when being transferred into large vessels.

The following techniques can be used to reduce or eliminate static electricity as an ignition source:

- Bonding or grounding (explained in the following section)
- Relaxation, a technique used when liquids from a pipe are discharged into the top of a vessel. The charge build-up can be reduced by enlarging the pipe diameter at the tip of the pipe, which reduces fluid velocity.
- Dip pipes reduce the static charge of non-conductive liquids in free-fall into a vessel. The pipe runs vertically down close to the bottom of the tank, minimizing free-fall.

Bonding and Grounding

Bonding and grounding are essential to electrical safety and used extensively in plants. **Bonding** means connecting two objects together with metal, usually a piece of copper wire. **Grounding** consists of connecting an object to earth with metal, usually copper wire. The connection to earth is usually made to a ground rod or underground water piping (see Figure 2-2). Electrical equipment is grounded first for protection of personnel, and second for the protection of equipment. In a refinery or petrochemical plant containing many large or tall metal vessels (flares, distillation towers, tanks), grounding is essential for protection from lightning.

Grounding serves two distinct purposes relating to safety. First, since the ordinary power circuit has one side grounded, a fault that results in electrical contact to the grounded enclosure will pass enough current to blow a fuse. Second, the possibility of shock hazard is minimized since the low-resistance path of a properly bonded and grounded system will significantly maintain all exposed surfaces at ground potential. Grounding is effective against the hazard of leakage currents. Electricity, like water in a pipe, is always looking for a way out. Electricity is contained by insulation, but if the insulation is worn or frayed electricity may leak out. All electrical insulation is subject to some electrical leakage which increases significantly as insulation deteriorates with age or as layers of conductive dust accumulate in the presence of high humidity. A proper grounding system with low electrical resistance will conduct leakage currents to ground without developing a significant potential on exposed surfaces.

SUMMARY

Pressurized vessels should be stored in locations away from heat sources, including the sun. Pressure in vessels should be released before working on equipment and the vessels checked with gauges for signs of pressure. Negative pressures (vacuums) are pressures below atmospheric level. Vacuums can cause closed vessels to collapse.

Figure 2-2   Grounded Pump

Air is a hazard in the petroleum and refining industry because it is a part of the fire triangle. All hydrocarbons are hazardous when their vapors are mixed with air in flammable proportions.

Steam is used as an inert gas to purge air from vessels and lines prior to startup of a unit. If a vessel has been steam purged and is left full of steam with valves closed, condensation of the steam can produce a vacuum great enough to collapse the vessel.

Light ends will vaporize enough at room temperature to form combustible mixtures with air. Light ends are very hazardous because they will evaporate rapidly at room temperature and pressure, plus they may also form explosive mixtures within process systems and vessels.

The big danger from water that accidentally enters a refining unit is that it can flash to steam and create pressures inside a vessel that will cause internal damage or actually cause it to rupture (explode).

Electric sparks and arcs occur in the normal operation of certain electrical equipment. Many sparks are caused by static electricity. Bonding and grounding are essential to electrical safety and used extensively in plants. *Bonding* means connecting two objects together with metal, usually a piece of copper wire. *Grounding* consists of connecting an object to earth with metal, usually copper wire.

REVIEW QUESTIONS

1. List four types of hazards associated with high-pressure systems.

2. Explain how a pressure gauge failure can injure a worker.

3. Explain how an operator should stop a whipping hose.

4. Describe one way vacuums are created in vessels.

5. List four ways to detect gas leaks.

6. Why is steam used as an inert gas for purging vessels and lines?

7. Explain one way to determine if steam purging has removed all air from a vessel.

8. Explain why steam purged vessels must be backfilled with fuel gas or any suitable gas.

9. Explain how steam heating of a blocked-in heat exchanger full of a liquid can be hazardous.

10. List the five compounds that make up light ends.

11. Describe three hazards of light ends.

12. How much does water expand at its boiling point at normal atmospheric pressure?

13. List two hazards of water in a process unit.

14. Explain how a tank foamover occurs.

15. Explain the danger of electrical sparks to the processing industry.

16. Describe the two dangers of an arc flash.

17. Define the terms *bonding* and *grounding*.

CHAPTER 3

Safety II:
The Permit System

Learning Objectives

After completing this chapter, you should be able to

- *List four types of work permits.*

- *Explain the function of the permit system.*

- *Explain the function of the hot work permit system.*

- *Explain the purpose of the lockout/tagout permit.*

- *Briefly explain how lockout of a piece of equipment is accomplished.*

- *Explain the purpose of the confined space permit system.*

INTRODUCTION

The chemical processing industry uses different permitting systems at its numerous sites. Some permits are developed by a particular plant and apply only to that plant. Each plant may have its own permit system that addresses routine work and maintenance. Some permit systems, however, are required by regulatory agencies such as the Occupational Safety and Health Administration (OSHA).

A permit system requires a special document (permit) before certain types of work can be done. Usually, a permit system is designed to transfer custody of equipment from one group (operations) to another group (maintenance or a contractor). The first group is charged with making the equipment and environment safe for the receiving group. As long as transfer of custody is involved a work permit is required. Personnel involved in the hazardous work must fill out a permit and the permit must be inspected and verified as complete before work can begin. The function of the permit system is to force personnel involved in a hazardous task to take the time to review all the steps, personal protective equipment (PPE), hazards, and additional equipment required to perform the task safely. The permit system places equal responsibility on the issuer of the permit and the recipient of the permit. In essence, a permit system is an extra step in the direction of safety and accident prevention. Some of the more common permits are

- Confined spaces
- Lockout/tagout (LOTO)
- Hot work
- Cold work
- Opening/blinding
- Radiation
- Critical lifts
- Electrical

There should be detailed work-permit procedures covering vessel entry, welding and cutting, repairs of rotating equipment, and any tasks that present a potential hazard. If work is interrupted for any reason the original permit should be canceled and a new one issued after the area has been inspected again and found to be safe. Permits should not normally be written to cover more than one operating shift. The incoming shift supervisor should inspect the work areas and carry out the proper tests before authorizing permit work to resume.

There are three permit systems very common to the petrochemical and refining industries because OSHA mandates them: (1) *hot work*, (2) *lockout/tagout*, and (3) *confined space*. These permits will be examined in detail in this chapter.

HOT WORK PERMIT (29 CFR 1910.119)

This regulation governing hot work permits is considered part of the PSM Standard by reference. Paragraph (k) of 29 CFR 1910.119, the PSM hot work permit element, refers to 29 CFR 1910.252(a), the welding and brazing fire protection standard. The purpose of a hot work permit is to protect personnel and equipment from explosions and fires that might occur from hot work performed in an operational area. **Hot work** is defined as any maintenance procedure that produces a spark, excessive heat, or requires welding or burning. Examples of work considered to be hot work include grinding, welding, internal combustion engines, soldering, and dry sandblasting.

The hot work permit has multiple layers of protection. A processing industry site might involve several people during the issue of a hot work permit. One person might wear two hats and assume the tasks and responsibilities of two people involved in the permitting process. The responsibilities of those involved are as follows:

- Process technician—Inspects area and ensures housekeeping, blinds, isolates and clears equipment, vessels, tanks and piping, immobilizes power driven equipment (LOTO), determines PPE required, fills out the permit, and posts it at the job site.
- Process supervisor—Delegates responsibilities to the process technician and ensures that all established procedures are completed.
- Maintenance supervisor—Inspects area and ensures that it is ready for the safety inspector; ensures that equipment, vessels, and piping are cleared; ensures that safety equipment is located near job site; reviews procedure with person performing the work; confirms PPE required; and signs permit.
- Person performing the work—Inspects the job site; gathers information from process representative and mechanical supervisor about potential hazards, special procedures, or conditions; and selects and dons appropriate safety equipment before beginning work.
- Safety permit inspector—Inspects the area and ensures that it is safe; performs gas test and determines oxygen level; ensures that equipment, vessels, and piping are cleared; confirms required PPE; signs permit; and sets time limit. Process technicians at many locations perform this function.
- Standby—is a person that ensures that the person performing the work is safe, wears the PPE required to perform the job, warns the person performing the work if a hazardous condition develops, and calls for help, if needed. The standby is sometimes called a "hole watch," buddy, or attendant.

The hot work permit must be filled out and signed before a safety inspector is called. When the system is ready, the mechanical supervisor and safety inspector will show up to inspect the area. Chemical concentrations and potential hazards are assessed and the need for a standby will be determined by the mechanical supervisor and the operator. If everything is in order, the safety inspector will cosign and post the permit and the work can be started.

The permit must be displayed at the work site until the hot work operation is complete. The hot work permit must indicate

- That fire prevention and protection measures will be in place before the hot work is initiated.
- The date(s) the permit is approved for.
- The location and equipment in use where the hot work is performed.
- That a fire watch will be posted and in place during the procedure and 30 minutes after the work is complete.

The reason the fire watch remains on post 30 minutes after the work is complete is just in case an unseen spark from the hot work should ignite material in the area.

CONTROL OF HAZARDOUS ENERGY (LOCKOUT/TAGOUT) 29 CFR 1910.147

The most effective way to control hazardous energy is to put it under lock and key. Most facilities covered by OSHA's general industry standards (Part 1910) must implement a lockout/tagout (LOTO) program. LOTO is a standard (Section 147 of Part 1910) designed to isolate a piece of equipment from its energy source in order to protect employees from the hazards associated with the accidental release of uncontrollable energy. In most cases, OSHA requires employers to have a written energy isolation program and to provide training to new employees upon initial assignment and every two years thereafter. Equipment modifications and new unit startups require the existing workforce to have additional LOTO training.

Despite the regulations, many employees continue to be severely or fatally injured each year as the result of getting caught in machines and equipment. In fiscal year 2000, OSHA issued citations for 4,149 alleged violations of LOTO standard 29 CFR 1910.147. Approximately one-third of these violations were issued for the lack of an energy control procedure or program. Another OSHA statistic illustrates that 6 percent of all workplace fatalities are caused by the unexpected activation of machines while they are being serviced, cleaned, or otherwise maintained.

Year after year, most LOTO injuries can be traced to one or more of these five causes, which have been given the name "The Fatal Five":

1. Failure to stop equipment
2. Failure to disconnect from power source
3. Failure to dissipate residual energy
4. Accidental restarting of equipment
5. Failure to clear work areas before reactivation of equipment

In order to avoid these types of violations and injuries, employers should take responsibility for protecting their employees from recognized hazards. One way employers can protect their employees is by implementing general machine guards and a LOTO program with emphasis on procedures, training, and periodic inspections.

In a lockout system, a padlock is placed through a gate or hasp covering the activating mechanism of an energy source, or is applied in some other manner to prevent energy from being turned on. The lock is often color coded to indicate the division (operations, maintenance, etc.) that applied it. Often, locks from more than one division are on a lockout hasp.

A tagout system is just like a lockout system except a tag is included with the lock, if the equipment can accept a lock. Tags alone should be used only in cases where a lock is not feasible. The tags should be sturdy, waterproof, and large enough to catch the eye. They should also have a string or wire for attaching to the equipment or device to be tagged out and a place for the person doing the tagout to sign their name and the date.

Lockout/Tagout Procedure

The chemical processing industry harnesses energy from seven basic forms: electrical, pneumatic, hydraulic, compressed gases, liquids, gravity, and spring tension. In order to avoid hazards and injuries, a LOTO procedure should be followed.

Two types of employees are involved in a LOTO procedure. They are the *affected employee* and the *authorized employee*.

1. An **affected employee** is an employee whose job requires them to operate or use a machine or equipment on which servicing or maintenance is being performed under LOTO, or whose job requires them to work an area in which such servicing or maintenance is being performed.
2. An **authorized employee** is a person who locks out or tags out machine or equipment in order to perform servicing or maintenance on the machine or equipment (see Figure 3-1).

OSHA has established a six-step procedure for locking out a piece of equipment.

1. Preparation for shutdown. During this phase, the type of energy being isolated must be identified and the specific hazards controlled. Authorized employees must prepare for shutdown by reviewing information such as the type and magnitude of the energy, controls, and hazards.
2. Shutting down the equipment.
3. Isolation, which involves the closing of valves, shutting down main disconnects and circuit breakers, and disconnecting pneumatic, electric, hydraulic, compressed gas, and liquid lines.
4. Application of LOTO devices to breakers and disconnect switches, valves, and energy isolating devices.

Figure 3-1   Equipment Lockedout and Taggedout

5. Control of stored energy takes place by relieving pressure, grounding cables connected, supporting elevated equipment, and moving parts stopped.
6. Verification that all energy hazards have been locked out. The term *lock-tag-try* is applied when the electrically disconnected equipment is checked by attempting to start the equipment at the local start-stop switch. If the procedure has been performed correctly the equipment will not start.

All of this information should be recorded in a lockout logbook.

The LOTO procedure must contain clear instructions covering procedures for re-energizing machines and equipment after completion of maintenance or repairs. The equipment must be inspected to verify that components are properly fastened and jumpers or grounds, if applied, are removed. Employees performing the procedure must be safely positioned before removing the LOTO devices. Affected employees not participating in the procedure must be notified that the devices have been removed.

CONFINED SPACE STANDARD (29 CFR 1910.146)

Carbon monoxide, hydrogen sulfide, too little oxygen? What is in the atmosphere of the confined space the worker is about to enter? Workers can't see or feel an improper oxygen level or toxic or flammable gas levels. How are they supposed to know what's in a confined space? Approximately 50 percent of the time a confined space is hazardous because of an oxygen deficiency. Two out of three deaths in a confined space occur to persons ill-equipped or poorly trained attempting rescue.

Breathing low levels of oxygen can give a person a feeling of euphoria and well-being. Normal atmospheres where we work, play, and live contain approximately 20.9 percent oxygen, 79 percent nitrogen, and 0.1 percent other gases and vapors. The good news is that 20.9 percent oxygen content is exactly the amount the body needs to function most efficiently. The bad news is that at levels below 19.5 percent oxygen, a person begins to lose their sense of judgment, motor skills, and consciousness. Without sufficient oxygen a worker can pass-out and die within minutes. If a worker enters a space that has been inundated with nitrogen and the oxygen level displaced below 10 percent, death can occur very quickly. Yet, since nitrogen is a major part of our everyday environment, the victim will have no way of naturally sensing their danger.

Determining atmospheric conditions is an essential part of the hazard identification and evaluation process. Air contaminants can enter the space from conduits and piping systems, or through seams and cracks in walls. Also, heavier-than-air gases can simply drop in through openings. Sometimes fluids and their associated vapors enter the space, like gasoline floating in with some water, for example. Some toxic gases like hydrogen sulfide can be formed by the decomposition of organic material already in the space. There can also be numerous sources of hazardous vapors and gases in the space itself, the surrounding work facilities, or in nearby operations or processes. The initial confined space hazard survey should account for both current and potential hazards.

On April 15, 1993, the OSHA rule on Confined Space Entry (CSE) 29 CFR 1910.146 became effective. OSHA's intent was to protect workers from exposure to toxic, flammable, explosive, or asphyxiating atmospheres and also from potential engulfment (burial) from powders or other free-flowing solids.

The reason OSHA has created the confined space standard is that working in a confined space (CS) is especially hazardous. A confined space makes it hard to get help to a worker in trouble, plus, it is hard for the worker to get himself out quickly. Thus, working in confined spaces requires extra precautions and these precautions are exercised through permit requirements for entry.

There are two important definitions regarding confined spaces in the United States:

1. In 29 CFR 1910.146 a **confined space** is an enclosed area which (1) is large enough to enable an employee to enter and perform assigned work, (2) has limited or restricted means for entry or exit, and (3) is not designed for continuous employee occupancy.
2. The second definition is a subset of the first. A *permit-required confined space* is a confined space with one or more of the following characteristics: contains or may contain a hazardous atmosphere, contains a material with the potential for "engulfment" of an entrant (such as sawdust, sand, grain, or earth), has an internal configuration or shape such that an entrant could be trapped or asphyxiated, or contains any other recognized serious safety or health hazard.

A confined space may contain a toxic atmosphere, chemicals, an oxygen deficient atmosphere, flammable materials, power-driven equipment, or other hazardous conditions. Some examples of petrochemical and refining equipment that are considered confined spaces are

- Storage tanks
- Large pipes
- Silos
- Distillation towers
- Underground cable ducts

Permit-Required Confined Space

The criterion for permit-required confined space is that it contains or has a known potential to contain a hazardous atmosphere, including chemicals, sludge, or sewage. *Hazardous atmosphere* means an atmosphere that may expose employees to the risk of death, incapacitation, impairment of ability to self-rescue (escape unaided), injury, or acute illness. Examples of hazardous atmospheres are a flammable gas or vapor or airborne combustible dust at a concentration that meets or exceeds its lowest concentration that will create a flame, also known as lowest flammable limit (LFL); unacceptable oxygen levels; or any atmospheric condition recognized as immediately dangerous to life or health (IDLH).

If a confined space contains a material that has the potential for engulfing an entrant, it is a permit-required confined space. *Engulfment* means the surrounding and effective capture of a person by a liquid or finely divided (flowable) solid substance that can be aspirated (breathed in) causing death by filling or plugging the respiratory system or exerting enough force on the body to cause death by strangulation, constriction, or crushing.

Monitoring of the atmosphere inside a confined space involves three elements: (1) properly calibrated instruments, (2) established procedures, and (3) critical evaluation of results by a qualified person. The order of testing for specific atmospheric elements is important. Testing for oxygen content must be performed first, followed by tests for flammable atmospheres, and then for toxic gases or vapors. Testing is done in this order because if there is not enough oxygen present, the lower explosive limit (LEL) test function will not work properly since oxygen is required for that part of the instrument's operation. It may reveal an atmosphere to be safe when it is really explosive.

Normally, some benchmark, such as 10 percent of LEL, is used to indicate that a potential fire hazard exists in any area when this concentration is reached. Monitoring should be continuous during all operations within the space. Concentrations of air contaminants may increase due to leaking or percolation from the adjoining environment even though initial tests revealed them to be low or non-existent.

CONFINED SPACE ENTRY PROGRAM

What steps are required before sending two men into a drained kerosene storage tank to do an inspection? They should include:

1. Test and monitor the space for hazards.
2. Provide a means of communication between the worker(s) in the confined space (CS) and an outside hole watch so that help can be summoned quickly.
3. Develop plans for rescue and transportation to medical facilities.
4. Alert the medical facility at your site.

The purpose of the confined space entry (CSE) program is to systematically carry out the CSE process. The major objectives of a CSE program are to:

● Keep unauthorized people from entering the CS (see Figure 3-2).
● Provide tools and instrumentation necessary to ensure that entry into a confined space is safe.
● Create a confined space permit system.
● Establish a buddy system.
● Generate a list of authorized workers.
● Form a notification system to medical services.
● Perform an annual review.
● Ensure proper training.

Confined space entry (CSE) is in many ways like an expanded safe work permit. Before issuing a permit for such areas, you must check for the presence of toxicants, flammables, and other hazards in the space. This can be deceptively simple. Imagine you are trying to measure the LEL inside a large storage tank. At one location, the LEL shows 1 percent. You might think that the tank has no flammability danger and a permit can be issued. Well, the problem with this is that in large vessels, the LEL may be very different at different locations. For a large vessel, you must check multiple locations (top, middle, and bottom). Also, the chemicals used in a process will influence the LEL readout or differences in LEL. Viscous chemicals or slurries require extreme caution before issuing a permit. As slurries

Figure 3-2   Confined Space and Warning Placard

dry up they leave a powder behind that can slowly release toxic or flammable materials. This can keep the LEL readout at a high value for a long time.

Many organizations create a CSE permit as a part of their safe work permit system. Obviously, the CSE doesn't stand by itself. Additional permits may be required for hot work, cold work, opening/blinding, and energy isolation. However, unlike many other permits, a CSE permit must list the names of all the people (entrants) working in the confined space. Employers are required to develop a written CSE program, identify all permit places in their plant, post warning signs, develop procedures, and provide training.

Confined Space Team
Safe work in a confined space requires teamwork. The confined space entry team is made up of the entry supervisor, the attendant, the entrant, and the rescue team.

Entry Supervisor.   An entry supervisor is responsible to coordinate all activities related to the CS before issuing a CSE permit. They plan the entry and develop rescue plans. They check the LEL or toxic components of the confined space and the list of all LOTO items and signatures. They also assign the hole watch and entry workers. The entry supervisor has the authority to withhold issuing a permit if unsafe conditions have been detected. Since no permit remains valid beyond the duration of a shift, they have the responsibility to initiate work stoppage as the permit nears its duration.

Entrant.   Confined space training is required for authorized entrants. They must be thoroughly familiar with the space and its hazards, and should be able to detect warning signs of overexposure. They should be physically fit to be able to get out of the vessel on their own. This may sound contradictory to the main idea about the buddy system and rescue services. But, if you think about it, the idea is that a physically able person is less likely to get trapped in the CS than a physically challenged person.

Attendant. The *attendant* (also called *buddy, standby,* or *hole watch*) has the primary responsibility of monitoring the safety of the persons working in CS. They review the permit before any entry and keep unauthorized personnel out of the area. The attendant ensures ventilation equipment is working, monitors the atmosphere in the CS, maintains constant communication with the worker(s) in the CS, and tends to the lifeline of the entrant(s). The attendant does not perform any rescue function and should not enter the CS in the event of a problem. Instead, the attendant should have the authority to stop work at once if unsafe conditions develop and contact rescue services, if needed.

Emergency Rescue Services. Rescue services may be provided by outside contractors specializing in that specific line of work. Generally, plant rescue personnel or outside contractors are selected in advance and are appraised of the hazards of the CS. They should be given all the pertinent information: location, name of the vessel, permit procedures, list of hazardous chemicals, material safety data sheet (MSDS), etc. In many instances, companies decide to provide their own in-house rescue services. These employees will have received extensive training in CS rescues.

Entry into a confined space during an emergency is sometimes unavoidable. Pre-entry testing or ventilating may not be possible in situations involving a rescue. In such instances, the atmosphere must be considered immediately dangerous to life and health. A positive pressure self-contained breathing apparatus (SCBA) or airline respirator with emergency escape bottle must be used. The entrant must be equipped with a full-body harness and lifeline that can be attached to a pulley and winch or hand-operated hoist which allows the hole attendant to begin rescue. Continuous communication must be maintained with the person attempting the rescue.

CSE Hazards

Numerous hazards may exist in a confined space. Flammable and/or toxic liquids or gases may be encountered, as well as engulfment hazards, oxygen deficient atmospheres and various physical hazards such as drop-offs, sharp metal objects, and bare electrical wiring.

- *Flammable* and *toxic hazards* are the most common CS hazards. Monitoring large vessels requires care. They should be monitored at various locations since most vessels do not have well circulated air, and a gas, depending on its density, may be concentrated at the top of the vessel or the bottom.
- *Engulfment hazards* are a real possibility in silos and excavations. Workers can be buried alive. The most effective way to prevent this from occurring is to issue a permit only after establishing that there is no material in the silo and all blinds and lockouts are in place and that adequate shoring of excavations is in place.
- *Oxygen deficient atmospheres* are another CS hazard. The use of a SCBA is mandatory, especially if there is a chance of encountering or developing oxygen deficient atmospheres. OSHA regards less than 19.5 percent oxygen as oxygen deficient. See Table 3-1 for a further breakdown of oxygen levels and their corresponding symptoms. A number of factors can contribute to lowering the oxygen level, some of which are poor mixing of the air in the vessel, a process leak, corrosion or other activities like welding.

Table 3-1 Oxygen Atmospheres

Vol% Oxygen	Symptom or Comment
23.5	Maximum OSHA permissible level
20.9	Normal air
19.5	Minimum OSHA permissible level
16.0	Impaired judgment and breathing
14.0	Faulty judgment, rapid bodily failure
6.0	Difficult breathing, death in minutes

- *Physical hazards* includes piping that may trip the entrant or cause the entrant to slip and fall. There could be holes or drop-offs, electric wiring, steam lines, and vessel internals that present hazards and may lead to injury.

Testing the Work Environment

CSE is a very critical operation. Issuing a permit places total reliability on gas-testing devices. The instrument (gas detector) must be accurate and reliable. All the observations and adjustments during calibrations must be recorded for regulatory compliance, and maintaining instrument calibration is critical. The person performing the testing must also be just as reliable and accurate. They should be trained and knowledgeable and perform the testing and calibration according to manufacturer specifications. The name of that person must be recorded along with test data and dates. It is important to be thorough and patient when issuing a permit since lives are at stake (see Figure 3-3).

The following are some things to keep in mind about gas testing:

- Keep a spare gas detector on hand.
- Ensure both detectors have been properly calibrated.
- Periodically communicate the instrument readings to the workers inside the CS.
- When the work is in progress, frequently monitor the work area in the vicinity of the worker.

Ventilation of Confined Spaces

Mechanical ventilation and purging are key entry preparations for vessels. If pre-entry monitoring indicates oxygen deficiency or the presence of flammable or toxic materials, the space must not be entered. First, it must be purged with forced mechanical ventilation and/or cleaned to remove all identifiable hazards.

Generally, ventilation consists of portable blowers. In some situations, it may be desirable to use portable air conditioner units especially in summer months. Obviously, the compressor and the ductwork must be checked carefully for integrity and the air should be free of oil and moisture. There is a hidden danger involved in using ventilation in a vessel that once contained flammable vapors. Air can dilute a concentration of flammable vapor that is too rich to burn and lower it into the flammability (explosive) range.

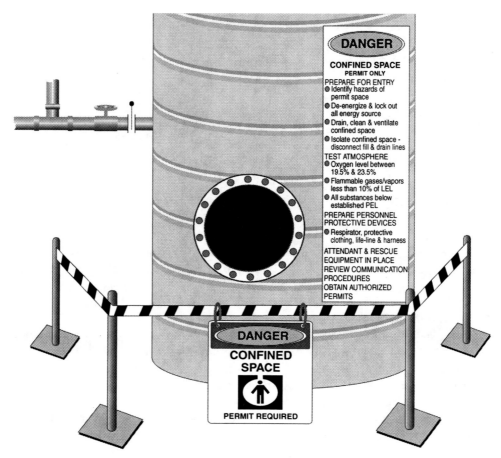

Figure 3-3   Confined Space Barricade Tape and Permit

A Final Note of Caution

Confined spaces are deadly, extremely deadly if they contain toxic vapors or gases, which can knock down entrants quickly. Often, fatality numbers for confined space indicate that for each entrant killed in a confined space two rescuers are killed attempting to rescue the downed entrant. This is because the rescuers attempted rescue using only excitement and adrenaline and did not bother with the proper personal protective equipment. Instead, they joined the ranks of dead heroes.

SUMMARY

Permit systems are used throughout the processing and manufacturing industry. They are mandated by federal law and their intent is to protect the worker. A permit system requires a permit before certain types of work can be done. Usually, the work involves some type of hazard. The function of the permit system is to force personnel involved in a hazardous task to take the time to review all the steps, PPE, hazards, and equipment required to perform the task safely.

The purpose of a hot work permit is to protect personnel and equipment from explosions and fires that might occur from hot work performed in an operational area. Hot work is defined as any maintenance procedure that produces a spark, excessive heat, or requires welding or burning, Examples of work considered to be hot work include grinding, welding, use of internal combustion engines, soldering irons, and dry sandblasting.

The purpose of the hazardous energy standard is to protect employees from the hazards associated with the accidental release of uncontrolled energy. The lockout/tagout procedure, also referred to as LOTO, is a standard designed to isolate a piece of equipment from its energy source.

Confined spaces present an array of hazards. By implementing a comprehensive planning process that encompasses atmospheric testing, hazard assessment, and protective equipment selection, many of these hazards can be addressed before entry activities begin. In order to avoid serious or fatal injuries, CSE requires a lot of caution, attention to details, and good planning before issuing a permit.

REVIEW QUESTIONS

1. Four types of permits are the _____ , _____ , _____ and _____ .

2. Explain the function of the permit system.

3. Explain the purpose of the hot work permit.

4. List five people who might be involved in a permit system.

5. Explain the responsibility of the process technician involved in a hot work permit.

6. Workers are protected from hazardous energy by the _____ permit system.

7. What is the percent of workplace fatalities due to failure to follow or have an energy isolation program?

8. Explain how to lockout a centrifugal pump with an electric motor driver.

9. Briefly explain how tagout of an energy source is done.

10. Explain why tagout is used instead of the lockout procedure.

11. Define *confined space* according to OSHA.

12. Three hazards that might exist in a confined space are _____ , _____ , and _____ .

13. List two criteria that necessitate a permit-required confined space.

14. Discuss some considerations involving the gas detector and its operator in monitoring the atmosphere of a confined space.

15. Explain how you would check the atmosphere in a very large gasoline storage tank in a refinery.

16. Describe the job of the attendant (hole watch).

17. Explain what is meant by an *engulfment hazard*.

18. Describe two symptoms of a 16 percent oxygen level.

CHAPTER 4

Quality as a Competitive Tool

Learning Objectives

After completing this chapter, you should be able to

- *Explain why quality is not only the job of the plant's quality department.*

- *Discuss how a worker can become involved in quality.*

- *Explain the importance of employee empowerment.*

- *List the costs of quality.*

- *Discuss why control charts are an important quality tool.*

- *Explain what is meant by the statement, "Quality is a function of the process."*

INTRODUCTION

Many workers do not realize that quality is an all-embracing concept. Quality does not involve just the finished product of a processing unit, such as styrene monomer or polyethylene. Quality involves every department and every person in a company and even people not part of the company, such as the company's raw material suppliers. Quality involves administrative clerks, shipping clerks, company sales representatives, warehouse personnel, engineers, safety and health personnel, corporate management, suppliers—everyone and everything. A customer that buys tens of millions of dollars of styrene monomer product a year from a processing unit may switch to another supplier, even if the styrene they have been purchasing is the best on the market. They quit buying because the

Figure 4-1 Key Relationships that Cause Quality

orders were frequently late (shipping department) or the prices and quantities were misquoted (sales department) or the quantity was insufficient due to unplanned unit downtime (maintenance department). The styrene process unit technicians may have done everything they were required to do in a quality manner but other parts of the company failed in the performance of their duties. Quality is a result of how your business is managed and quality improvement results from changes to business practices (see Figure 4-1).

WHAT IS QUALITY?

There are numerous definitions of quality, some of which are conformance to specifications, fitness for use, whatever the customer says it is, and inexpensive but reliable. In the petrochemical and refining industry, **quality** is usually defined as conformance to specifications. Specifications can be the acceptable level of impurities in a final product, product color, density, pour point, corrosiveness, etc. Also, on the bill of lading, further specifications may be spelled out, such as the carrier type, quantity of material, and date of delivery.

Quality is a cornerstone of the competitive strategy of companies and corporations that hope to survive far into the twenty-first century. It has a profound effect on the way companies are managed. Businesses must be involved in a quality process because of the ever-growing expectations of customers around the world who are demanding high-quality products and services at low prices. Quality is a necessary competitive tool for survival that increases market share and profits. A real commitment to creating quality and then improving on that quality also results in higher productivity and lowers costs.

A small group of quality experts had been saying for years that quality was a cost-effective and necessary business strategy. Today, a growing number of U.S. firms are following the coaching of these experts as they seek to survive, compete more efficiently, and dominate their markets. Five of the most respected quality coaches are Philip B. Crosby, W. Edwards Deming, Kaoru Ishikawa, Joseph M. Juran, and Genichi Taguchi. They all recognize that there are no short cuts to quality and that the improvement process is a never-ending cycle requiring the full support and participation of individual workers and, most importantly, the top level of management. Beyond that, the five coaches disagree about how to improve quality. All of their quality systems work, they just have different systems and theories.

QUALITY SYSTEMS

As previously mentioned, there are numerous quality systems. A company could adopt Juran's system, or Demings, Crosby's, the International Organization for Standardization's (ISO-9001:2000), Total Quality Management, a combination of these, or any other system available. Quality doesn't just happen. A plant manager can not just say that starting tomorrow everyone in the plant will do quality work and produce quality material. There has to be a system that will cause quality to happen. What do we mean by *system*? A system is the materials, methods, manpower, and equipment that produce a product or service. The plant must develop quality policies and procedures, train workers on quality concepts and quality tools (charts, graphs, data analysis, etc.), create a way for workers to present their ideas and suggestions for improvement to management, set performance standards, and establish a reward and recognition system.

CONTINUOUS IMPROVEMENT

The output of a quality system is a product or service desired by customers. Let's think about that. The technician on a styrene unit could say they were producing an on-specification (on-spec) styrene before the initiation of a plant quality system, so why go to the trouble and expense of installing a quality system? The answer is because no system is perfect. Every system can be improved upon and tweaked until it runs a little better. *Continuous improvement* is a goal of all quality systems. The technician on the styrene unit can claim they made on-spec styrene 361 days out of 365, which is 98.9 percent of the time. The four days they weren't making on-spec styrene were due to equipment problems that required shutting the unit down or caused off-specification production. Technicians can claim that problems like these are bound to happen because all equipment will eventually fail; however, although the technician's statement is true, failing equipment doesn't have to affect a process unit. Predictive maintenance can detect equipment at the beginning of its failure mode and operations can plan to prevent a production interruption.

The first step in continuous improvement is to begin thinking about managing quality from a personal perspective. Forget about the entire company or process unit and focus on just your responsibilities. You must strive to make improvements in the trenches where your work is done, the ideas are born, and the bottom line is supported. The concepts at the heart of most quality philosophies can be applied to *any* scope of work, whether loading railcars or managing a process unit. They can be applied whether you only manage yourself or oversee five technicians.

Remember the old axiom, "If it ain't broke, don't fix it"? A system that promotes continuous improvement fixes things before they break and cause production problems and increased expenses. Predictive maintenance programs that monitor the vibration, temperature, and amperage of rotating equipment alert technicians to the reliability status of their equipment. It allows them to fix things before they break.

IMPORTANT QUALITY CONCEPTS

Quality is built around several other concepts, such as employee empowerment, customer focus, process focus, the cost of quality, and management by data and facts. Depending on the quality philosophy adopted, other concepts can be added, but we will discuss only these five in the following paragraphs.

Employee Empowerment

Employee empowerment is critical because employees are embedded in the system and know how it operates. If management wants to understand the real nut and bolts of process unit (system), they go to the technician who is putting in 40 hours a week in the system year after year. The technician has a greater probability of supplying an answer than the engineer who is putting 10 to 15 hours a week in the system for three years before they are promoted and move on to another plant or process unit.

A quality program is concerned with more than just the mechanical aspects of a process. It should focus on improving the indirect value characteristics of the organization, such as trust, responsibility, participation, harmony, and group affiliation. Employees must be empowered to make the necessary organizational changes. The concept of empowerment is based upon the belief that employees need the organization as much as the organization needs them and that management understands that employees are the most valuable assets in the business. It is important that management recognizes the potential of employees to identify and to derive corrective actions to quality problems. Empowered personnel have responsibility, a sense of ownership, satisfaction in accomplishments, power over what and how things are done, recognition for their ideas, and the knowledge that they are important to the organization. Organizations with empowered workers that accept empowerment will be more efficient, productive, and profitable.

Customer Focus

Quality is all about the customer. Few process technicians interact with the external customer, the person or company buying the produced product. Instead, they interact with other technicians, engineers, and personnel in the plant that accept an output from them (information, intermediate product, a service, etc.). These are *internal customers*. Customer focus requires that you strive to understand their specifications or needs, strive to develop a cooperative attitude, and definitely strive to avoid an adversarial attitude. Attitude is key! When you understand that your job is to deliver the most accurate and timely rendition of what the customer desires—locked and tagged out pump for maintenance, proper bill of lading, courteous service to contractors—then working relationships become more positive, intragroup collaboration improves, and barriers between groups begin to break down because goals start to merge. Quality calls for collaboration to expose processes fully. Only through collaboration can workers determine if their piece of a larger process is meeting customer requirements. A technician willing to collaborate with internal customers and suppliers will find these people have many ideas that can add value to their core processes.

Process Focus

Technicians should ensure that the processes that fall within their scope are documented, communicated, measured, and continuously improved as time goes on. Do they have all the procedures required for their tasks? Are they current and available? Do they have the right equipment or supplies? If supplies are short or occasionally late, why is this happening? Is their equipment maintained and operating efficiently? In order to focus on process improvement, technicians should identify the core processes for which they are responsible. Basically, they should ask what do they get paid to do? The easiest way to identify their core processes is to identify their outputs or deliverables. Within the scope of their work the outputs are defined as anything that they are responsible to deliver to someone else (the internal customer). When technicians begin to think this way they will reap the benefits of removing inefficiencies they had been accepting while improving quality.

Cost of Quality

The cost of quality is all the business costs incurred in achieving a quality product or service. These include the following:

- Prevention costs
- Appraisal costs
- Internal failure costs
- External failure costs

Figure 4-2   Cost of Quality

- The cost of exceeding customer requirements
- The cost of lost opportunities

Taken together these costs can drain a company with a poor quality system of 20 to 30 percent of its revenue. Key areas of waste in a company include material, capital, and time, of which time is the biggest cost.

Quality must be measured in order to manage it. We must measure the results of how every part of our company is doing at providing internal and external customers with what they want. Whatever measure we choose should work for each department so that we can understand and assess the impact of individual decisions on the company as a whole and on our customers. One way to do this is to measure the **cost of quality (COQ)**. COQ is a dollar amount that represents the cost of avoiding, finding, making, or repairing defects in products or services. It is *prevention costs*, *appraisal costs*, and *failure costs* (see Figure 4-2).

Prevention costs are costs to prepare for an activity and perform that activity error-free. Examples of such costs are

- Training
- Process design reviews
- Measurement of processes on an ongoing basis

Appraisal costs are the costs of evaluating an output to ensure it is error free. Appraisals are usually done after the product is produced but before it is released to the next recipient or shipped to the customer. Examples of appraisals are

- Inspections
- Operations and equipment tests

Failure costs are the monies spent correcting defects after they are discovered. There are two types of failure costs: internal and external. Internal costs are detecting an error after a product is made but not shipped. External failure costs are the costs accrued after the product is shipped. Examples of both are listed here.

Internal failures

- Scrapping
- Rework
- Redesign

External failures

- Warranties
- Recalls
- Complaint resolution

COQ can be an effective management tool because it helps you compare the prevention, appraisal, and failure costs of one business process to another. It is an indicator of the effectiveness of your business. The higher the COQ, the less effective your business is. Most American companies spend about 20 to 25 percent of their operating costs on the cost of quality. Steadily, they are bringing this number down by increasing prevention costs and reducing appraisal and failure costs.

Management by Data and Facts

Managing by data and facts is important because it helps to understand how a process works and what improvements can be made. Since decisions are based on data and facts, it is essential to collect and evaluate appropriate metrics. Any measurement that helps managers and workers understand their processes and operations is a potential business metric. Some examples are number of units completed per hour, percent of defects or errors from a process, and hours required to deliver a certain number of units.

The core of quality improvement methods can be summed up in two words: *scientific approach*. It isn't complicated. A scientific approach is really just a systematic way for individuals and teams to learn about processes. It means agreeing to make decisions based on data rather than hunches, to look for root causes of problems rather than react to superficial symptoms, and to seek permanent solutions rather than quick fixes.

Companies within the process industry use quality tools that utilize statistical thinking all the time. **Quality tools** help individuals and teams to continuously improve their work processes. A quality tool can be a chart, graph, statistics, or a method of organizing or looking at things. There are many quality tools commonly used by process industries and they are simple to use and reveal important information without requiring much time or user effort.

Quality tools can help the technician to:

- Identify, agree upon, and satisfy customer requirements.
- Identify the internal and external customers that make up the customer-supplier chain.
- Recognize that improving business processes is everyone's job.
- Base troubleshooting decisions on factual information.
- Develop and track a scorecard of measures to monitor performance and guide improvement.

Some common quality tools are Pareto charts, histograms, SPC charts, process models, check sheets, flow charts, scatter diagrams, and run charts. These are all easy to create and understand.

STRIVE TO INCREASE UNDERSTANDING

Workers cannot be effective and productive if they do not understand their process unit's equipment and processes, and the big picture of how their unit is capable of affecting other units downstream. We cannot manage what we do not understand. This holds true for being responsible for and interrelating with people, projects, and processes. Process technology is high tech. The technology upgrades that make a processing unit more productive require the technician to keep learning. The more knowledge you have the more valuable you become to the company. A technician who just knows how to punch buttons and open valves and who truly doesn't understand their process is useless when things go wrong. They can only stand there and wait for someone to tell them what to do.

GET INVOLVED IN QUALITY

Quality is not the job of the quality department. In fact, many companies don't have a quality department. They have one or two people responsible for ensuring the company has a quality system in place and continues to improve on it. They are quality leaders in addition to the other jobs they are responsible for. But they can't do it all and can't afford to hire a bunch of people to staff a quality department. One or two quality leaders cannot make quality happen; it is up to the dozens or hundreds of workers in the plant to make quality happen.

How does a worker in any business get involved in quality? First management must train them in the quality system the company is using and then train them on the quality tools they will use to make incremental improvements. Then management must not only encourage them to use the tools but demand that they use the tools. Company quality metrics must be posted on bulletin boards so workers can see the categories and quantities of mistakes and the cost of making mistakes. When a worker realizes that their production site lost $1.2 million last year due to nonconformances (failures in meeting specifications), they become aware of the importance of quality. That is a lot of money to waste. Can a company stay profitable making mistakes that costly? Will waste like that affect pay raises?

Workers must learn to use quality tools to improve their processes. They must be alert to smarter work methods to save the process unit money, to cut waste, and to document their proof. Quality is firmly embedded in the scientific method and statistics. If the worker won't gather the data and interpret it with one of the quality tools, management will as a rule ignore the potentially fallible data. Management will only take the time to look at prepared "evidence" (data expressed with a quality tool). Management's attitude is, "Show me the proof."

The process technician intimately knows the nuts and bolts of their process. They know how their equipment works, how it is supposed to work, and what changes would make things work better. They know this because they spend an enormous amount of time on the unit compared to process managers and engineers. *Engineers and managers know how things are supposed to work but not how they actually perform on a daily basis.* The process technician is the person of greatest value to use the quality improvement system to make incremental changes on a regular basis. For example, in one year a technician documents data that is acted on by management and saves the processing unit $8,000. Another technician's suggestions are acted on and saves the unit $11,000. This is $19,000 saved by just two workers actively involved in the quality improvement process. Imagine the savings to a plant if 200 of its workers were all submitting data for quality improvements!

CONTROL CHARTS FOR QUALITY

Quality doesn't just happen. A system(s) makes quality happen. Prevention is one system that makes quality happen. If the technician can prevent things from going wrong, they can make on-spec product at minimal cost. Control charts based on **statistical process control (SPC)** are one of the most important prevention tools a technician has at their service. And reading and understanding control charts is one of the most important skills a technician can develop. Control charts are valuable tools for improving product quality and making processes predictable. The unique feature of the control chart is its ability to form data into patterns, which when tested statistically can lead to information about the process. Process technicians have the responsibility to monitor control charts and determine what the charts are revealing about the process. Some of the advantages of SPC are

- Improved product quality
- Increased quality consciousness
- Cost reduction
- Data-based decisions
- Predictable processes

Statistical process control (SPC) is a quality tool used to continuously monitor process performance with charts and graphs. If a company wants to make a product on specification all the time then it must have a process capable of consistently meeting those specifications. It must use SPC to define nonconformances and assist in eliminating them. For quality improvement to occur these concepts should be clearly understood and a part of each individual's job. Most everyone agrees that a policy of defect prevention is a sensible goal, but that goal cannot be reached without quantitative measures of quality to effectively identify variability in a process. SPC and its associated problem-solving techniques contain the quantitative tools that allow an objective approach to quality.

SPC recognizes that variations will always be in a process, however, it is acceptable only as long as the variations are minor. SPC divides variations into two major groups: *normal variation* and *abnormal variation*. Normal variation is a minor variation that is natural to the process. Since it is to be expected, it is *normal*. Abnormal variation is a large variation in a process variable or in product quality not normal to the unit; hence it is called *abnormal* variation. Abnormal variation can be caused by a malfunctioning pump, failed temperature controller, bad feed quality, etc. The cause doesn't have to be something major. For example, during a reactor startup to make polyethylene just a minor amount of impurity in the reactants can cause major variations in product quality.

SPC charts and associated problem-solving techniques provide a picture of performance for a process. This picture can be analyzed to detect an incipient problem and make a correction before a process produces an off-specification product. Further, the chart can be analyzed to help identify the root cause of a problem. SPC enables processes to run more consistently and increases the percentage of product that meets customer expectations (improved quality). It enables the technician to understand their process and to answer the following questions: Am I in control of the process and is there room for improvement?

SPC is an important tool for controlling all kinds of processes, however it can only work if those actually running the process use control charts and understand them. It is also necessary to have control over outside factors influencing the process, such as raw materials, equipment, or operator skills. Control charts reveal when a process is unstable and alerts the technician to take corrective action (see Figure 4-3). They reveal when the process variation becomes unacceptable. At that point a diagnosis of the cause of the variation and corrective action is required. Control charts are worthless unless corrective action is taken when evidence of excessive

48

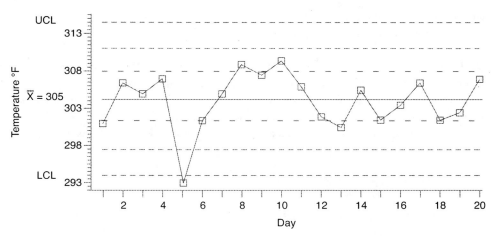

Figure 4-3 Example of a Control Chart

changes and variation occurs. The concept of statistical process control is based on the idea that someone will take action to improve the situation if process variation is unacceptable.

QUALITY IS A FUNCTION OF THE PROCESS

The process technician is just a worker in a process, such as the process for making styrene or diesel fuel. The process, also called the *system*, consists of tools, procedures, training, equipment, policies, etc. The process, not the worker, determines quality and management designs the process. Therefore, management, according to Juran and other quality gurus, is responsible for 80 percent or more of quality problems. Employees making products and dealing with customers do indeed have the greatest influence on quality. However, they are only as good as the system in which they work. Most people want to do a good job but if a good worker is thrown into a bad system, the system will win every time. Management is responsible for building two types of quality into the processing system:

1. Operational quality—Processes must be driven by prevention and efficiently promote the production and delivery of high-quality products and services.
2. Environmental quality—Individuals should work in a setting that supports and enhances quality performance.

With these environmental conditions, plus sufficient skills and logical work systems, individuals can do what they prefer to do—take pride in their work, produce high-quality products and services, and make their company efficient and competitive.

SUMMARY

Quality does not involve just the finished product of a processing unit, such as styrene monomer or polyethylene. Quality involves every department and every person in a company and even people not part of the company, such as the company's raw material suppliers. Quality involves administrative clerks, shipping clerks, company sales representatives, warehouse personnel, engineers, safety and health personnel, corporate management—everyone and everything. Today, businesses constantly analyze the *efficiency* of their operations. You can't be efficient without quality, which depends on systems, which to operate efficiently, depend on a quality improvement process. And a quality improvement process will be ineffective without an understanding of some basic quality concepts, such as worker empowerment, continuous improvement, the cost of quality, etc.

Quality doesn't just happen. A system(s) makes quality happen. Prevention is one system that makes quality happen. If the technician can prevent things from going wrong they can make on-spec product at minimal cost. Control charts are one of the most important prevention tools a technician has at their service and reading and understanding control charts is one of the most important skills a technician can develop. Control charts are valuable tools for improving product quality and making processes predictable.

REVIEW QUESTIONS

1. Explain why quality is dependent on all operational aspects of a company.

2. Explain why quality is not the responsibility of the quality department or quality team.

3. How does a worker in any business get involved in quality?

4. Why is the process technician most responsible for making quality happen on their process unit?

5. Explain why it is important to empower employees.

6. Explain why process focus is important to a technician.

7. List the costs of quality.

8. The _____ is one of the most important prevention tools a technician has at his service.

9. List three advantages of using control charts.

10. What is statistical process control (SPC)?

11. List four things that make up a process or system.

12. Describe the two types of quality management is responsible for incorporating into a system.

CHAPTER 5

Process Economics

Learning Objectives

After completing this chapter, you should be able to

- *Explain how economics affects the productivity of a process unit.*

- *Explain why it is important that technicians understand the economics of their unit.*

- *Describe the characteristics of a productive technician.*

- *Discuss several ways to increase boiler efficiency.*

- *Describe how technicians can increase furnace efficiency.*

- *List several ways that steam can be wasted on a process unit.*

- *Discuss why it is important for a process unit to avoid heat losses.*

- *Calculate the heat loss from uninsulated piping.*

INTRODUCTION

Economics is the production, distribution, and consumption of goods and services. In this chapter we will look at the production side of economics and how a process technician can add to the bottom line of their process unit. If John Smith is a process technician his standard of living, either short term or long term, will be determined by the productivity of his process unit. Competition in the processing industry is global. American workers are competing against workers overseas who make $7 per hour and receive few, if any medical benefits. Today, more than ever, it is very important for John Smith to understand the economics of his unit and become a productivity asset.

It is a well-known fact that process technicians have good paying jobs with great benefits. How much technicians are paid determines their standard of living. Whether they get a raise or how big their raise will be is determined by two factors: (1) the supply and demand for their product and (2) the productivity of their process unit. The skills and initiative of the technicians that run the unit will determine the unit's productivity. If productivity doesn't increase, pay raises are minimal, and benefits may be cut. If productivity remains stagnant over several years, there is the chance that the process unit will be shutdown and its technicians laid off, or the plant or unit may be sold.

THE PRODUCTIVE PROCESS TECHNICIAN

It is important for technicians today to realize that petrochemicals, refining and several other process industries are in a commodity market—a market of mass production for mass consumption. Commodity market items, in most instances, have marginal profit margins, not high profit margins. The fact that profits are relatively marginal makes it necessary for technicians to understand the importance of sticking to a unit's budget in order for it to be profitable. Put simply, a processing unit is like a store. A store stays open only as long as it makes money. Many corporations today require unit managers to give presentations of unit business goals to their technicians for the coming year and tie in incentive pay and bonuses based on attaining those goals. These goals are directly related to the economics and productivity of the process unit.

Technicians are expected to be assets, not liabilities. Technicians are productivity assets if they are conscientious; understand their roles; know how their equipment works; make good routine checks of all their vessels, lines, and equipment; and know how to handle emergencies and problems. Alert technicians produce on-specification product. Technicians understand that off-specification materials wastes time, energy, and often product, which all cuts into unit profitability. They understand that they can prevent waste in their area by the things they do and by the things they do not do. Productive technicians run their equipment so as not to waste

- Fuel
- Steam
- Compressed air
- Electricity and water
- Product by running off specification
- Raw materials
- Manufactured supplies
- Time and money that would be required for repair of carelessly handled equipment (see Figure 5-1)

Saving $10 or $20 a day may not seem like much in the operation of a large unit, but one dollar saved by cutting costs may be worth approximately $24 generated by sale of product. The $24 figure will vary from plant to plant, being higher in some plants and lower in others. Using the $24 value, cutting costs at the rate of $20 a day is the same as

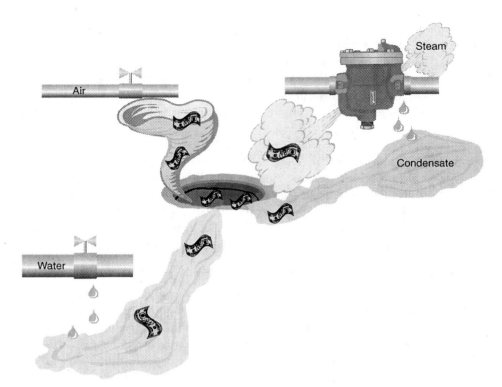

Figure 5-1 Wasting Utilities

increasing sales by $480 a day, or $175,200 a year. Think of how much money could be saved in a large plant with 250 technicians if each technician made an improvement that saved $20 a day!

It costs less money to maintain good equipment than to repair it. The average repair cost of a large centrifugal pump, of which there are hundreds on a large processing unit, is in the thousands of dollars. Good technicians monitor their equipment closely and perform preventive maintenance regularly to keep it in good condition so that it operates efficiently. Technicians should realize when their equipment is not running right and either fix it themselves or write a maintenance work order to fix it before the damage becomes more extensive and thus more expensive.

Technician alertness and knowledge about their equipment prevent trouble from occurring but the pay-off is that trouble-free equipment (1) makes the technician's job easier and (2) adds to the unit bottom line. Checking equipment daily doesn't take nearly as much time or effort as does babysitting defective equipment. Key instrumentation on their equipment informs an alert technician when equipment is having problems.

FUEL CONSERVATION

Huge amounts of fuel are consumed firing furnaces, producing steam in boilers, and producing power with pumps and electric motors. Conserving heat, steam, or electric power means burning less fuel. The following paragraphs will explore how technicians should seek to conserve fuel and improve their unit's profitability.

Furnace Economics

Energy is probably the greatest expense for the process industry. Furnaces burn large amounts of fuel. Assume natural gas costs $5.00 per one million British Thermal Units, or MBTU. A typical furnace yielding 60 million British Thermal Units (BTU) an hour might burn over $300 worth of fuel in an hour, which amounts to $7,200 in a 24-hour day. A fuel cost of $7,200 a day bills out at approximately $219,600 a month. A 1 percent saving of fuel is a saving of about $2,196 a month.

A productive technician understands how their furnace works and keeps furnace efficiency high to keep fuel costs low. Furnaces mix fuel with air and burn the mixture. Complete combustion of this mixture requires a proper balance of air to fuel. Unburned fuel exiting the furnace stack is wasted and reduces efficiency. To ensure that all fuel is burned, the furnace must admit a slight excess of air but if too much excess air is admitted, the excess air results in a higher fuel usage. Fuel is wasted heating excess air that is dumped to the environment.

A knowledgeable technician uses instruments and analyzers to determine and control the smallest amount of excess air required to burn all of the fuel. An oxygen analyzer in the stack can analyze the amount of oxygen in the flue gas. If there is too much oxygen, there is too much excess air. On the other hand, if there is no oxygen in the flue gas, all the oxygen must have been used burning the fuel. Without excess air in the furnace, some of the fuel is probably not being burned, which wastes fuel and may lead to environmental compliance problems. The operating directions for a furnace usually indicate the percentage of oxygen that must be in the flue gas to show that all the fuel is being burned, which is usually between 1 and 5 percent.

Let's look at the chemistry of a burning fuel, such as methane. The chemical formula for the burning of methane is

$$CH_4 + 2O_2 \rightarrow CO_2 + 2H_2O + \text{heat (BTU)}$$

As we can see, the fuel and oxygen react to form carbon dioxide and water and heat. To form carbon dioxide each atom of carbon reacts with two atoms of oxygen. If there is not enough oxygen, the carbon combines with just one atom of oxygen and makes carbon monoxide instead of carbon dioxide. Carbon monoxide comes from a flame starved for oxygen, which results in incomplete combustion. The heat released by the combustion of fuel is measured in **British thermal units (BTU)**, the amount of heat required to raise the temperature of one pound of water 1°F. The complete combustion of one pound of carbon releases 14,100 BTU. When one pound of carbon is oxidized to carbon monoxide, only 4,000 BTU are released. So, incomplete combustion severely reduces the production of heat (BTU) in the preceding formula by about 70 percent. In addition, incomplete combustion can result in unburned fuel that poses a fire or explosion hazard in the burner box.

Furnace Air Control. Controlling the flow of air through the furnace is also important in controlling furnace pressure. If the stack damper is closed too much so that flue gas can't enter the stack fast enough, the pressure in the furnace becomes higher than the pressure of the outside atmosphere. Now the furnace is operating at a positive pressure, which may damage the furnace roof or walls. To minimize furnace repairs, the dampers and air inlets must be adjusted to prevent the occurrence of positive pressure. The furnace should instead be operated at a slightly negative pressure.

Another consideration in furnace air control is airflow affecting the burner flame. If air is flowing fast (a strong draft), it may change the direction of the flame and cause flame impingement on furnace tubes. **Flame impingement** makes hot spots where the flame touches the tubes and the tube may be oxidized at that point. Oxidation eventually results in a hole in the tube and reduces the normal life of the tube.

Flame impingement can cause the following expensive results:

- Tube oxidation (destruction)
- Product breakdown
- Lowered product flow rates
- Increased fuel requirements
- Reduced production
- Costly shutdown for decoking (removal of internal fouling of the tubes)

Besides damaging tubes, flame impingement may cause localized overheating of the material flowing through the tubes. This breaks chemical bonds and creates substances not normal to the fluid in the stream. It may also cause black carbonaceous solids in the fluid and/or a product that doesn't meet specifications. Over heating also causes a coating of coke or carbon to form on the inside of the tube. This coating reduces the rate of flow through the tubes and reduces the efficiency of the furnace by acting as a layer of insulation that gradually

- Reduces the transfer heat through the tube wall.
- Increases the amount of fuel needed to heat the product to the correct final temperature.
- Reduces the efficiency of the furnace.

Assume that the furnace has been limited by flame impingement. A 10 percent reduced rate of flow cuts back the operating capacity of the entire unit by 10 percent. If the unit earns $50,000 a day when operating at full capacity, a 10 percent reduced rate of flow results in a loss of at least $5,000 a day. No refinery can afford daily losses of $5,000 from reduced operating capacity for very long. However, shutdown of the unit to repair the damaged tubes and restore 100 percent production and efficiency may result in a loss of $100,000 or $200,000 in production. Management must balance the high cost of unit shutdown for repairs against the accumulating loss from reduced throughput.

Fuels and Heat Values. Different fuels release different amounts of heat when burned. Table 5-1 reveals the heat values of four hydrocarbon fuels and the air required for complete combustion. As you look down the table you see the fuels that release more BTU also require more air to burn.

It is not uncommon during furnace operation for the fuel gas composition to change. When this happens the amount of heat generated by the furnace will change and it becomes necessary for the technician to adjust the fuel flow to maintain the furnace at the desired temperature. This, in turn, will require adjusting the air flow for complete combustion or

Table 5-1   Heat Values of Hydrocarbon Fuels

Fuel	BTUs/Ft.^{3}	Air Required (Ft^3) per Cu. Ft. Fuel
Methane	909	9.54
Ethane	1617	16.70
Propane	2316	23.86
Butane	3025	30.54

to prevent wasting heat on excess air. Assume the furnace technician receives a call that 15 percent more propane is being added to the plant fuel gas system because enough methane is not available. This means the heating value of the fuel gas (BTU) will increase and the technician will have to cut back on the fuel gas sent to the burners. A good technician will make proper furnace adjustments when their fuel or fuel composition changes.

Boiler Economics

If properly cared for, steam boilers provide safe and efficient heat transfer for a wide variety of processes. However, a poorly maintained boiler gobbles up energy dollars at an astounding rate and results in production, safety, and maintenance problems. Energy is almost always the biggest expense associated with steam production (see Figure 5-2). Consider a boiler system that produces 25,000 to 40,000 pounds of steam per hour. Using natural gas for fuel ($5/MBTU), a conservative fuel bill for the boiler would be $100,000 to $150,000 a month. A steam boiler system essentially consists of converting fuel to heat energy followed by a series of heat exchange processes. Anything that interferes with these steps results in an energy loss. To minimize boiler expenses, operators can do several things, such as the following:

- Ensure burners are tuned
- Eliminate scale
- Minimize corrosion
- Maximize condensate return

A typical boiler that generates 40,000 pounds of steam an hour may cost $150,000 a month to operate. A gain of just 1 percent in efficiency saves 1 percent of $150,000 or $1,500 a month. It is possible for a conscientious technician to produce a 3 percent increase in boiler efficiency, which saves $4,500 a month. Flame impingement in a boiler oxidizes tubes just as it does in a furnace. Efficiency can be improved in boilers as well as furnaces by coke removal and by proper control of fuel and excess air.

Boiler Feed Water and Scale. *Boiler feed water (BFW)* added to the boiler always has some

solids dissolved in it. When the water is evaporated to make steam, the dissolved solids do not evaporate but remain in the water and gain in proportion as more water is evaporated. Scale deposits are formed on the insides of the tubes when the concentration of dissolved solid material gets too high. Scale, like coke in the furnace tubes, insulates the tubes and retards heat transfer, which increases fuel usage. Depending on boiler design, a 1/4-inch thick accumulation of calcium scale or iron oxide can reduce a boiler's efficiency by 40 percent. (See Figure 5-3 for an example of the effect of heat reduction by an increasing level of

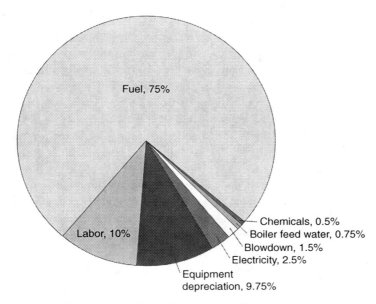

Figure 5-2 Typical Cost Factors for 1,000 pounds of Steam

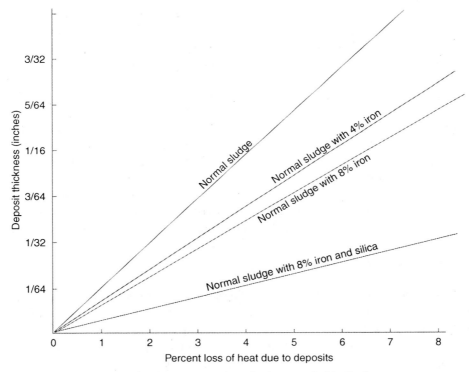

Figure 5-3 Percent Reduction of Heat Transfer by Iron Oxide Scale

iron oxide scale.) In addition, the scale keeps the BFW from carrying the heat away from the metal surfaces and metal deterioration occurs more quickly. The scale accumulation also plugs piping connections to level switches, gauge glasses, and other safety devices that could lead to equipment damage and/or safety hazards.

If scale deposits get thick enough to seriously interfere with the flow of the water and the efficiency of the boiler, the boiler must be shut down for scale removal. Shutting down a boiler costs time and money. In rare cases where there is only one boiler for a plant, the whole plant may shut down or expensive package boilers be brought in while the plant boiler is repaired. Scale formation must be kept as low as possible to reduce boiler down time. Performing **blowdown**, removing water from the boiler, when the concentration of solids becomes too high retards scale formation. The water that is blown down is replaced with make-up water. The efficiency of the boiler is lowered by blowdown because water and heat are wasted.

The only way to reduce blowdown and boiler shutdown to a minimum is to keep the concentration of solids in BFW as low as possible. Boiler feed water is usually treated before use to remove dissolved solids. It is the boiler technician's job to ensure that the concentration of solids in the feed water entering the boiler is not too high.

Air Control in Steam Systems. Where possible, technicians should eliminate air in their steam systems. Air, in varying amounts, will always be present in steam. Air is one of the best insulators known. The fiberglass in fiberglass insulation has virtually no R-value (the resistance a material has to heat flow). Instead, it is the air films trapped by the spun fiberglass threads that create the insulation value. Just as in fiberglass, air in steam systems creates blankets or films inside heat exchangers that result in significant reductions in exchanger efficiency. As little as 0.10 percent air in steam inside a heat exchanger will reduce heat transfer efficiency by 50 percent. Air enters steam systems when the steam condenses and forms a vacuum, which draws air into packing glands, pipe threads, gauge glass washers, etc. To avoid this problem. air vents should be installed at appropriate locations in the steam distribution system to remove air.

Air in steam lowers the surface temperatures in heat exchangers and less heat than expected will be transferred. The temperatures on heat exchanger surfaces are lower than expected due to the saturation temperatures in the **steam tables**, documents that provide information about the physical properties of steam. This situation can be described with Dalton's Law of Partial Pressure. Dalton's Law states that the total pressure exerted by a mixture of gases is made up of the sum of the partial pressures of the components in the mixture. The partial pressure is the pressure exerted by each component as if it was occupying the same volume of the mixture. Reducing the volume of steam reduces the effective steam pressure, and hence, heat. Here is an example of Dalton's Law.

The pressure in a steam/air mixture made up of 3 parts steam and 1 part air is 75.7 pounds per square inch absolute (psia). Steam is responsible for three-fourths of the pressure or 56.7 psia. What is important is that since the steam has an effective pressure of only 56.7 psia, the steam/air mixture will have a temperature of 282°F rather than the expected saturation temperature of 305°F for 75-pound steam. This has a major effect on the heat transfer capability of a heat exchanger since the exchanger is running about 25°F cooler than its normal operating condition because of air in the steam system.

Steam Condensate. *Condensate* is water of almost distilled water quality that contains a substantial amount of heat, both latent and sensible, which was added to it by burning fuel.

For these reasons, as much as possible condensate should be returned to the boiler for reuse. An increase in the amount of condensate return still containing sensible and latent heat will result in fuel savings. In addition, clean, hot condensate contains practically no scale-forming impurities and only small quantities of oxygen, thus boiler feed water treatment chemical usage is reduced, saving more money.

Corrosion Control in Steam and Condensate Lines. Corrosion of steam and condensate lines is a costly problem facing processing plants and costs about $1 billion annually in the United States. In afterboiler corrosion, the return of corrosion products (metal oxides) to the boiler water via the condensate line is often overlooked. These metal oxides have low solubilities and can deposit on boiler heat transfer surfaces. The two major causes of corrosion in steam and condensate lines is (1) oxygen, which results in pitting, and (2) low pH, which results in the general thinning of piping. There are various monitoring methods for corrosion control based on BFW treatment. Some of the monitoring methods are a hydrazine test for oxygen concentration, pH test for neutralizing inhibitors, and corrosion coupons to determine the efficiency of filming inhibitors.

Electric Motor and Pump Economics
Electrical motors account for two-thirds of the U.S. industrial usage of electricity. Pumping systems account for an estimated 25 percent of that electrical usage. When the economics of using an electric motor are considered it becomes apparent that purchase price is miniscule compared to overall operating cost. For a typical motor operated at full load 24 hours a day the purchase price is 2 percent, rewind costs 0.7 percent, and electricity consumption is 97.3 percent of the total lifetime cost. A typical 50 horsepower motor costs $25,000 to operate continuously for one year at full load ($0.07 per kilowatt hour [kWh]). Technicians should discuss with engineering and maintenance personnel to determine at what load the motors for their area of responsibility can operate and do their job most efficiently.

Poor maintenance of equipment wastes money for repairs and replacements. Equipment that limps along consumes too much of a technician's attention and time. If a pump's bearings are not getting enough lubrication, they may freeze up and the pump must be shut down for maintenance. A centrifugal pump is partially lubricated by the liquid it pumps. A centrifugal pump that loses its prime so that no liquid is running through it may be seriously damaged by lack of lubrication, overheating, and vibration requiring costly repairs of hundreds or thousands of dollars or even requiring a new pump altogether. A new general-purpose pump costs $10,000 to $20,000; specialized pumps for hot oil cost up to $50,000. If a pump has to be primed too often, it should be taken out of service for repair before it is seriously damaged. At the first sign of trouble, a conscientious technician either fixes a problem or notifies someone who can fix it.

STEAM ISN'T CHEAP
It is easy to waste many thousands of pounds of steam in a few seconds because of its many uses in a processing plant. Wasting steam wastes fuel and wasting fuel wastes money. Fuel costs may vary from $5 to $7 to produce one million BTU of heat. If fuel cost is $5 per million BTU, steam costs about $6 per 1,000 pounds. No matter the cost of 1,000 pounds of steam, wasting steam wastes fuel, and wasting fuel wastes money. The man who is in the best position to ensure that steam is not wasted is the man using the steam—the technician.

Steam Turbines

A technician can reduce steam waste in steam turbines. A steam turbine produces power from high-pressure steam. Steam flows from areas of high pressure to areas of low pressure but if the exhaust pressure of a steam turbine is the same as the inlet pressure, steam won't flow. If steam does not flow through the turbine, no power is produced. Lowering the turbine's exhaust pressure increases the amount of work a turbine can produce from a given amount of steam. The more the pressure drops as steam passes through the turbine, the more power the turbine produces. For this reason, the pressure of steam leaving the turbine should be as low as possible.

On some steam turbines a condenser cools and condenses the steam at the turbine exhaust. This condensation decreases the volume of the steam and reduces the exhaust pressure to a point below atmospheric pressure (a partial vacuum). Thus, condensers increase turbine efficiency by creating a vacuum in the turbine exhaust line. However, when the exhaust line is under vacuum, if air leaks into it a loss of vacuum will occur and increase the exhaust pressure. Air leaks in the exhaust line prevent the production of a partial vacuum as the steam passes through the turbine and the power output of the turbine is reduced. Air in the steam cannot be condensed. A large air leak can flood the vacuum system with air, reduce the efficiency of the condenser, and increase the exhaust pressure, which will lower the power output of the turbine still more. To bring the power output back up, more high-pressure steam must be fed into the turbine, wasting energy.

Plugs, Steam Tracing, and Steam Traps

Some liquids in a line or tank may become too thick to flow smoothly if its temperature gets too cool. If the liquid gets cold enough, it may harden and plug the line because dissolved waxes drop out of solution and form solids. A plug in a process line may affect the operation of the whole unit. Steam is often fed temporarily into process lines to purge heavy oils or wax that may form plugs or to steam out plugs that have already formed.

Tanks and other large pieces of equipment are often steam-jacketed and lines are often steam-traced to keep them warm. A line is **steam-traced** by running a separate steam line along side it or around it. Materials that can be harmed by direct contact with steam can be safely warmed by steam tracing. If the tracing steam rate is too low, the material may not be warmed enough and a plug may still form in the line. If the tracing-steam rate is too high and the material is warmed more than necessary, steam will be wasted. It takes skill and experience to maintain an exact balance of effectiveness and economy in setting the steam for tracing. During winter when freezes are expected many lines may be steam traced; however, when the weather becomes warm enough that there is no danger of a freeze, tracer steam should be turned off to conserve steam.

Steam traps, devices used to remove condensate from a steam system, can also waste steam. If the valve seat in a steam trap is worn, or if the closing mechanism is broken, the trap may leak steam continuously. Leakage from steam traps, leaky valves, and piping adds up. An alert technician watches for steam leaks and has them repaired as soon as possible. A few steam leaks can waste as much as 1,000 pounds of steam an hour. No matter where steam is used or for what purpose, it is important to use no more than is necessary to do the job.

AVOID HEAT LOSS

Suppose we consider the boiler unit to be like a warehouse. The material stored there is heat energy and the piping leaving the boiler (steam header) is like a conveyer belt that carries material (heat energy) to points of use throughout the plant. Now, if material kept falling off this conveyer belt, someone would do something about it because they were losing expensive material. Once this material (heat energy) falls off, you can't get it back. It's gone. Therefore, you must keep it from falling off. This requires insulation for steam and condensate piping. Heat is energy. To save energy, processes must minimize heat losses. Whenever process or storage equipment is hotter than the air around it, it gives off heat. The hotter the equipment, the more heat it gives off. Equipment must be kept just hot enough to give good liquid flow and to operate correctly. Keeping the temperature higher than needed is wasteful.

Suppose an uninsulated 120,000-barrel tank is normally kept at 1,500°F. Letting the temperature of the tank increase to 1,750°F increases the heating cost about $1,200 per month. When equipment must be kept at temperatures higher than the air around it, it may be possible to save money by insulating it to decrease the heat loss. Whenever the heat loss costs more than insulating the equipment, the equipment should be insulated to save money.

Insulation can typically reduce line and valve energy losses by 90 percent and help insure proper steam pressures and temperatures of plant equipment. In general, any surface over 120°F should be insulated. At times when insulation on some lines becomes damaged or is removed for repairs, it is often never replaced. To prevent situations like these, it is important to include action plans to manage steam and condensate line insulation as part of a unit's plan to control energy usage. In Table 5-2 are some examples of uninsulated line heat losses and costs associated with these losses at a natural gas price of $4/MBTU.

Remember losses in Table 5-2 are stated in *per foot of pipe*. For example, if there is a total of 10 feet of uninsulated, 12-inch, 875-pound steam line, the heat loss is 2,708 MBTU per year, which amounts to $5,040. In addition, lines are not the only thing an alert technician should be monitoring. A typical 6-inch gate valve may have over 6 square feet of surface area from which to radiate heat to the environment. Technicians should conduct a survey of their area's steam distribution and condensate lines and valves and look for ways to save energy.

Table 5-2   Heat Loss/Cost for Uninsulated Pipe

Pound Steam	Line Size	Heat Loss	Cost
875	6″	67.7 MBTU/yr	$270
875	12″	270.8 MBTU/yr	$504
165	6″	18.3 MBTU/yr	$ 73
165	12″	33.3 MBTU/yr	$133
65	18″	27.8 MBTU/yr	$111
65	30″	53.5 MBTU/yr	$214

Heat losses are in MBTU a year per foot of uninsulated pipe.
Cost of natural gas is based on a price of $4/MBTU.

Table 5-3   Energy Saving Resulting from a Steam Trap Survey by Yarway

Leak Source	# of Leaks	Lb/Hr	$/Day
Piping leaks	5	—	12.50
Valves	2	260	62.40
Traps	20	34	81.60
		Total	156.50
	52 weeks × $156.50 = $40,690 a year based on a 5 day week		

Yarway, a company that specializes in the installation and repair of steam traps conducted several surveys on the savings obtained by reducing the loss of steam to the environment. The company assumed some typical condition, in this case, steam costs of $5.00 per 1,000 pounds of steam. The potential energy savings was estimated based on the following:

- Leaks—gaskets, pinched tubes, etc.
- Valves—isolation, bypass, etc.
- Traps—for an installed population of 150 steam traps

The energy savings revealed by the surveys is detailed in Table 5-3. The savings is based on an average number of piping leaks, valve leaks, and defective steam traps per week and those leaks/defects being detected and corrected.

RELIEF VALVE LEAKS

If pressure in any part of a process gets too high, a relief valve automatically opens to allow the excess material to flow out before high pressure damages equipment. When pressure gets too high, process materials that vaporize easily may be sent to a *flare*, equipment designed to safely burn excess hydrocarbons. The heat from gases burning in the flare is not recovered for use in the process. Both the material and the heat are completely wasted but this is a deliberate choice made for the sake of safety.

It is important to note that relief valves occasionally leak resulting in additional wasted materials that are accidentally vented to a flare or blowdown pit. To prevent this waste, relief valves should frequently be checked for leaks. As long as nothing is flowing through a relief line, the line has the same temperature as the air around it. Generally, leakage of products from the process through the relief valve can make the relief line hotter or colder than the air around it. A technician can detect this temperature difference. Some leaky valves chatter or whistle and can be detected by these sounds. Changes in the appearance of the flare may also indicate a leaking valve. An alert technician checks for leaks by frequently noticing the appearance of the flare, listening for noisy valves, and feeling the temperature of the relief piping.

AVOID WASTE OF UTILITIES

All our life we have heard that air is free. This is not true in processing industries. The cost of compressed air is the cost of running the compressor. It costs about 13.5 cents to compress 1,000 standard cubic feet of air. In one month, over 4 million cubic feet of air can leak through a 1/4-inch opening. At 13.5 cents for 1000 cubic feet, such a leak wastes $540 worth of air in one month. Water leaks can also be expensive. In one month, 400,000 gallons of water can leak through a 1/4-inch opening. At $1.50 for 1,000 gallons, a water leak of this size wastes about $600 per month.

Observant technicians should monitor all the water and air piping in their area of responsibility, note any leaks, and repair them or have them repaired. True, both air and water are relatively inexpensive, but small leaks can accumulate large costs over time. Utilities should be kept on only while they are needed. Cooling water and other water lines should be turned off when not in use. Electricity should be turned off when not in use. There are hundreds of ways in which air, water, electricity and other utilities can be wasted in a refinery. A good rule is "Turn it off if you're not using it."

SUMMARY

Today, efficient process economics depends on technicians learning the skills and knowledge needed to improve productivity and reduce waste. Technicians are productivity assets if they are conscientious; understand their roles; knows how their equipment works; make good routine checks of all their vessels, lines, and equipment; and know how to handle emergencies and problems.

Huge amounts of fuel are consumed firing furnaces, producing steam, and producing power. Conserving heat, steam, or electric power means burning less fuel. Technicians should seek to conserve fuel to improve their unit's profitability. Energy is a big expense associated with production, thus saving energy increases efficiency.

Good preventive maintenance adds to the bottom line of a process unit. Poor maintenance of equipment wastes money for repairs and replacements. Equipment that limps along consumes too much of a technician's attention and time. The trouble-free operation of equipment depends primarily on process technicians. They pay attention to preventive maintenance and follow correct startup, shutdown, and lubrication procedures to keep their equipment running at peak efficiency.

REVIEW QUESTIONS

1. Write the definition of *economics*.

2. Explain why it is important that technicians understand the economics of their unit.

3. The productivity of a process unit is dependent upon what four factors?

4. Explain how excess air in a furnace wastes fuel.

5. Describe two ways that incomplete combustion is detrimental to furnace operations.

6. How does operating under positive pressure affect a furnace?

7. Discuss how changes in furnace fuel gas affect furnace efficiency.

8. List three ways flame impingement affects the economics of a furnace.

9. The biggest expense associated with steam production is

_____ .

10. _____ is one of the best insulators known, which is why it is important to keep it out of steam systems.

11. List the two major causes of corrosion in steam and condensate lines.

12. State the percentage of an electric motor's total lifetime cost represented by electricity consumption.

13. Explain how lowering the exhaust pressure of a steam turbine affects efficiency.

14. List two purposes of steam tracing.

15. Discuss why it is important for a process unit to avoid heat loss.

16. Describe how a technician could check for leaking relief valves.

CHAPTER 6

Communication

Learning Objectives

After completing this chapter, you should be able to

- *Discuss some problems that hinder good communications.*

- *Describe the importance of feedback in oral communication.*

- *Describe the various written communication documents used by process technicians.*

- *Discuss some of the ways that nonverbal communications are possible.*

- *Explain why good communications is important in the chemical process industry.*

- *Write a clear and concise operating procedure.*

- *List several important guidelines for writing procedures.*

- *Explain the importance of procedure validation.*

INTRODUCTION

The art of getting your message across effectively is a vital part of individual success, plus good communication is the lifeblood of organizations. It takes many forms, such as speaking, writing, and listening, and its purpose is always to convey a message. Good communication or poor communication directly affects productivity, safety, and quality. Correct instructions, procedures, bills of laden, and exchanges of information between shifts are critical for safe and efficient process operations.

The most frequent complaint at any place of business is the *lack of communication*. The complaint is heard so often that it seems to imply that people don't talk to each other, don't e-mail each other, don't write things down, and don't pass on memos. You would think everyone was mute or illiterate but we know that isn't the case. Instead, *everyone is communicating but they are doing a poor job of it*. They are saying things but what they are saying and what is being perceived is not the same thing; they are writing memos and emails but the *message perceived is not the message meant*. Or their body language conveys a different message. Or their rapport is so bad with their coworkers that nothing they say will ever be believed or accepted. Good communication is extremely important and extremely difficult. Maintaining good communication is the greatest frustration of all workers in any business.

This frustration is not uncommon in the processing industry, whether the industry is petrochemicals, mining, or pharmaceuticals. Opportunities abound for poor communication and miscommunication. Following is a partial list of verbal and written communication responsibilities of a process technician. Failure to clearly and precisely communicate on any one of these duties has the potential to create one or more of several outcomes, such as a dangerous situation, loss of profit, customer dissatisfaction, work stoppage, or bad feelings that result in divisiveness and workplace tensions.

- Maintain logbooks
- Order replacement materials and supplies
- Write clear instructions
- Complete and route forms and bills of laden
- Communicate with maintenance, lab, or technical services
- Fill out checklists
- Receive shift change information when making relief

THE ELEMENTS OF GOOD COMMUNICATION

Communication requires three elements: a message sender, the message, and a message receiver. It seems so simple. You (the sender) say something (the message) to someone (the message receiver) and communication is complete. These three elements can create an amazing number of communication problems. Ask anyone who has been married for 20 years and they will readily admit that communication problems in their marriage still exist.

There are several reasons for poor communication. The person delivering the message may be someone we don't like so we focus on our dislike instead of the message. The form of the message may hinder communication, such as the sender's handwriting is difficult to read or the sender leaves a long-winded, recorded telephone message. If the receiver has little patience for such things, then little or no communication has taken place. The receiver may erect barriers that block or filter the message. If the receiver feels strongly about a topic or is biased in one way or another, these feelings will affect their understanding of the message. For example, a rancher and a vegetarian will express very different meanings about the nutritional facts on a package of sausage.

COMMUNICATION METHODS FOR THE PROCESS TECHNICIAN

Good communications are essential to the safe and efficient operation of a processing unit. Technicians may be involved in scheduling for turnarounds or other non-routine duties, arranging for equipment repairs with maintenance, and discussing safety and environmental issues, quality issues, or process improvement issues. A technician who cannot communicate effectively will eventually find him or herself politely ignored by other technicians and management. Good communication, though seemingly simple, is not simple at all because too much is taken for granted. Good communication requires as much work and concentration as being a good ball player.

The three methods of communication constantly used by process technicians are *oral*, *written*, and *nonverbal*. Each of these methods is appropriate depending on the situation at hand.

Oral Communication

People often deliver a spoken message that is frequently misunderstood because

- They are not clear about what they want to say.
- They are vague in how they say it.
- Their body language contradicts their message.
- The listener has decided in advance what the message is regardless of what the speaker is saying.

A useful way to avoid misinterpretation is to rehearse your message aloud or get the listener to restate your message after you have given it. In the latter way, you use their feedback to avoid misinterpretations and errors.

Feedback. Feedback is essential to good communication because it (1) verifies the other person has understood your message and (2) allows you to react to what the listener has said and done, yielding further information to the speaker.

A few examples of important daily oral communications include the following:

- Shift meetings and safety meetings
- Shift change discussions that convey current operating information to the relief technician
- Telephone conversations with the quality control laboratory, maintenance or utilities

It is impossible to run a processing unit without direct verbal communication by everyone involved. Poor verbal communication is frequently a contributing cause to the development of serious problems.

Listening. The two-way nature of communication that ensures both sides understand each other is widely ignored. People generally want to be heard, but they don't like to listen. Listening is important since how you listen not only conveys meaning to the speaker but also ensures you have correctly received their message. If you listen carefully you will be able to note the feelings behind the other person's response. Listening is just as important as speaking. Listening is 50 percent of any oral communication. To facilitate listening skills you should test your understanding by rephrasing statements and repeating them back to the speaker to ensure you understand the speaker's point. Some important points to remember about listening are

- Listening intently inspires confidence in the speaker.
- Misunderstandings are caused by hearing only what you want to hear.
- Constant interruptions make it difficult for the speaker to clearly present their point of view.
- What you are told should be regarded as trustworthy until proven otherwise.

Written Communication

Written communication involves the printed or written word. It is more formal than oral communication and may include anything from a brief note left for the oncoming shift to detailed logs and reports completed on a daily basis. Because written communication does not lend itself to immediate feedback, some follow up may be needed to ensure that the message receiver understands the message.

Documents that are well written and easy to understand are composed by people who have clarified their thoughts and probably rewritten the document several times. Your document should not be a work of literature. Its purpose is to convey information. The key to writing any document whether it is a business letter, a work procedure, or a training method is to write clearly and concisely. Use simple words and write straight to the point. If you are concise, you will reinforce the clarity of your report. Never use two words where one will do. Keep your audience in mind and write to the level of understanding of your audience. As an example, don't use engineering terminology on operators. Use the following guidelines:

- Short words (*said* rather than *announced*)
- Short sentences
- Active verbs (*open* versus *should be opened*)
- Avoid jargon
- Write the way you speak
- Revise when you have finished
- Cut unneeded words and sentences

Since your document is going to be read by others it should be accurate and well formatted so that it is easy to follow. Depending on the type of document you are writing—for instance an operating procedure—you may want to use numbered paragraphs to make following it easier. That also keeps important points separate. Use headings for changes of subject and include subheadings if necessary. When you are done, get several colleagues to proof read it for clarity and accuracy.

The process technician uses written communications in many applications, some of which are

- Memos and notes
- Charts and graphs
- Meeting minutes

- Reports
- Training manuals
- Written instructions (daily operating instructions and night orders)
- E-mail

Nonverbal Communication

Nonverbal communication includes body language, gestures, signals and alarms, and the examples that we set for others. The signals that the process technician might use to communicate with a crane operator while moving heavy equipment or the engineer running the switch engine used to move railroad cars around are examples of nonverbal communication techniques frequently used by the technician.

Body language is a common form of communication between many species. If you have dogs or cats, you are probably familiar with their body language: the cocked ear, the wagging tail, and facial expression. Humans have a body language with a range of unconscious physical movements that can either strengthen or damage communication. Even sitting completely still, people may be communicating their feelings. We have all evidenced this form of communication with family members or a boy or girlfriend. They don't have to say a word to let us know they are angry. Sitting very still with their arms rigid across their chest and lips pressed tight speaks just as effectively and loudly as do words.

Because of its subtlety body language can be difficult to read and control, but it is still recognized as a reliable communicator of feelings. A few key points to remember about body language are that whenever possible when standing with people, leave a personal space of three feet between you and them because crowding too close creates an uncomfortable feeling. Also, stances convey feelings. Standing with hands on hips and a big smile while directly facing the person you are talking to conveys determination, interest, and confidence. A direct unsmiling gaze and hands at the side indicate you are paying attention but have neutral feelings. If your body is turned slightly away and your gaze is evasive, this indicates a negative attitude.

Gestures, together with other nonverbal communication, such as posture and facial expressions, are an important part of body language. All skilled public speakers use gestures for emphasis. Supportive gestures, such as making eye contact and nodding while someone else is speaking, are a way of showing support or empathy. Other familiar gestures that connote feeling are

- Raised eyebrows indicate interest.
- Making eye contact and leaning toward the speaker show readiness to assist the speaker.
- A hand around the throat and/or the around the body indicate the listener needs reassurance.
- Closed eyes and nose pinching reveal inner confusion and conflict.

Communication Filters

Communication filters are perceptions that the receiver uses to interpret the message received. Filters may exist because the individual receiving the message has a lower knowledge level, bias about the subject, or special points of view. Such filtering can change a message in ways that the sender never intended or even block communications completely. A good example of biases and filters affecting communication is a democrat trying to convince a republican to become a democrat.

IMPORTANT PROCESS UNIT DOCUMENTS

Most process plants will have a collection of documents related to the technology and operations of their processes that will include technical information, operating guidelines, safety concerns, and product requirements. They also include data and records of day-to-day plant operation. In some cases, this information is required for regulatory compliance or for quality certification programs such as the International Organization for Standardization's standards (ISO 9001-2000). The following paragraphs list the names and descriptions of some of these documents that operating technicians will become familiar with during their career (see also Table 6-1). The documents or manuals should be available both as hard copies and on-line electronic copies.

Daily operating instructions (DOI) are normally written by the unit superintendent or operating engineer and are located in the control room. The instructions include information that technicians need to know about product scheduling, daily operating objectives or targets, special operating instructions, maintenance schedules, etc. The DOI book should be read by all operators at the beginning of each shift or read to the crew by the lead supervisor. Another term sometimes used for this book is the *daily order book (DOB)*. The DOI is an important document that may be used in legal situations (discipline, injury, etc.) and should be protected and maintained in good order.

The *operator logbook* is located in the control room and maintained by process technicians. There may be more than one. There may be a board operator logbook and a logbook for each outside area, a maintenance logbook, and a permit logbook. Most logbooks are the written passed-down information for an area of responsibility and include all information pertinent to the safe and efficient operation of that area. Safety hazards, equipment problems,

Table 6-1   Important Process Unit Documents

Process Unit Documents
Daily operating instructions (DOI)
Operator logbooks
Product specification sheets
Standard operating conditions (SOCs)
Operating procedures manual
Unit technical manual
Safety manual
Material safety data sheets (MSDS)
Daily operation records
Work requests
Standard operating procedures (SOPs)

operational problems, and operational adjustments made during the shift are entered into the operator logbook. The information entered during a 12-hour shift should be an accurate account of what happened during the shift. The operator logbook should be read by all relieving operators at the beginning of each shift. Information recorded in the log should be accurate and detailed since these records provide a history of operations, which may be useful when troubleshooting problems. They could also become legal documents in the event of an accident investigation, an Occupational Safety and Health Administration (OSHA) audit, or lawsuit.

Product specification sheets provide information about the products produced by the unit. If the unit produces different grades of material, or completely different products (as many batch units do), each grade or product will have its specification sheet. Specification sheets list the properties used to judge product quality and the acceptable values for on-specification product. Specification sheets are contained in the specification book in the control room. If the unit is an ISO-9001-2000 certified unit, each specification sheet is numbered and treated as a controlled document. See Table 6-2 for an example of a specification sheet for diesel fuel oils. Specification sheets are referred to when filling out a ***certificate of analysis*** for a unit product. For example, if the unit loads a truck with styrene monomer, the laboratory analysis of the truck sample is compared to the specification sheet and then recorded on the certificate of analysis. This is a way to prevent off-specification product from being shipped.

Standard operating conditions (SOCs) are the goals and acceptable ranges for the controllable variables (temperature, pressure, flow, level, composition) of the processing unit. Operating within these ranges is necessary to produce on-specification product. A complete set of SOCs should give the goal and ranges for the process variables, plus they should summarize the consequences of not running the unit at these conditions. The SOCs are kept in the control room and may exist as a separate document or be incorporated in the operating procedures.

The ***operating procedures manual*** contains detailed instructions for properly operating the unit. It will include startup, shutdown, and emergency procedures, plus normal operating procedures. It may also include safety information. A copy of this manual should be available in the control room. Several OSHA standards require accurate and up-to-date operating procedures.

The ***unit technical manual*** contains detailed information about the unit process and equipment. A copy will be located in the control room. It will include a process description, the process chemistry, unit limitations, information about major equipment (construction material, stress rating, dimensions, etc.), and a summary of the unit's instrument control strategy.

The ***safety manual*** will contain detailed information about unit safety procedures, site safety policies, and safety standard operating procedures (SOPs). The operator should be familiar with all safety procedures that apply to their operating area.

Material safety data sheets (MSDS) provide information concerning the hazards of the chemicals present in the unit. These are a collection of information sheets provided by the suppliers of the chemicals to the unit, plus sheets created by the unit to describe the hazards of its feed, intermediate, and final product(s). They will provide data concerning

Table 6-2   Partial Specification Sheet for Diesel Fuel Oils

ASTM D 975 Requirements for Diesel Fuel Oils						
Property	**Test Method**	**Low Sulfur No. 1.D**	**No. 1.D**	**Low Sulfur No. 2.D**	**No. 2.D**	**No. 4.D**
Flash point, °C, min	D93	38	38	52	52	55
Water and sediment, % vol, max	D 2709 D 1796	0.05	0.05	0.05	0.05	0.50
Distillation temperature, °C, 90% vol recovered min max	D86	288	288	282 338	282 338	
Kinematic viscosity, 40°C, cSt min max	D445	1.3 2.4	1.3 2.4	1.9 4.1	1.9 4.1	5.5 24.0
Ash, % mass, max	D482	0.01	0.01	0.01	0.01	0.1
Sulfur, % mass, max	D2622	0.05	0.50	0.05	0.50	2.00
Copper strip corrosion, 3 hr at 50°C, max rating	D130	No. 3	No. 3	No. 3	No. 3	
Cetane number, min	D613	40	40	40	40	30
One of the following: 1) Cetane index, min 2) Aromaticity, % vol, max	D976 D1319	40 35		40 35		
Ramsbottom carbon residue on 10% distillation residue.	D524	0.15	0.15	0.35	0.35	

the toxic, flammable, and reactive characteristics of the chemical materials. The sheets are also a source of information on safety equipment that should be used and special precautions that should be followed when handling these materials. A copy of the MSDS should be available in a binder in the control room.

Daily operation records generated by technicians as they work their shift are also important communication tools. These include routine checklists completed during the shift, data sheets filled in as the technician completes area rounds, and any shift reports that are generated. These are historical records that can prove useful when troubleshooting and updating procedures. These are usually maintained in a file for some period of time before going to storage.

Work requests are used to request service and repair of process equipment. The work request system also provides a record of the completed maintenance work on each unit. Operating personnel are responsible for writing the work requests and prioritizing the maintenance work on a daily basis. Each outside operator is responsible for writing his or her own work requests. All work requests should provide specific written instructions about the work requested, the equipment to be repaired, the priority of the work, and the safety implications of the job. Incomplete information could result in the job not being accepted and scheduled. Maintenance supervisors review work requests and if they consider the request to be incomplete, they may bounce the request back to operations, delaying repairs. Work requests that deal with a safety hazard or potential safety problem should be given a high priority, and as a rule, they are acted on promptly.

Standard operating procedures (SOPs) are procedures that may not relate to the running of the unit to make product. They are plant or unit policies that should be followed without deviation. These may be SOPs that are plant-wide in their scope, such as a hurricane, freeze protection, housekeeping, and security procedures.

MANAGEMENT OF CHANGE

We have already seen that there is a tremendous amount of documentation required on a processing unit. Keeping those documents up to date is critical. Management of changes in processes or procedures is a highly vital element in any process safety program. Many of the catastrophic accidents over the past few decades can be traced, in large part, to a management of change system that was not in place or was dysfunctional.

Too many serious accidents happen because workers were not made aware of the changes in procedures, chemicals, or equipment in their unit. As changes are made to the unit, they must be incorporated in the documents and workers made aware of the changes. Site management must establish and implement written procedures to manage changes except "replacements in kind." The OSHA standard, ***Process Safety Management (PSM)*** requires the work site employer and contract employers to inform and train their affected employees on the changes prior to unit start-up. Process safety information and operating procedures must be updated as necessary.

WRITING OPERATING PROCEDURES

Why is there a procedure manual? Why is management concerned that all procedures be kept up to date? Why has OSHA created regulations requiring procedures to be up-to-date and easily accessible to process technicians? What is the big deal about procedures? Procedures are critical to operations because they

- Reduce the risk of technician errors.
- Reduce the frequency of errors.
- Comply with OSHA's Process Safety Management regulation, 29 CFR 1910.119.

In the past, the task of writing operating procedures was frequently delegated to the shift supervisor or the process engineer, but today the task is often assigned to senior process technicians since they have the operating experience and knowledge to understand and explain the operation of their unit.

A procedure is a list of steps that direct you to start or stop a piece of equipment or a unit. It is a list that says "start here and end here," and when you are done you will have accomplished a certain task. In essence, a procedure tells you the correct way to perform a task. Good procedures must be clearly written, contain accurate information, and include enough detail to be useful to technicians, but not so much detail that they confuse or irritate more experienced technicians. Workers avoid using procedures they consider wordy and confusing. A procedure is not a history of pieces of equipment in a pretentious language. It is a brief document written at a ninth-grade reading level that tells you in very short sentences how to start or shutdown a piece of equipment or process unit. Procedures should be written to eliminate the two basic human error categories: (1) errors of omission (failing to perform a task) and (2) errors of commission (performing a task incorrectly). Procedures can prevent both types of errors through careful wording and visual clues.

Procedures should have a consistent format. A format establishes the major sections of a procedure, the indentation, the font, the use of text boxes, warnings, etc. The same format should be used for all procedures. The advantages of a standardized format are that the familiarity bred by standardization makes it easier to follow the procedure and also easier to find information faster. A generic format might have the following major sections listed in Table 6-3.

Many different formats exist for writing procedures. There is *play script*, *narrative*, *paragraph*, and the most common, *outline* format. By an outline format we mean we use only

Table 6-3   Format Sections

Procedure Format Categories
Title (what this procedure is about)
Scope (where the procedure starts and where it ends)
Responsibilities (who modifies, updates, and verifies the procedure)
Brief description of major pieces of equipment
Personal protective equipment required to perform the procedure
Body of text (the actual steps performed)
References (documents containing information helpful to the procedure user)
Management of change (revision number and brief statement of changes from last revision)

the minimum words and write in short brief sentences. The following is an example of the outline format:

Start feed pump P-105A.

Notice the action verb *start* is bolded. That is deliberate because it draws the user's attention as to what they are required to do. Another thing to do to make following instructions easier is to number each step. Thus the example above becomes:

1. **Start** feed pump P-105A.

Now, if the user has a question about a procedure they can say, "I don't understand step number one". This is much easier than counting bullets or paragraphs.

Guidelines for Procedure Writing

Keep in mind when writing procedures that you are writing for clarity. You want to be 100 percent understood with no questions or confusion remaining after the procedure is written. Good luck, because that is a hard target to hit. To accomplish it, do the following:

1. Use descriptive procedure titles. Good titles tell the technician what equipment or process the procedure covers and what phase of the operation is covered. An example title is *Reformate Unit Start-up* or *Paraxylene Unit No. 2 Emergency Shutdown.*
2. Use the minimum number of words per task and bold or capitalize the action verbs and the receiver of the action. Bolding or caps jump out at us and draw our attention to what must be done and to what piece of equipment.

    Example: **START** pump P-105C.

3. Always write in the active tense. Do this by phrasing each step as a command. (See Table 6-4 for a list of action verbs for procedure writing.)

    Passive tense: The Operator should open valve V-203 two full turns.

    Active tense: **Open** valve **V-203** two (2) full turns.

4. Minimize the number of actions per step. For highly stressful procedures, especially, use only one action per step. The following example (an actual example from a processing unit) has too many actions to be just one step and could be confusing or lead to a mistake in a high-stress situation. The example should be broken down into at least three steps.

    PLACE control valve FIC-203 in manual mode and OPEN the valve 30%. Begin throttling bypass valve V-203 until bypass valve is completely closed. ADJUST FIC-203 setpoint to match manual mode, then PLACE controller in AUTOMATIC mode.

5. Insert text boxes for safety statements or warnings. This is another way of drawing attention to an important piece of information.

Table 6-4 Action Verbs for Procedure Writing

Activate	Count	Insert	Plug	Run	Throttle
Add	Decrease	Inspect	Prepare	Sample	Tighten
Adjust	Deenergize	Install	Pressurize	Select	Transfer
Align	Depress	Isolate	Prevent	Send	Trip
Analyze	Discharge	Label	Pull	Send	Turn
Assign	Disconnect	Latch	Purge	Shift	Turn on
Attach	Drain	Limit	Raise	Shut down	Turn off
Block	Energize	Lock	Realign	Silence	Unlock
Bypass	Ensure	Log	Record	Start	Unplug
Calculate	Enter	Maintain	Reduce	Stop	Update
Calibrate	Evacuate	Monitor	Regulate	Store	Vent
Change	Fill	Notify	Release	Submit	Verify
Check	Flush	Observe	Remove	Supply	Wait
Clear	Guide	Obtain	Repair	Survey	
Clean	Identify	Open	Report	Switch	
Close	Increase	Perform	Resume	Tag	
Confirm	Indicate	Place	Review	Terminate	
Connect	Inject	Plot	Rotate	Test	

> Danger: Contact with this chemical can cause severe chemical burns. Check the MSDS for the proper PPE.

6. Where possible, insert drawings or diagrams. A picture is worth a thousand words.

7. Always identify the vessel or piece of equipment you are writing about. If you begin with "START feed pump P-105A" there is no mistaking which feed pump we are talking about. While still referring to that pump, you can say "ADJUST feed pump flow to 250 GPM." To write "START the pump." without previous reference to a specific pump causes confusion. An operator working with flammable or toxic liquids cannot afford to guess what a particular step in a procedure is telling him or her to do. The procedure should be written so that the operator clearly understands what they are to do.

8. Use simple terms that carry the same meaning throughout the procedure. For instance, *verify*, *check*, *ensure*, *confirm*, and *assure* all mean the same thing. They mean "Look at it and if it isn't right, make it right." If you start the procedure using *verify* or *confirm* then use that word consistently throughout the procedure.

Table 6-5   Visual Cues for Written Procedures

Use all caps to indicate WARNINGS and CAUTIONS
Use bold text to highlight the action verbs in a step.
Provide check-off boxes in front of each step in the instructions. The box would be checked as the step is completed.
Use colored text to highlight special instructions, such as those required by safety procedures or federal regulations.

9. The level of detail at which a procedure is written should be based on the expected skill level of a new worker who may be asked to perform the task. Writing for a new hire who has no operating experience is very different than writing for a new hire who has eight years previous operating experience.
10. As a rule, most plant procedures are written at a ninth-grade level of comprehension. This is not meant as an insult, rather, it is meant as a reminder to *keep things simple*. Your procedure is not meant to be a scholarly dissertation. Use very simple language that avoids large words, slang, jargon, and complex sentences.
11. A document control section helps the technician determine that the procedure they are using is the most recent revision. This section should include the procedure number, revision number, date of revision, copy number, number of pages, the author(s), reviewer(s), and if the procedure is a controlled document.
12. Use visual cues. Visual cues make use of spacing, underlines, colored text, indentations, font styles, text bullets, bold text, and other eye-catching devices to draw the reader's attention to specific sections of the procedure. Table 6-5 presents a list of visual cues.

Procedure Data and Contents

A procedure will only be as useful as the information it contains. Before any writing starts, the needed data should be assembled. This task should involve technicians across two crews on a unit because no crews operate identically, much less all five operators on a single crew. What operators will discover as they sit down and begin drafting a procedure is that most of them perform a task slightly differently from their coworkers. If the writing team is updating a procedure because of equipment modifications several of them should do a dry run or actually perform the procedure, taking notes as they go through the process. Another way to start a procedure for an existing operation is to list the steps required to perform the task.

It is not always obvious which tasks need written procedures. In general, a procedure should be written for any operation or task that could adversely affect safety, production, quality, or the environment if performed incorrectly. Furthermore, if a procedure is needed, it should cover all phases of that operation, including start-up, shutdown, normal operations, and emergency operations. However, all tasks do not require a written procedure. Some are so simple that a ***job aid*** will suffice. A job aid is essentially a memory jogger or short list. An example of a job aid would be a five-line list of steps for operating a piece of equipment.

> **Preparing FCO-305 for Laboratory Analysis**
>
> 1. Collect the sample in a wide-mouth pint jar with a Teflon-lined lid.
> 2. Fill the pint jar two-thirds full and screw cap down tightly.
> 3. Place the jar in a small ice chest and cover the sample with ice.
> 4. Attach sample label to the ice chest handle.
> 5. Call the laboratory at extension 1284 and tell them the sample is ready for pickup.
>   John Smith,
>   Unit Supervisor 9/5/2006

Figure 6-1   Example of a Job Aid

The job aid would be taped onto the equipment and dated and signed by an approving authority. An example of a job aid is shown in Figure 6-1.

Once an operation that needs a procedure is identified, it should be analyzed to develop the data needed to write the procedures. This analysis should consider the following:

- What skills, knowledge, and equipment are required?
- What special tools or equipment are required?
- What hazards are involved?
- What are the major steps in doing this operation properly and the consequences of missing a step?
- What existing documentation is available (vendor literature, existing equipment and operation files, process hazards studies)?
- What level of detail is needed? The level of detail should consider the skill levels of the technicians who will use the procedure, the hazardous nature of the operation, and the complexity of the process and equipment.

The procedure might include the following:

- Warnings and cautions about hazards
- Equipment description and location
- Simple diagrams to clarify discussion
- Step-by-step check lists for complicated procedures
- Calculation aids
- Acceptance criteria that indicate when a step was performed correctly

PROCEDURE VALIDATION

After a procedure has been written, it should be validated. Validation ensures the procedure does what it is supposed to do. Validation ensures that it is accurate, complete, and written so that employees are willing and able to follow its instructions. Validation also considers safety, quality, and productivity factors. One approach to validation uses a team of employees familiar with all phases of the operation but who were not involved in writing the procedure. This team uses the procedure to carry out the operation and then evaluates the results. Their evaluation should consider the following:

- Is sufficient detail provided for an inexperienced employee to do the job properly?
- Are the steps of the operation listed in the proper order?

- Is sufficient guidance provided to do the work safely?
- Does the procedure maintain product quality and unit productivity?

Once the procedure has been written and verified, it is ready for final approval and implementation. Examples of two different styles of operating procedures are presented at the end of this chapter. Neither is right or wrong, they are styles used by different plants. Exhibits 6-1 and 6-2 at the end of this chapter are examples of operating procedures.

SUMMARY

Good communications are essential to the safe and efficient operation of a processing unit. Technicians may be involved in scheduling for turnarounds or other non-routine duties, arranging for equipment repairs with maintenance, discussing safety and environmental issues, quality issues, or process improvement issues. Opportunities abound for poor communication and miscommunication. Failure to clearly and precisely communicate on any one of these duties has the potential to create one or more of several outcomes, such as a dangerous situation, loss of profit, customer dissatisfaction, work stoppage, or bad feelings. The three methods of communication constantly used by process technicians are *oral*, *written*, and *nonverbal*. Each of these methods is appropriate depending on the situation at hand.

Most process plants will have a collection of documents related to the technology and operations of their processes that will include technical information, operating guidelines, safety concerns, and product requirements. They also include data and records of day-to-day plant operation. In some cases, this information is required for regulatory compliance or for quality certification programs such as ISO 9001-2000. The documents or manuals should be available both as hard copies and on-line electronic copies.

A procedure is a list of steps that direct you to start or stop a piece of equipment or a unit. It is a list that says "start here and end here," and when you are done you will have accomplished a certain task. In essence, a procedure tells you the correct way to perform a task. Good procedures must be clearly written, contain accurate information, and include enough detail to be useful to technicians, but not so much detail that they confuse or irritate more experienced technicians. After a procedure has been written, it should be validated. Validation ensures the procedure does what it is supposed to do. Validation ensures that it is accurate, complete, and written so that employees are willing and able to follow its instructions.

REVIEW QUESTIONS

1. Describe several adverse outcomes that may occur due to a failure to clearly and precisely communicate information.

2. List five communication responsibilities of technicians.

3. List the three methods of communication most often used by technicians.

4. Explain why feedback is essential to good communication.

5. Listening is _____ percent of communication.

6. Give three examples of communication filters.

7. List five applications of written communication used by technicians.

8. Discuss how nonverbal communication aids in communicating.

9. List five important process unit documents.

10. What information is contained in the daily operating instructions?

11. What information is contained in the operator logbook?

12. What information is contained in product specification sheets?

13. What information can you find in the operating procedures manual?

14. What information can you find in the unit technical manual?

15. What are the advantages of a standardized format for procedures?

16. What type of tasks require written procedures?

17. Write the definition of a job aid?

18. What is the purpose of procedure validation?

Exhibit 6-1

PROCEDURE FOR SODIUM TRANSFER FROM RAILCAR TO STOCK TANK T-101

NOTE: Refer to Personal Protective Equipment Section for the minimum PPE required for handling liquid sodium. Liquid sodium is very toxic and ignites on contact with air. The equipment is stored in the cabinet on the 2nd floor of the Sodium Building.

1. The inside technician will **inform** the outside technician of the need **to transfer** sodium to sodium storage tank T-101.

> **WARNING:** If this is the first transfer from the railcar, verify that pressure testing has been completed.

2. The inside technician will **check** the weight of **T -101 contents** to ensure there is room for the transfer and record the starting weight on the Tanker Transfer Sheet.
3. The inside technician will **verify** that automatic shutdowns 1-2, 1-3, and T-101 fire system **are not tripped**.
4. The inside technician will **verify** that **T-101 pressure** (PIC-302) is below .07 kg/cm^2.
5. The outside technician will use nitrogen to **raise the pressure** on the railcar by **increasing the setpoint** on PIC-317 to 30 psig.
6. The outside technician will **line up valves** in the sodium transfer line from T-101 to the spool piece.
7. The outside technician will **open** the **product valve** in railcar dome.
8. The outside technician will **check for a weight increase** in T-101 on the local mounted weight indicator (IWI-900).
9. The outside technician will **monitor** the transfer line **for leakage**.

> **NOTE:** DOT rules require that the outside technician remain within sight of railcar for the duration of sodium transfer. We have an approval exemption to monitor the railcar using the realtime video camera. Therefore, we are permitted to leave the railcar connected and unattended after the transfer line has been blown back and isolated at the railcar. We do not have an approved exemption for leaving the railcar unattended during a sodium transfer.

10. **When** the **weight** inside T-101 reaches 10,000 to 10,450 kgs, the outside technician will **close the sodium valve** in the railcar dome.
11. The outside technician will **lower** the nitrogen **pressure setpoint** on the railcar pressure system by changing the setpoint on PIC-317 to 10 psig.
12. The outside technician will **vent** the railcar through the automatic valve in the railcar vent line by **turning HS-822**, located on the west platform, to the **open position**.
13. The inside technician will **raise the setpoint** on T-101 **pressure controller** (PIC-302) to 1.8 kg/cm^2. (The outside technician may use the nitrogen by-pass valve on T-101 to speed up the pressure increase.)
14. **When** the railcar pressure is **less than 10 psig**, the outside technician will **close** the **automatic vent valve** on the railcar pressure system by **turning HS-822** to the closed position.

> **WARNING:** It is important that the automatic vent valve is closed before beginning blow-back of the sodium transfer line. If the railcar is venting while the sodium transfer line is blown back, the nitrogen from the blow-back will bubble up sodium and carry over entrained sodium into the vent header creating a hazardous situation.

Exhibit 6-1 (Continued)

15. **When** the railcar **pressure is less than 10 psig and T-101 pressure is greater than 1.5 kg/cm^2**, the outside technician will **open the sodium valve** in the railcar dome to clear the sodium from the transfer line.
16. The outside technician will **monitor** the **local pressure gauge** for a slight pressure increase usually occurring in 5 to 10 seconds. (A rumbling sound indicates sodium/nitrogen is forced back into the railcar.)
17. The outside technician will **close** the **sodium valve** in the railcar dome.
18. The outside technician will **close** the **first out valve** in the sodium transfer line in System 1 (CA-l001).
19. The outside technician will **inform** the inside technician that the **blow-back** has been completed.
20. The inside technician will **lower the setpoint** on T-101 pressure controller PIC-302 to **0.07 kg/cm^2**.
21. The inside technician will record transfer information on the Tanker Transfer Sheet.

Sodium Rail Car Transfer | **Page 2 of 2** | **Revised 4-3-2001**
Issued: 7-11-1996 | Revision 4

Exhibit 6-2

PROCEDURE FOR H-3 FEED PREHEATER STARTUP

	OPER.	Time

1. WHEN a good plume of steam exits the stack,
   PRESS PURGE button on Heater panel.   _____   _____

> **NOTE:** The heater purge will last about 5 minutes.

2. WHEN PURGE COMPLETE light is illuminated,
   CHECK the LEL concentration using a portable meter.   _____   _____

3. IF the LEL concentration is greater than 1%,
   PURGE firebox UNTIL LEL concentration is less than 1%.   _____   _____

<u>LIGHT PILOTS</u>

> **NOTE:** When the reset button is depressed, the solenoids on the pilot gas isolation and bleed valves will latch. The vent valve will close and isolation valve will open.

4. PRESS RESET button on Heater panel.   _____   _____

5. Board Operator and Outside Operator VERIFY that all
   ESDs are cleared on Heater Panel.   _____   _____

6. TEST the igniter to verify proper operation.   _____   _____

7. NOTIFY Board Operator that pilots are ready to be lit.   _____   _____

8. WHILE depressing igniter switch, SLOWLY OPEN 1"
   pilot valve selected by Board Operator   _____   _____

> **WARNING:** A face shield must be worn when viewing through the firebox inspection windows for flame verification.

9. VERIFY that pilot is lit.   _____   _____

CHAPTER 7

Process Physics

Learning Objectives

After completing this chapter, you should be able to

■ *Explain how a temperature increase or decrease affects vapor pressure.*

■ *Explain how an increase or decrease in pressure affects vapor pressure.*

■ *Explain how specific heat affects the operation of heat exchangers.*

■ *List four thermal properties of substances.*

■ *Discuss how linear expansion affects piping and equipment.*

■ *Explain why liquids sold by volume have their volumes adjusted to 60°F.*

■ *Explain how failure to take into effect volume expansion of liquids during transportation or storage is a safety hazard.*

INTRODUCTION

This chapter is concerned with some of the physics involved on a process unit and the thermal properties of materials. Physics is the study of energy. Process units contain enormous amounts of energy. Energy, which if not understood, can cause serious and expensive accidents and injuries. The energy contained in a process unit can be in the form of heat energy, kinetic energy (rotating equipment, flow), potential energy (vessels full of chemicals), and pressure (steam, bottled gases).

Process technicians use the principles of heat (energy) and material balances (mass) to control and follow the performance of their process unit. Their tools for monitoring and understanding energy and mass balance of their unit are their operating manuals and instrumentation readings. Process technicians should understand how heat is being transferred in each vessel or piece of equipment in order to follow the performance of their equipment and understand the control logic that supports their standard operating procedures. For heat to be transferred there must be a donor of heat and a receiver of the heat. Heat will not move if it has no place to go. It will build up until something ruptures.

Heat energy is used to do useful work in steam generation, refrigeration, distillation, chemical reactions, and other unit operations, such as the heat exchanged in reboilers, heaters, coolers, and condensers. The technology involved to both add and remove heat from a process stream is identical. The issue in most processes is how effectively the transfer of the desired amount of heat is controlled. Thermal properties, such as specific heat, are used to quantify the heat content of materials and are involved in most of the processes in a chemical plant.

Two factors important to a process technician involving heat energy are (1) conservation and (2) safe usage. Conservation of heat energy is a significant economic factor in unit operations because energy is expensive. Also, heat energy not properly used is a major contributor to hazardous conditions and accidents that result in damaged equipment and personnel injuries.

HEAT MEASUREMENT AND BRITISH THERMAL UNITS

The British thermal unit (BTU) is the most common expression of heat in the processing industries. The amount of heat in a substance depends on both its temperature and its mass. To define a unit for measuring heat, both mass and temperature must be specified. If two samples of the same substance have the same mass, then the amount of heat required to raise the temperature of each object by a specified amount should be the same. Units of heat are defined as the amount of heat that will cause a specified temperature change in a specific amount of material. The British thermal unit (BTU) is defined as the amount of heat that will increase the temperature of one pound of water by 18F. It would take 1 BTU to increase the temperature of one pound of water from 508F to 518F. To heat 10 pounds of water from 508F to 518F, it would take 10 BTU. To heat 100 pounds of water from 508F to 518F, it would take 100 BTU.

Latent heat is the heat required for a phase change. It is the heat needed to convert ice into water or water into steam. The quantity of latent heat energy is large per unit weight and is always absorbed at a constant temperature as long as the pressure is constant and the two phases are present. It is heat absorbed but not sensed by a thermometer or other heat sensing device. When a solid such as ice gains enough latent heat, it melts. Another word for this melting is *fusion*. If enough latent heat is added to ice, it will fuse (change state and become a liquid). The amount of latent heat required to fuse a substance is called the heat of fusion. One pound of ice at 32°F must gain 144 BTU of heat to fuse into water. Remember, heat of fusion involves a change in state from solid to liquid or vice versa. Therefore, heat of fusion is latent heat and a temperature gauge will show no temperature change. The heat of fusion of one pound of ice is 144 BTU. So, 10 pounds of ice would have to gain 1,440 BTU to change to water.

Now, suppose we have one pound of water at 32°F and want to determine how much energy is required to change all the water to steam. Before we can begin to change the water to steam, we must heat it up to its boiling point (212°F). Our first step will be to determine the amount of heat required to raise the temperature of the water to 212°F. This will be sensible heat since it causes a temperature increase. First, you will be heating the water to 180°F (from 32°F to 212°F). It takes 1 BTU to heat one pound of water 1°F. To heat one pound of water to 180°F it will take 180 BTU. Before we can go further we need to know that it takes 970 BTU to vaporize (convert to steam) one pound of boiling water. So, by adding that figure (970 BTU) to the sensible heat required to bring the water to the boiling point (180 BTU) the total heat energy required to change one pound of water at 32°F to steam is 1,150 BTU. Table 7-1 presents the BTU values for the states of water.

Now, determine how much heat energy is required to melt 10 pounds of ice to water and then evaporate (vaporize) the water completely.

1. The heat of fusion to change 10 pounds of ice to water is
   $10 \times 144 = 1,440$ BTU.
2. The heat required to raise the temperature of 10 pounds of water from 32°F to 212°F is $10 \times (212°F - 32°F) = 1,800$ BTU.
3. The **heat of vaporization** for 10 pounds of water at 212°F is
   $10 \times 970 = 9,700$ BTU.
4. The total energy required is $= 12,940$ BTU.

The point being made is that a large amount of energy contained in latent heat is stored in heated materials. This is the reason why steam burns to operators can be very serious burns. Technicians should be aware of the serious hazards of steam burns. Table 7-2 reveals some interesting BTU values.

VAPOR PRESSURE

Increasing the temperature of a liquid will mean that a greater proportion of the liquid's molecules will gain enough energy to escape from the liquid as a vapor. When the temperature of a liquid is increased, its rate of evaporation will increase. When a hydrocarbon is placed in a closed container, the molecules that evaporate cannot get far away from the liquid. The evaporating molecules move randomly in all directions until eventually some of the molecules slow down and return to the liquid. A point will be reached where the number of molecules returning to the liquid equals the number that are evaporating. At this point, the liquid and its vapor (the gas molecules) are in *equilibrium.*

Table 7-1   BTU Values for the States of Water

BTU Values for the States of Water	
Heat of Fusion—One pound of ice at 32°F to one pound of water at 32°F	144 BTU
Sensible Heat—One pound of water at 32°F to one pound of water at 212°F	180 BTU
Heat of Vaporization—One pound of water at 212°F to steam	970 BTU

Table 7-2   BTU Values of Substances or Actions

Substance or Action	BTU
A match	1
An apple	400
Stick of dynamite	2,000
One pound of wood	6,000
Running a TV for 100 hours	28,000
One pound of coal	13,200
One gallon of gasoline	125,000
One barrel of oil (42 gallons)	5,800,000
20 days cooking on gas stove	1,000,000
Appollo 17's trip to the moon	5,600,000,000
Hiroshima atomic bomb	80,000,000,000

Vapor Pressure and Trayed Distillation Towers

At a substance's equilibrium point, the rate of evaporation is equal to the rate of condensation. In a closed container, the liquid and the gas will eventually come to equilibrium at all temperatures. If the temperature of the liquid in a closed container is increased, its molecules will move faster, and because they are moving faster, more will evaporate. When the temperature of the liquid is increased, the rate of evaporation becomes temporarily greater than the rate of condensation but eventually the liquid and the vapor establish a new equilibrium point for this higher temperature. The gas molecules (vapor) moving inside the container exert a force on the sides of the container and on the liquid below. This force is called *vapor pressure*. The greater the energy possessed by the vapor molecules, the greater the vapor pressure. Heating a liquid increases its vapor pressure.

The general effect of pressure is to maintain a substance in the liquid phase. A common process in petroleum processing is fractionation (distillation). The *light* fractions in the tower are hydrocarbons with small molecules, low boiling points, and high vapor pressures. The higher the vapor pressure of a substance, the easier it vaporizes. The light fractions vaporize before the heavier fractions in the tower. The trays in a trayed distillation tower all represent zones of varying vapor pressure created by (1) a different temperature for each tray and (2) the molecular weight of the compounds vaporizing from that tray.

Vapor Pressure and Cooling Towers

The more water that can be evaporated in a cooling tower, the greater the amount of cooling the tower accomplishes. The lattice construction/fill in a cooling tower is designed to increase the evaporation rate by breaking up the water into a spray of droplets and increasing the exposed surface area of the water. The fans in a mechanical draft cooling tower provide a constant circulation of air to remove the vapor from the water. By exposing the maximum water surface to the air and by constant removal of water vapor, the cooling tower evaporates water rapidly, which lowers the temperature of the remaining liquid.

The heat of vaporization of water is 970 BTU per pound, which means that for each pound of water that evaporates, the remaining water in the basin will lose 970 BTU of heat. The removal of 970 BTU of heat from 970 pounds of water would result in a temperature decrease of 1.0°F. If 10 pounds of water were evaporated the temperature of 970 pounds of water would decrease by 10°F.

STEAM AND HEAT ENERGY

The temperature at which a liquid boils depends upon the external pressure exerted on the liquid. The boiling point of a liquid can be defined as the temperature at which its vapor pressure is equal to the external pressure on the liquid. Water at a normal atmospheric pressure of 14.7 psia boils at a temperature of 212°F, which is when its vapor pressure reaches 14.7 psia. The higher the pressure exerted on a liquid, the higher the boiling point of that liquid. If water in a refinery boiler is operating under a pressure of 300 psia, the water will boil when its vapor pressure reaches 300 psia, which occurs at 417.5°F. At that point, the boiler will also produce steam with a temperature of 417.5°F. The heat of vaporization of water is 970 BTU per pound. It would take 970 BTU of latent heat to change one pound of boiling water into steam. So, if 970 BTU of heat were removed from the steam at this pressure, one pound of the steam would condense into steam condensate.

Steam as a Source of Latent Heat

Steam is an excellent source of latent heat in a refinery. Heat flows from the hot condensing steam to the process liquid. When the steam is allowed to condense to water, a pound of saturated steam at 360°F can yield approximately 1,100 BTU of heat. Most of this heat is latent heat. Look at the latent heat of vaporization for the three fluids below.

	Heat of Vaporization	**Mid-Boiling Point**
Water	970 BTU/lb.	212°F
Gasoline	140 BTU/lb.	240°F
Butane	165 BTU/lb.	31°F

Water (steam) has the highest latent heat of vaporization (970 BTU/lb.) when compared to gasoline (140 BTU/lb.) and butane (165 BTU/lb.). Although steam is not the only fluid used in heat exchangers, the high latent heat of steam makes it a good fluid for heating. And because of its high specific heat (discussed later in this chapter), water is a good fluid for use in cooling systems. A cooling liquid must absorb heat from a hot process fluid. A liquid absorbs more heat when it does not evaporate easily.

Superheated Steam

If the steam produced by water boiling at 417.5°F and 300 psia was heated further until its temperature reached 467.5°F, the steam would be 50° hotter than the water that produced it. The steam would contain 50°F of superheat. **Superheated steam** is steam that contains more heat than what is required to maintain it above its dew point. If steam's dew point at 300 psia is 417.5°F, any steam hotter than 417.5°F is superheated. To condense the superheated steam, its temperature would first have to be lowered to 417.5°F, the boiling point of the water (at 300 psia pressure) that produced it.

Before the condensation of one pound of the superheated steam could begin, the 50°F of superheat has to be removed. The specific heat of superheated steam varies with the absolute pressure and the degrees of superheat. Specific heat is the amount of heat it takes to increase the temperature of one pound of a substance by 1°F. For our purposes, we can use an average value of 0.5 for the specific heat of superheated steam. Thus, to remove 50°F of superheat from one pound of superheated steam, we would have to remove 25 BTU of heat (0.5 × 50). Therefore, before it would condense, the one pound of steam would have to lose a total of 995 BTU:

$$970 \text{ (heat of vaporization)} + 25 \text{ (superheat)} = 995 \text{ BTU}$$

Steam loses heat as it travels down the steam header. Assume that 1,000 BTU of heat will be lost from 40 pounds of steam during its transmission from the boiler to the process unit. If the steam is not superheated, about one pound of the steam will condense in the header on the way to the process unit and be lost as a useful heating medium. If the steam is produced at 417.5°F and superheated to 467.5°F before transmission, each pound of the steam will contain 50 degrees of superheat that can be given up before any steam is lost through condensation. If, during transmission, the 40 pounds of steam loses 1,000 BTU, the temperature of the 40 pounds of steam will decrease by 50°F. This happens in the following way:

1. 1,000 BTU / 0.5 (average specific heat of superheated steam) = 2,000.
2. 2,000 / 40 lbs. = 50.
3. Thus, the steam's temperature will be lowered to 417.5°F (467.5 − 50) and none of the steam will be lost by condensation.

Superheated steam is important in many operations because it results in less condensation in steam headers as it is transmitted from the boiler house to the process units.

Steam is important as a heating medium in many processes because of its high latent heat of vaporization and condensation. The heat of condensation of steam is 6.9 times as great as that of gasoline. This means that the heat from one pound of steam as it condenses will vaporize 6.9 pounds of gasoline. Butane has a heat of vaporization of 156 BTU per pound. The heat of condensation of steam is about 6.2 times as great. Therefore, the condensation of one pound of steam would release enough heat to vaporize about 6.2 pounds of butane.

SPECIFIC HEAT

Our earlier chapters revealed that one of the greatest expenses of these industries is energy, and that heat was a form of energy. Processing industries seek to conserve energy to cut production costs. One way to do this is with heat exchangers. Heat exchangers have a specific surface area, metals of construction, and calculated flow rates designed to transfer heat (save energy). The two fluids exchanging the heat and the amount of heat are critical to an efficient heat transfer process. Why the fluids are critical is explained in the following paragraphs.

Specific heat is the quantity of heat required to increase the temperature of one pound of a substance by 1°F. Different substances heat up at different rates. A substance which requires only a small amount of heat to raise its temperature 1°F has a low specific heat. A substance that requires a large amount of heat to raise its temperature by 1°F has a high specific heat. An aluminum pan heats up quickly because aluminum requires only a small amount of heat to raise its temperature 1°F. The specific heat of aluminum is low. The wooden handle of the aluminum pan heats up slowly and requires a large amount of heat to change its temperature 1°F. Thus, wood has a high specific heat.

Water has a specific heat of 1. Thus, if 1 BTU of heat is added to one pound of water, the temperature of the water will increase by 1°F. To increase the temperature of 10 pounds of water by 10°F, 100 BTU of heat would have to be added. To raise the temperature of one pound of gasoline by 1°F takes only 0.5 BTU of heat, which is why the specific heat of gasoline is 0.5. Therefore, if 1 BTU is added to one pound of gasoline, its temperature will increase by 2°F. How many BTU would be required to raise the temperature of ten pounds of gasoline at 0°F to 100°F? The math is 10 lbs. $\times$ 0.5 $\times$ 100°F = 500 BTU.

There are two easy steps to follow to figure out how much heat is gained or lost when the temperature of a substance increases or decreases: (1) determine how much heat would be required if the substance were water, then (2) multiply this result by the specific heat of the substance. Often process fluids are too hot to go into immediate storage. They have to be cooled down. Suppose you wanted to figure out how much heat would have to be removed from eight pounds of crude oil to lower its temperature from 80°F to 60°F. This is a temperature reduction of 20°F. If the crude oil were water, you would have to remove 8 $\times$ 20 = 160 BTU of heat. Since crude oil has a different specific heat than water, you must multiply the 160 BTU by the specific heat of crude oil, which is 0.5 (see Table 7-3). Since the specific heat of crude oil is 0.5, the amount of heat you would have to remove from eight pounds of crude oil to bring it from 80°F to 60°F is (8 $\times$ 20) $\times$ 0.5 = 80 BTU.

Table 7-3   Specific Heat of Substances

Substance	Specific Heat
Water	1.000
Gasoline	0.500
Toluene	0.400
Kerosene	0.500
Crude oil	0.500
Sulfuric acid	0.336
Hydrochloric acid	0.600
Coke	0.203
Aluminum	0.224
Carbon	0.165
Copper	0.092
Iron	0.122
Lead	0.030
Mercury	0.088
Sulfur	0.175

How much heat must be removed to lower the temperature of 700 pounds of kerosene from 680°F to 650°F? The calculation is outlined here:

1. Temperature change × weight of sample: 30 × 700 = 21,000
2. Answer above × specific heat of kerosene: 21,000 × 0.5 = 10,500 BTU

Determine how much heat must be added to 1,000 pounds of toluene to raise its temperature by 5°F. First, this 5°F temperature change would require 5,000 BTU for water. For toluene with a specific heat of 0.4, the temperature increase can be accomplished by adding 2,000 BTU of heat (0.4 × 5,000). Specific heat is taken into consideration in the design specifications of heat exchangers.

THERMAL PROPERTIES OF SUBSTANCES

As we have mentioned, a major goal in process operations is the economic and efficient use of heat energy. Technicians can attain this goal only if they understand the basic thermal properties of the materials involved in their processes. Cooling water is a good example of thermal properties because water has a high specific heat. The basic thermal principle behind its use is the fact that the higher the specific heat of a substance, the more heat required to increase its temperature. In the case of water, this means it can remove more heat from product streams before the cooling water becomes too hot for use and must be returned to the cooling tower for cooling. See Table 7-4 for the thermal properties of other hydrocarbons.

Distillation utilizes the basic thermal property of boiling point. Differences in boiling points are used in distillation to separate a mixture into its components. If a selection had to be made between kerosene and toluene for use as a cooling medium for one of the distillates, ignoring cost, the best choice would be kerosene because it has the highest specific heat (see Table 7-4).

Table 7-4   Thermal Properties of Some Hydrocarbons

Substance	Heat of Vaporization	Specific Heat	Melting Point	Boiling Point
Water	970 BTU/lb.	1.000	32°F	212°F
Kerosene	108	0.50	—	425°F
Gasoline	140	0.50	—	220°F
Benzene	172	0.450	42°F	176°F
Decane	110	0.436	−22°F	345°F
Octane	128	0.420	−70°F	258°F
Heptane	133	0.415	−131°F	209°F
Hexane	156	0.406	−140°F	156°F
Pentane	158	0.402	−202°F	97°F
Toluene	151	0.40	−139°F	231°F
Butane	156	0.396	−217°F	31°F

The heat of vaporization of water is 970 BTU per pound. So, one pound of steam would have to lose 970 BTU before it would condense. Pentane has a heat of vaporization of only 158 BTU per pound. Therefore, the loss of 158 BTU of heat from a sample of pentane vapor would cause one pound to condense.

If the water used in a condenser to condense gasoline is heated from 90°F to 120°F, each pound of water will remove 30 BTU of heat from the gasoline. The heat of vaporization (condensation) of gasoline is 140 BTU per pound. One pound of cooling water will remove 30 BTU, so 4.67 pounds will remove about 140 BTU. Therefore, for every 4.67 pounds of cooling water passing through the heat exchanger about one pound of gasoline will condense.

Remember that the heat removed in condensing the gasoline is *latent* heat, thus the temperature of the condensed gasoline will be equal to its boiling point, 220°F. To cool the gasoline to 100°F after it has condensed, some of the *sensible* heat (heat that raises or lowers the temperature of a substance without causing a change of state) would have to be removed. The temperature reduction from 220°F to 100°F is a reduction of 120°F. If the substance were water, this would require the removal of 120 BTU of heat for each pound. But the specific heat of gasoline is only 0.5. Therefore, 60 BTU of heat would have to be removed to lower the temperature of one pound of gasoline by 120°F (0.5 × 120 = 60 BTU).

LINEAR EXPANSION

Various metals have different rates of linear expansion. **Linear expansion** is the lengthening of a pipe or metal bar caused by an increase in the temperature of the metal. The coefficients of expansion for several metals are listed in Table 7-5. These coefficients tell how much a 1-foot bar of the material will expand in length when its temperature is increased by 1°F.

If each of the materials listed in Table 7-5 were heated by 10°F, aluminum would show the greatest linear expansion. If the temperature of a 1-foot steel bar were increased by 1°F, its length would increase by 0.0000072 of a foot. If the temperature of the 1-foot steel bar were increased by 1,000°F, its length would increase 1,000 × 0.0000072, or 0.0072 feet. The coefficient of linear expansion tells how much a 1-foot bar of a material will lengthen when its temperature is increased by 1°F. Suppose the steel bar was 100 feet long and its temperature was increased by 10,000°F. Then its length would increase 100 times as much as the 1-foot bar that was heated to 1,000°F. The bar would lengthen by 100 × 0.0072, or 0.72 feet.

Total linear expansion = coefficient of expansion × length of bar × temperature increase

Table 7-5   Coefficients of Linear Expansion

Metal	Coefficient
Aluminum	0.0000122
Copper	0.0000094
Steel	0.0000072
Iron	0.0000067

Suppose a 1,000-foot steel pipe in the refinery were cooled from 200°F to 100°F. The length of the pipe would decrease by 0.0000072 × 1,000 × 100, or 0.72 foot. This is over half a foot! You can see how important expansion joints are in refinery piping. These expansion joints compensate for the linear expansion and contraction of piping so that it will not break when heated and cooled. When pipe is installed on stanchions, movable hangers and shoes are used to allow for the piping's movement due to the expansion and contraction of the pipe.

A hot oil pump is shimmed when it is set cold on its foundation so that the pump will be level when the pump is in hot service and expansion takes place. Because of linear expansion, heat exchangers are manufactured with a fixed tube sheet and a floating tube sheet and head to allow for movement caused by expansion and contraction. Also, because of linear expansion and contraction, rapid changes in operating temperatures should be avoided to minimize damage to equipment due to thermal expansion or contraction.

Although our examples of linear expansion seem very small, the concept is very important in any application that involves using different materials in an environment where they are heated and cooled. For example, if a rivet of one metal is used inside a hole in a different material, it can become too tight or too loose if the thermal expansion of the two materials is very different.

VOLUME EXPANSION

When a liquid is heated and is free to expand, it will expand in all directions. For this reason, liquid expansion is called **volume expansion**. In Chapter 13 you will learn that volume expansion or contraction must be taken into account during custody transfers of bulk liquids and for inventory control. To standardize the measurement of many hydrocarbons, its volume is always computed for a standard temperature of 60°F. The volume expansion of hydrocarbons varies for different American Petroleum Institute (API) gravity ranges. These various gravity ranges have been assigned group numbers for identification (see Table 7-6).

Table 7-6   Volume Expansion of Hydrocarbons

Group Number	Coefficient of Expansion at 90°F.	Gravity Range of Group (Degrees AFI or 60°F.)
0	0.00085	up to 14.9
1	0.00040	15.0 to 34.9
2	0.00050	35.0 to 50.9
3	0.00060	51.0 to 63.9
4	0.00070	64.0 to 78.9
5	0.00080	79.0 to 88.9
6	0.00085	89.0 to 93.9
7	0.00090	94.0 to 100.0

Just as a liquid will expand when heated, it will also contract slightly when cooled. Because of the expansion and contraction of hydrocarbons due to temperature changes, a standardized temperature for measuring the volume of hydrocarbon is important in custody transfers. This is explained in the following examples.

Assume you must determine the volume decrease of 1,000 gallons of oil at 75°F if it were cooled to 60°F. If the oil (hydrocarbon) is in group 3 (gasoline), its volume would contract by 0.00060 for each degree it is cooled. Cooling from 75°F to 60°F is a reduction of 15°F, so the coefficient of expansion must be multiplied by 15.

$$15 \times 0.00060 = 0.0090 \text{ per gallon}$$

Thus, the 1,000 gallons of oil would reduce in volume by 9.0 gallons if cooled from 75°F to 60°F. For more exact volume corrections, the National Bureau of Standards has prepared tables for each gravity range group of a substance (see Table 7-7).

Table 7-7   Volume Correction Table

Temperature	Multiplier
50	1.0061
51	1.0054
52	1.0048
53	1.0042
54	1.0036
55	1.0030
56	1.0024
57	1.0018
58	1.0012
59	1.0006
60	1.0000
61	0.9994
62	0.9988
63	0.9982
64	0.9976
65	0.9970
66	0.9964
67	0.9958
68	0.9951
69	0.9945

Data from "Petroleum Measurement Table, vol 2, Table 5B," reproduced courtesy of the American Petroleum Institute.

Referring to Table 7-7, the multiplier for 69°F is 0.9945. This means that one gallon of oil at 69°F is equal to only 0.9945 gallons of oil at the standard temperature of 60°F. To convert 1,000 gallons of oil at 68°F to its new volume at standard temperature, use the multiplier 0.9951 from Table 7-7.

1,000 gallons × 0.9951 = 995.1 gallons at standard temperature

Compensation for Volume Expansion

Technicians have experienced catching a sample of gasoline when it is cold in a bottle with a cork stopper and then having the cork pop off the bottle when the gasoline has warmed to room temperature. But the effects of volume expansion can be much more dangerous in the case of sample containers of propane or butane under pressure. If a pressurized vessel is filled completely, the volume expansion of the material could rupture the vessel. Since the content is a flammable liquid under pressure, immediate vaporization would occur and may result in an explosion.

The Natural Gas Processors Association has specified a 20 percent outage when filling pressure containers at temperatures above 0°F. This means that technicians collecting samples should leave at least a 20 percent outage in the propane or butane sample container to allow for the volume expansion that occurs when the sample is brought into a warm room. Compensation for volume expansion must also be made when filling liquid petroleum gas (LPG) pressure tank cars. The Interstate Commerce Commission has set specifications on how full a tank car can be loaded under specific conditions. Three conditions are taken into account are

1. The specific gravity of the LPG being loaded
2. The season of the year for transport
3. The temperature of the LPG being loaded

In the summer, a tank car loaded with LPG (specific gravity 0.51) at a temperature of 66°F can be filled to 92.6 percent of its capacity. If the same LPG were loaded under the same conditions for transport in the winter, the tank car could be loaded to slightly more than 92.6 percent capacity because winter temperatures are lower.

SUMMARY

Process units contain enormous amounts of energy, which if not understood, can cause serious and expensive accidents and injuries. The energy contained in a process unit can be in the form of heat energy, kinetic energy (rotating equipment, flow), potential energy (vessels full of chemicals), and pressure (steam, bottled gases). Large amounts of energy are contained in the latent heat of process materials.

The British Thermal Unit (BTU) is the most common expression of heat in the petrochemical and refining industries. The amount of heat in a substance depends on both its temperature and its mass. A BTU is defined as the amount of heat that will increase the temperature of one pound of water by 1°F.

Gas molecules (vapor) moving inside a container exert a force on the sides of the container and on the liquid below. This force is called *vapor pressure*. The greater the energy possessed by the vapor molecules, the greater the vapor pressure. Heating a liquid increases its vapor pressure.

Steam is an excellent source of latent heat in a refinery. Heat flows from the hot condensing steam to the process liquid. When the steam is allowed to condense to water it yields large amounts of heat energy. Most of this heat is latent heat.

Specific heat is the quantity of heat required to increase the temperature of one pound of a substance by 1°F. A substance which requires only a small amount of heat to raise its temperature 1°F has a low specific heat. A substance that requires a large amount of heat to raise its temperature by 1°F has a high specific heat.

Various metals have different rates of linear expansion. Linear expansion is the *lengthening* of a pipe or metal bar caused by an increase in the temperature of the metal.

When a liquid is heated and is free to expand, it will expand in all directions. Liquid expansion is called *volume* expansion. To standardize the measurement of many hydrocarbons, its volume is always computed for a standard temperature of 60°F.

REVIEW QUESTIONS

1. Define *British thermal unit*, *latent heat*, *heat of fusion*, *sensible heat*, and *heat of vaporization*.

2. List two factors important to process technicians involving energy.

3. Explain how an increase or decrease in temperature affects vapor pressure.

4. Explain how an increase or decrease in pressure affects vapor pressure.

5. Define *superheated steam*, *dew point*, and *specific heat*.

6. Explain how specific heat affects the operation of heat exchangers.

7. List four thermal properties of substances.

8. A section of a steam header constructed of iron is 300 feet long at ambient temperature (75°F). What will be its length when superheated steam with a temperature of 600°F is introduced into the header?

9. What will protect the steam header piping from distorting or fracturing when it heats up?

10. Explain why liquids sold by volume have their volumes adjusted at 60°F.

11. A rail car is filled with 10,000 gallons of a liquid at 80°F. The liquid belongs to group 4. Referring back to Table 7-6, what is the volume of the liquid when it reaches its destination where the temperature is 97°F?

12. Explain how failure to take into effect volume expansion of liquids during transportation or storage is a safety hazard.

CHAPTER 8

Process Samples

Learning Objectives

After completing this chapter, you should be able to

- *List the categories of unit samples.*

- *Discuss the hazards associated with collecting samples.*

- *Discuss some safety precautions taken when collecting samples.*

- *Explain the importance of using the correct sample container.*

- *Discuss how non-representative samples can be collected.*

- *Explain the importance of sample identification.*

- *Discuss why a technician would use a gas detector.*

INTRODUCTION

Process technicians are responsible for the operation and control of their processing units so that they produce on-specification product consistently in the most economical manner. Technicians cannot determine the quality of their finished product or the chemical properties of the various process streams of their unit by just sight or smell, which could be hazardous to their health. For the technicians to know the chemical composition of their finished product and process streams they refer to analyzers, if they have them, or collect a small amount of each stream or product for analysis at a site quality control laboratory. Technicians must understand the importance of their unit samples, the proper way to collect, the

Figure 8-1 Technician Collecting a Sample

types of samples containers, and the hazards of collecting samples. They will also sample the air around certain vessels and piping using a gas detector as they check for leaks or hazards. Figure 8-1 shows a technician in PPE collecting a sample from a unit vessel.

Unit samples are collected for

- Verification of process stream composition
- Verification of on-specification product(s)
- Troubleshooting process problems
- Comparison with on-line analyzers

UNIT SAMPLES

A processing unit can have several categories of samples, some of which might be

- Feedstock
- Finished product(s)
- Process streams
- Auxiliary systems and utilities
- Environmental

Feedstock

Feedstock may be raw materials brought into the plant by pipeline, tank cars, rail cars, ship, and barge. Feedstock (or feed) can also be the finished product of an upstream unit. Though the upstream unit's finished product is not a salable product for external customers, it is a finished product (feedstock) for the receiving processing unit (the internal customer). Feedstock should be tested to ensure it meets the receiving unit's specifications. If the feedstock is from an outside supplier, it is tested before it is accepted and unloaded. The quality control laboratory usually performs the tests. At some sites a benzene feed sample might be collected for analysis twice a day. At other sites, based on excellent supplier performance, feedstock is accepted upon receipt of a certificate of analysis and does not require analytical tests. For instance, a refinery might ship benzene to a chemical plant via pipeline if they fax a certificate of analysis that verifies the benzene tank being pumped out meets the chemical plant's specifications.

Finished Product(s)

Technicians collect samples of finished products to be tested for comparison with the manufacturing specifications that are located in the specification manual (also called *spec book*). Finished products go to the rundown tank, and when the rundown tank reaches a certain level, its contents are tested. If the finished product is on-specification, it is pumped to product storage tanks. If the finished product is a solid, such as powders or pellets, the same process is followed except that product will be stored in hoppers or silos, or go directly to the bag house to be bagged. See Figure 8-2 for a normal sampling scheme for a process unit.

Process Streams

Process streams, also called intermediate streams, are streams of process materials making their way through the unit before they become a finished product or consumed in the process. An example might be crude styrene stream coming out of a reactor. It is an important process stream because it informs the control room that the reactor is, or is not, operating correctly. Process streams must be tested for the parameter(s) that indicate that section of the process is still within standard operating conditions (SOCs). If not, the unit will be unable to make a final product that meets specifications. Process streams are tested as necessary to aid in controlling the unit. Such samples represent the process stream only at the time it was sampled because the unit is in continuous production and material is constantly flowing past the sample point.

Often, if a unit is complex, on-line analyzers are used in conjunction with unit samples. The analyzers report results much faster than the quality control laboratory (15 minutes versus about 1 to 2 hours); however, whenever there is a conflict between an analyzer and laboratory

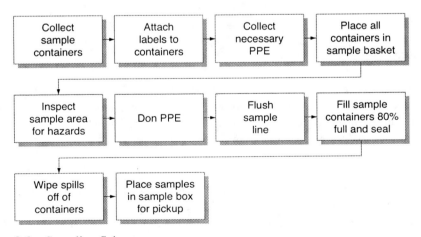

Figure 8-2 Sampling Scheme

result, the general rule is that the processing unit accepts the laboratory result. Also, unit analyzers, as a rule, do not give the detailed analysis that the quality control laboratory does. The analyzer might be looking at one or two key components while a laboratory analysis, which analytically takes much longer, may take about an hour but will yield information on 15 or more components.

The process technician is responsible for collecting an uncontaminated sample that is representative of the process stream or product, properly labeling the sample, and either taking it to the quality control laboratory or placing it in the unit sample box for pickup by the laboratory. The quality control laboratory will analyze every sample representing a shipment and create a certificate of analysis that lists the required analytical tests, the product specifications, and the test values for each test. A copy of the certificate of analysis is sent to the customer. Because of the possibility of a final product shipment becoming contaminated after loading or degrading under certain conditions, tested samples are often retained by the laboratory for months in storage. If an external customer complains of receiving off-specification product, the retained sample can be pulled and retested to verify product quality.

Auxiliary Systems and Utilities

Systems that support the processing unit may also require sampling and analysis because they may be the cause of unit problems. For example, de-ionized water can be contaminated or have a break through and is no longer pure enough for a reactor. Unit nitrogen can be contaminated with air and interfere with processes. Cooling water can be contaminated with hydrocarbons from leaking heat exchangers. Auxiliary systems can include steam and cooling water supply systems, waste treatment and disposal systems, equipment lubricating oil, etc.

Environmental

The processing industry is subject to compliance with state and federal environmental regulations. Failure to do so may result in severe penalties. Some of the most common environmental samples are of plant water released to ditches and streams. These samples are called **outfall** samples because they represent water allowed to fall back out into the environment. Rainfall and the use of process water create outfall samples. Before the water can be released back to the environment it must be analyzed for pH, total organic carbon (TOC), oil, and grease. Other tests may be specified as regulations change. Outfall samples may be collected from within storage tank dikes, stormwater basins, or the site's wastewater treatment plant.

Another very common type of environmental sample is scrubber gas samples. Analyzers usually monitor these samples. A **scrubber** is a piece of equipment designed to remove regulated air pollutants from stack gases, such as sulfur dioxide, sulfur trioxide, and hydrogen sulfide, usually by countercurrent contact with a liquid.

SAMPLE SCHEDULES

Each unit has a sample schedule. The **sample schedule** lists the samples operators are to collect, the date and time to collect them, what sample container to use, and what hazard labels to use. Sample schedules may be included in a technician's daily routine checklist. Some sites have computerized preprinted sample labels that list all of the above information;

at other sites technicians write the information on blank sample tags that are attached to the sample containers.

SAMPLE COLLECTION AND TECHNICIAN HEALTH

Technicians might wonder what is the big concern about catching a 4-, 16-, or 32-ounce sample. It is a big concern if they work in the refining or petrochemical industry. Their sample might contain hazardous compounds that fall into one of the following categories:

- Mutagen
- Teratogen
- Neurotoxic
- Flammable
- Toxic
- Carcinogenic
- Corrosive
- Allergen

Process technicians working a 12-hour shift may collect samples four different times during their shift. Assume the technicians are collecting, on average, 10 samples each time they make their rounds. Then, on a normal workday the process technicians are exposing themselves to physical or health hazards 40 times just from collecting samples. Carry the math (or health risk) out further. They work 52 weeks a year minus 4 weeks of vacation (on average), thus they collect 52×48 samples a year, which is 2,496 samples a year. Or, the prudent way to look at it is they expose themselves to a health or physical hazard 2,496 times a year in the process of sample collecting.

Besides collecting samples for process control reasons, technicians also use certain instrumentation to sample the air quality in certain areas or vessels. The instrumentation they use (gas detector) informs them of

- Oxygen levels in a vessel (tank, reactor, tower)
- Explosive levels of hydrocarbons in an area or vessel
- The presence of toxic levels of compounds in the air

A critical function of all process technicians is to be able to correctly operate and field calibrate a portable gas detector.

Hazards of Collecting Samples

Earlier, we mentioned that a process technician collected 2,496 samples a year on average. We also mentioned that some of these samples might possess different health or physical hazards. In fact, in a refinery or petrochemical plant it is a fairly safe bet that most samples will have several hazards associated with them. Why make a big safety issue about collecting samples? *Because of technician health and safety.*

Management does not want its workers hurt, sickened, or killed, if for no other reason than it is not good business. Humans being human tend to get careless about what they consider to be a non-threatening task. Why worry about collecting 16 ounces of para-xylene or gasoline? Go back to the math and see a reason to have a sincere concern about collecting samples in a safe manner. Assume a technician worked in a refinery or petrochemical plant 25 years and then retired. They will have collected 25 × 2,496 samples, or a total of 62,400 samples. Assume the technician's age is 60. They can expect to live at least another 20 years—especially if they collected the thousands of samples safely.

When a technician attaches the sample labels to the sample containers they fill with sample, they should notice the hazard warnings on the labels that are mandated by The Occupational Safety and Health Administration's (OSHA's) Hazard Communication Standard. The labels should read something like, "Warning: health hazard, benzene a known carcinogen." Or "Warning: flammable." The point being made is that every sample collected has the potential to harm a technician either internally or externally in an acute or chronic way. If a technician is careless and doesn't wear their respirator when collecting a benzene sample, and does this several hundred times over a 25-year career, will that amount of exposure be enough to initiate cancer after they retired? Or if they carelessly expose themselves hundreds of times to just a drop or two on their hands because they don't wear gloves to collect a chemical known to be neurotoxic, will that be enough to cause a nervous system dysfunction that will force them to walk with a walker? How many micro-insults can the human body take before they are cumulative enough to cause serious harm? No one knows.

Micro-insults is a term that means exposure to small or supposedly insignificant amounts of a harmful chemical, such as one inhalation of a few parts per million of butadiene or one drop of benzene on the skin, both of which are carcinogens. Do this often enough and they add up. Management supplies technicians with the required personal protective equipment (PPE) to protect themselves from skin contact and inhalation, but it is up to the technician to use the PPE and use it in a responsible manner.

Safety Precautions for Collecting Samples

A good operator always practices safe sample collecting. The following list includes some safety precautions that should be observed when collecting samples.

- Use the proper PPE for the job. There are dozens of chemically resistant gloves and they are not all equal. If certain chemicals penetrate latex gloves then a glove resistant to that chemical must be used. The sample room in the control house usually has several boxes of different gloves available because each glove has different protective abilities.
- Check the sample area for a source of ignition before sampling flammable compounds.
- Leave a vapor space above the liquid in the sample container to allow for thermal expansion. A good rule of thumb is to leave 20 percent of the container empty.
- Stand upwind when sampling to prevent breathing sample vapors. Some samples are so hazardous that the technician must wear a respirator.
- Label and identify all samples. Written labels should be legible and in pencil. Most hydrocarbons dissolve pen ink.
- Never use a leaking container or a container with faulty valves or other closures for sampling.
- Do not carry samples on your person, such as in your pockets. This contaminates clothing and risks exposure, especially if the sample container leaks or breaks. Carry samples in sample baskets.

- Do not use pliers or a wrench to tighten valves on sample cylinders. Excessive tightening damages the valve seat and causes leaks.
- Clean up any material spilled during sampling.

TYPES OF SAMPLES

Two types of samples are used in controlling processing operations, grab and composite samples. A **grab sample** is simply a sample collected from a sample point that represents the sample in that stream at that time. For example, if a technician fills a 6-ounce sample bottle with benzene for their 6:00 a.m. sample, that is a grab sample. Grab samples are principally used for unit control and verification of feed, final product and process stream composition, and are collected more often than composite samples. The operator uses grab sample test results to make routine adjustments in operating conditions as they become necessary. Samples are collected at regular time intervals, such as every four hours. Batch operations units collect samples whenever each processing step is completed to determine that the intermediate batch step has been completed properly.

A **composite sample** is a collection of samples from the same sample point over a period of time (usually 24 hours) mixed together in one container to make a composite that represents the material in that vessel or line. For example, to determine the average composition of a stream a pint of the stream's fluid might be poured into a one-gallon container once every four hours over a 24-hour period. This would be called a 24-hour composite sample and would represent the composition of the stream for that 24-hour period. Another example would be collecting polyethylene pellets for product quality certification from a railroad hopper car. An equal size sample would be collected from the top and from the bottom of each compartment in the hopper car. The composite sample produced by blending all of these samples together would represent the quality of the material in the rail car.

Some units might have automatic compositors built onto a process stream. These are mechanical samplers with programs that activate the device to withdraw a specific volume of sample at prescribed intervals and inject the sample into a container for collection by a technician.

Obviously, composite samples do not give information needed for monitoring and control of a process unit. They are used to develop material balances around the unit, determine the average composition of large storage tanks of material, and assign a dollar value to various stocks for interplant sales and purchases.

SAMPLE CONTAINERS

Sample containers come in all sizes and shapes and are made of varying materials. They can be

- Clear glass, wide-mouth jars varying in volume from 4 to 32 ounces
- Clear or amber bottles of varying volume
- Pint, quart, and gallon round paint cans
- Plastic bottles of varying sizes
- One gallon oblong cans
- Plastic bags
- Rubber football bladders
- Stainless steel cylinders called sample bombs

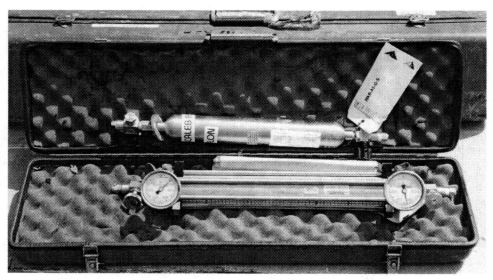

Figure 8-3   Sample Bombs

There are several reasons for such a variety of sample containers. The volume required for analysis is one of them. Process streams are eventually converted into salable product, so why collect a quart sample when only 2 ounces are needed? Some samples are gaseous, hence you need football bladders or sample bombs. Some samples are so hot they might crack the glass sample bottles, thus they are collected in cans. Some samples cannot be collected in cans because the solder used to make the cans releases metals into the sample and interferes with an analysis for metals. Other samples have components that migrate into the glass container, so they are collected in plastic containers. A few samples are affected by direct light and are collected in amber bottles.

The sample cylinder in Figure 8-3 is used to collect liquid and vapor samples under pressure. These cylinders are usually made of grade 304 stainless steel and are available in sizes ranging from 150 to 1,000 cubic centimeters. They have a working pressure rating of 1,800 pounds per square inch gauge (psig) and may have an overpressure safety device (rupture disk). A valve on each end of the cylinder provides control of flow into and out of the cylinder and is used to connect the cylinder to the sample collection point. One end of the cylinder may have an internal dip-tube to provide for liquid discharge when there is a vapor space at the top of the cylinder. If the cylinder has a dip-tube, the end with the dip-tube will be marked in some way. Sample cylinders provide safe containment for storage and transportation of both liquids and gases. These cylinders are DOT rated to 1,800 psig (124 bar) at 70°F (21°C).

SAMPLING METHODS

Collecting a sample in a container is not always as simple as it sounds. Sampling difficulties occur because of the sample's temperature, pressure, viscosity, or because the sample may fractionate (release vapors that should remain in the sample). Technicians must collect representative samples despite the previously mentioned conditions if the samples are to give valid information about the composition of the streams on their unit. Chemists and engineers will have determined the correct way to collect the unit's samples. Technicians must learn and follow the sample collection procedures of their unit if they are to collect a representative sample.

This chapter will not detail the various types of sampling methods. These are better learned through practice than description. However, two sampling practices will be mentioned.

First, it is a general safety rule that all samples containing liquids have an outtage (vapor space) of 20 percent in the sample container. This is a safety precaution against thermal expansion. Sample containers 100 percent full left in the sun develop pressure due to thermal expansion. Anyone opening that container may get splashed with the sample as it erupts from container, or debris and the sample hit the individual if the pressure is so great that it ruptures the container.

Second, there is a unique way to sample large liquid storage tanks or barge tanks that may have stratified into layers of light components at the top, slightly heavier components in the middle, and the heaviest compounds at the bottom. If there is no way to roll or mix the tank, how do you collect a representative sample of what the tank would look like if it were well mixed? This is how: A sample bottle is weighted with an attached weight and fitted into a rope harness or specially designed bottle holder. The stoppered sample bottle (with a line looped around the stopper) is lowered to the bottom of the tank. Then the stopper is removed by jerking the stopper line and the bottle raised at such a rate that it is not quite full when it reaches the top. As the bottle is pulled toward the surface it is collecting sample from the bottom, middle, and top layers. In the reverse scenario, separate bottles are lowered to collect samples from the top, middle, and bottom of the liquid level in the tank to see if a tank is well mixed.

Precautions for Safe and Accurate Sampling

The sample collected must have a composition that is representative of the stream being sampled if the test results are to have any value. If the sample is improperly taken and non-valid results reported back to the process unit, the technician will have to collect a **check sample**. Check samples are collected to investigate suspicious conditions or analytical results. Collecting a non-representative sample usually causes extra work for a technician because they will have to collect extra samples (check samples). It is rare that a process unit will make operational changes due to atypical analytical results from just one sample. A non-representative sample may result for any of the following reasons:

- The sample may be taken from a **dead spot** in a vessel where stagnant material not representative of the mainstream has collected. This is a spot out of the main flow within the vessel where material just collects in a quiet pool while the real sample stream bypasses the dead spot. Sample points should be relocated if they are located near dead spots.
- Long sample lines (8 or 10 feet long) trap a lot of sample. Often these sample lines are described as **dead legs**. If the line is not purged long enough to drain all the old material in the line the sample will have the same composition as the last sample taken from this point. Technicians become impatient collecting samples, especially if it is hot, cold, raining, or mosquitoes are bad and they may not wait long enough to purge the line.
- Samples may be collected in the wrong container and invalidate the sample. As an example, samples for a metals analysis are not collected in cans because

the solder used to seal the seams of the can may release lead and tin atoms into the sample. It is critical to use the correct sample container for each sample.

● A volatile material may be sampled while too hot and its lighter components may vaporize, fractionating the sample and leaving only the higher boiling components (heavier compounds) behind.

● Samples that have lighter components—compounds containing 1 to 6 carbon atoms—easily lose their lighter components through loose sample caps or cylinder valves. The lighter components vaporize and escape from the leaking sample container leaving the sample fractionated.

● Samples collected in dirty or contaminated containers can be expected to yield atypical results.

● The contents of many storage tanks stratify if the tank isn't constantly active. It is not uncommon to have three different zones in a storage tank with each zone having a different chemical composition. Lighter material will migrate to the top zone, heavier materials to the bottom, and the rest in the middle zone. To get a representative sample the tank will have to be mixed or samples collected from each zone and mixed together.

● Samples taken during a period of unit upset, or just before or after the upset, will not be representative. During these times operating conditions and product and process compositions are changing. Except for certain check samples to be used for guidance, all other samples may be useless.

SAMPLE IDENTIFICATION

Samples should always be tagged (labeled) immediately after collection. Many sites will have preprinted adhesive sample labels that list all information about the sample. Technicians take the sheet of sample labels for their samples and pulls and attaches the label to the sample. If blank string or wire sample tags are used, they are usually made of a tough material that resists moisture. They should be filled out with pencil or permanent marker ink so that the writing will not wash off when exposed to rain or unit sample material. The tag may include a National Fire Protection Association (NFPA) symbol for the sample hazards which the technician fills in. They will also include the following information on the tag: process unit name, sample name, date, time collected, and analytical tests required.

SAMPLING WITH GAS DETECTORS

Technicians will be required to operate gas detectors and sample various atmospheres on a processing unit. Atmosphere samples are not collected and submitted for analysis to a laboratory. Instead, the atmosphere within vessels, confined spaces, and in suspicious areas is sampled using portable gas detectors and the results recorded. Sampling of the ambient air is an integral part of any safety, environmental, or health program for several reasons, some of which are noted here:

● Assure compliance to environmental regulations
● Detect fugitive emissions
● Protect employee and contractor safety and health

Information from ambient air monitoring may also be useful in designing better control systems.

Air monitoring can be performed easily by many direct reading instruments, such as lower explosive limit meters (LEL), volatile organic compound meters (VOC), meters for detecting hazardous gases, oxygen monitors, and multiple function meters. The general trend in gas detector instrumentation is miniaturization, ease of use, digital read out, and the capability of uploading the data to a computer. Many of the instruments are *active*, meaning they use a pump to draw a specified quantity of air into the instrument. Others are passive and use no pump. Generally, the accuracy of the active instruments is higher than that of the passive instruments. Almost all of the detectors are small, compact, portable, and rugged. They have been designed for the rough conditions of the process industries.

Types of Gas Detectors

LEL Meters. *Lower Explosive Limit (LEL)* meters are probably the most common direct reading instruments. In a broad sense, the LEL instruments are based on the heat released by the sample of gases when they are burned under a controlled environment. Most instruments show the reading as the %LEL. There are several different designs but one of the most common is based on the heat of combustion of the burned hydrocarbon.

In the heat of combustion design, a catalyst-coated filament is kept heated at a certain temperature. When the gas sample passes over the filament it is burned and the heat of combustion changes the resistance in the filament. This change in resistance also changes the electrical conductivity of the filament. This conductivity is proportional to the concentration of the contaminant in the air. Since LEL meters are used for %LEL values of many different contaminants, the general practice for calibrating these instruments in the industry is to calibrate them with specific gases (for example, methane) and use a table of response factors for other gases. As a safety measure, analyzers with combustion sources have flashback arrestors to keep the flame from spreading out. Like many other instruments, LEL meters must be checked for zero reading before use. Calibrations on these meters are generally done with a known concentration of methane.

Like most other direct reading instruments, LEL meters are susceptible to steam or moisture. Moisture may extinguish the flame. Most vendors advise not to use their LEL meters in environments rich in steam or moisture. LEL meters based on combustion need oxygen to function and would not work in an atmosphere totally devoid of oxygen. The LEL meters have visual as well as audible alarms, digital displays, and self-diagnostics. Figure 8-4 shows a gas detector with modes for detecting LEL, oxygen, and several toxic gases.

Oxygen Meters. Often, LEL meters have an oxygen detection function built into them. Panametric analyzers use the paramagnetic property of oxygen to detect its concentration.

Detector Tubes. Probably, the simplest (and least accurate) of the direct reading instruments are the detector tubes. The detector tubes are often referred to by the name of their manufacturer, such as Drager tubes or Sensidyne tubes. The detector tube is hermetically sealed at both ends and filled with a granular or powdery reagent. Each tube is specific for a particular gas, hence tubes for ammonia detection are different from the tubes for carbon monoxide or methane detection. The chemical the tube tests for is printed on the tube.

Figure 8-4   LEL/Oxygen Meter

Detector tubes are operated in conjunction with a hand pump or battery operated pump that draws ambient air into the tubes. To sample the air, both ends of the tube are broken off in the special slots built into the pumps for that purpose, and the tube is inserted in the pump in the proper direction. An arrow on the tube indicates the end that goes into the pump and a known amount of the ambient air sample is drawn into the pump. Hand pumps require the diaphragm to be pumped a required number to times; for battery driven pumps, there is a time-limiting device. The reagent in the tube is sensitive to the compound being tested for and will change the color of the reagent in the tube on contact with the compound. If the compound is present in the air sample, the tube develops a band of color and the length of the discolored reagent is proportional to the concentration of the contaminant. The tube is calibrated and marked in parts per million (PPM) levels on its side and the compound's concentration can be read directly from the tube where the colored band stops.

Detector tubes are very convenient, low-cost items (a tube may cost $5 each) but they are estimation devices. Because of this, they cannot be used for some types of permitting applications, such as confined space entry.

Precautions must be taken when using detector tubes because moisture or particulates affect their performance. Although there is a guard section at the entrance to the tubes, moisture and/or particulate matter can quickly exceed the capacity of the guard section. Detector tubes should not be used in a rainy, damp, or dusty area. Before taking a reading from the tube, wait three to five minutes for the sample to diffuse completely through the tube. Although the detector tubes are specific for specific gases, they are not immune to interference from other chemicals. For instance, tubes for benzene may be affected by the presence of phenol or toluene.

SUMMARY

Technicians cannot determine the quality of their product by looking at it or smelling it. Nor can they determine the chemical properties of the various process streams of their unit that control the economics of the operation and predict the properties of the finished product(s). For the technician to know the chemical composition of their finished product and process streams they refer to analyzers, if they have them, or must collect a sample of each stream or product for analysis. Typical process unit samples are feedstock, finished product(s), intermediate streams, and environmental samples. The two most common types of samples are grab samples and composite samples.

Samples may be hazardous and initiate acute or chronic health problems. They may contain mutagens, teratogens, neurotoxins, toxins, carcinogens, corrosives, and allergens and may be flammable. Technicians should follow all sampling safety precautions prescribed by their unit SOP. Non-representative samples yield erroneous data. Non-representative samples can be collected from *dead spots*, *dead legs*, from using the wrong container, and from fractionating off the volatile components.

Technicians will be required to operate portable gas detectors and sample various atmospheres on a processing unit. Sampling of the ambient air is an integral part of any safety, environmental, or health program to assure compliance to environmental regulations, detect fugitive emissions, and protect employee and contractor safety and health.

REVIEW QUESTIONS

1. List four reasons for the collecting of unit samples.

2. List the five categories of samples.

3. What is the purpose of a *certificate of analysis*?

4. List several reasons why you would collect samples from auxiliary and utility systems.

5. What are *outfall* samples?

6. List five hazards associated with collecting samples.

7. List the types of containers used for sample collection.

8. Discuss three reasons why a technician would use a gas detector.

9. What does the term *micro-insult* mean?

10. Describe the two major types of samples.

11. The *outtage* (vapor space) of a sample container should be about _____ percent.

12. What is the hazard associated with collecting a liquid full sample cylinder of propane?

13. How would you collect an average liquid sample from a large storage tank?

14. What sampling technique would you use to determine if a storage tank is well mixed?

15. Why wouldn't you collect a sample from a dead leg in a piping system?

16. What do you need to do if you are collecting a sample from a point that has an extra long line between the sample vessel and the sample point?

17. What information should be included on the sample label or tag?

18. The most common gas detector used in process plants is the _____.

19. What precautions must be taken when using detector tubes?

CHAPTER 9

The Significance of Common Analytical Tests

Learning Objectives

After completing this chapter, you should be able to

- *Explain the usefulness of analytical tests to process technicians.*

- *Discuss the importance of collecting representative samples.*

- *List four tests that test for physical properties.*

- *Explain how a distillation test helps technicians troubleshoot a tower.*

- *Explain the information a flash point test yields.*

- *Explain why pH information is important to a process unit.*

- *Discuss why testing for sulfur is important.*

- *Explain why tests for color are important.*

- *Discuss why more online analyzers are being added to process units.*

INTRODUCTION

A styrene unit has a customer who buys 500,000 pounds of styrene monomer a month. The customer wants the same styrene monomer for his production process each month. The customer is after supplier consistency. Styrene monomer is a basic building block (raw material) the customer uses in their process to make a salable product. The customer wants their raw material to be consistent because they do not want to have to make adjustments to their process every time. To ensure the styrene unit produces quality styrene monomer all the time for its customers, the unit will have to test its feedstock, process streams, and product frequently because the processing unit equipment is subject to entropy, feedstock can be off specification, and technicians make mistakes. Ideally, a unit comes up after a turnaround, lines out, and produces on-specification product until the next turnaround. But in reality, samples must be collected routinely and analytical tests run on a regular schedule to assist a processing unit in remaining within standard operating conditions and producing products that consistently meet specifications.

This chapter looks at several very common analytical tests for refinery and petrochemical products. The tests chosen for this chapter were based on their simplicity of understanding and usefulness to technicians for controlling their units. One of the most important jobs of a technician is to catch valid samples and submit them to the quality control laboratory for analytical testing. Analytical testing of unit samples serves the following purposes:

- Verifies the unit is on specification.
- Alerts the technician to a developing problem.
- Helps troubleshoot unit problems.

THE USEFULNESS OF ANALYTICAL TESTS

Analytical tests are usually run on samples taken from a processing unit, loaded cargo tanks, and stock about to be released into a pipeline. Before a feedstock becomes a product it may have to go to a process unit where it undergoes a reaction and separation process, or it passes through several process units before it becomes a final product that meets customer specifications. Technicians should understand the basic chemistry of their unit and what analytical tests are revealing about their process. They do not have to be chemists to understand the information revealed by analytical test results.

Unit product specifications usually include only the properties that are controlled by the unit. Let's say a sample of a kerosene distillate meets distillate specifications. Since kerosene distillate is not a final product, it will be sent to process Unit X for further processing. Assume Unit X's treatment of the stream does not affect the API gravity of the kerosene during processing. Because it does not, an API gravity specification and test is not a requirement for Unit X. Usually the number of specifications for a finished product is more than the number for an intermediate product. On some units some products require special treating. For example, a unit may treat an oil with acid, which will necessitate an acidity test on the oil.

Here is an example of the value of analytical tests. A particular crude oil can be fractionated into naphtha, kerosene, gas oil, and reduced crude. Testing the sample will reveal how much kerosene the crude oil will yield. If a crude distillation unit is suddenly required to produce 20 percent more kerosene than normal, it will need to use a crude feed that contains a higher percentage of compounds that will make kerosene. To determine how much kerosene a crude oil can produce, a crude assay test is run. A crude oil assay test reveals what the components of a crude oil are and the percent of each component. In the assay comparison shown in Table 9-1, Crude B yields more kerosene indicating that it is the better feed for the process unit production goals.

Table 9-1   Crude Oil Assay

Crude A		Crude B	
Naphtha	20%	Naphtha	10%
Kerosene	15%	Kerosene	25%
Gas oil	55%	Gas oil	50%
Reduced crude	10%	Reduced crude	15%

THE IMPORTANCE OF VALID SAMPLES

Test results are no better than the sample submitted. A unit that submits bad samples (contaminated, dead-leg, fractionated) gets bad information. A valid (representative) sample is required to get useful data. The results are useless for unit control or product verification before shipment if the sample is contaminated or in any way non-representative.

Assume a sample was supposed to be from a side stream of a distillation tower but was really drawn from the overhead stream and labeled as a sidestream. A sidestream has a different composition from the overhead. It will contain a much larger percentage of heavier compounds. The overhead stream always contains the lighter components of the tower feed. No other tower stream contains such a high percentage of light molecular weight compounds (often called **lights** or **light ends**). Test results from the sidestream are not useful because the sample was drawn from the wrong sample point and yields misleading information. If a laboratory technician receives an overhead sample mislabeled as a sidestream, they will use the wrong equipment or analytical program (for automated equipment) to analyze the sample. If the technician were running a distillation or a flash point test, there is the possibility of a fire occurring in the test equipment because the sample will flash and distill at much lower temperatures. Equipment may be damaged or laboratory personnel hurt. Mislabeled samples are hazardous and are potential safety and health accidents.

The unit may also make unnecessary operational changes based on the results of the mislabeled sample if results from analytical tests performed on mislabeled samples are sent back to the unit. Though this is not normal, it can occur. Usually, if a sample yields abnormal information, a second sample (check sample) is caught and submitted for analysis. The quality control laboratory's job is to report sample results and not to argue with the unit about whether the sample is contaminated or correctly labeled. That is the unit's problem. What the laboratory supervisor will argue about is the fact that the sample submitted was not what the label said it was and caused a fire or exposed a laboratory technician to a serious safety hazard.

Assume your unit is producing naphtha. Gauges show that tower pressure has increased slightly. You want to be sure that the distillation end point of naphtha is still in the 440°F to 450°F range. You pick up the last sample bottle in your sample basket, one you had contaminated earlier with a few drops of kerosene. It was such a small amount of kerosene, you didn't want to walk back to the sample room and get a clean bottle. You poured the kerosene residue out and figured the thin film left in the bottle was so miniscule *it couldn't possibly* affect test results. After all, you are adding a quart of naphtha to the bottle. You collect the sample and send it in to be tested. The analytical results come back bad. Why? Naphtha boils at a lower temperature than kerosene. The little bit of kerosene residue pushed the endpoint beyond acceptable limits. Not only did you get unreliable results, now you've got to go back outside and collect another sample. The smart thing to do is to do it right the first time.

TEST RESULTS FOR PROCESS CONTROL

Usually samples are not collected when a process unit is upset unless the unit needs the test results to guide its way back to standard operating conditions. When the unit is upset, technicians know that the final product probably won't meet specifications. The following example illustrates using samples to get a unit back on standard operating conditions (SOCs).

A unit had a reboiler problem and is distilling naphtha at a higher temperature than normal, causing the naphtha end point to be off specification. The **end point (EP)** is the highest temperature attained during the laboratory distillation and contains the highest boiling compounds and highest temperature(s). The unit has been off specification for several hours and technicians still don't know how much farther they have to adjust the reboiler temperature to get back on spec. The end point specification for the unit's naphtha is 420°F to 460°F. A sample is sent to the quality control laboratory, analyzed, and reported with an EP of 475°F. This tells the unit technicians that the tower is still too hot. Heavier molecules that should be contained on lower trays have enough thermal energy to make it up to the naphtha sidedraw. The technicians adjust the reboiler temperature downward and wait long enough for the temperature change to make it to the top of the column before collecting and submitting another sample to find out if their naphtha product is back on specification.

Technicians should understand the chemistry of their unit and what the chemistry is telling them. Most technicians think only of four process variables—temperature, pressure, flow, and level—and forget there are others, and one important other is *composition*. **Composition** is the chemistry of their unit, the key chemical parameters that guide them toward making a salable product. It is the weight percent, the density, gravity, color, and other chemical characteristics of the solids and fluids in their piping and vessels. When off-specification product comes out of a reactor and a double-check of reactor SOCs confirms the SOCs for temperature, pressure, level, and flow as correct, what is left to investigate? *The unit chemistry—reactants and catalyst.*

Physical Versus Chemical Tests

Individuals have physical characteristics (height, color of hair, color of eyes) that distinguish them from their neighbor. Petroleum products also have distinguishing physical properties, some of which are vapor pressure, API gravity, color, viscosity, etc. Physical properties are measured by physical tests. **Physical tests** do not change the chemistry of the original sample. When kerosene is tested for its physical properties it still remains kerosene after the tests. It is unchanged. A **chemical test** involves a reaction that changes the characteristics of the original sample. If benzene is subjected to a chemical test it will cease to be benzene at the end of the test. The benzene molecule is changed into one or more different molecules.

TESTING FOR PHYSICAL PROPERTIES

The common tests for physical properties of chemicals we will discuss are distillation, vapor pressure, flash point, specific and API gravity, pH, and viscosity.

Distillation Test

Different liquids have different boiling points. Crude oil is a liquid mixture of thousands of chemical compounds. Look at the boiling range of the various cuts (groups of compounds) of crude oil X in Table 9-2.

Crude oils vary greatly. They have names such as West Texas Intermediate, the North Sea's Brent, Indonesia's Minas, Nigeria's Bonny Light, Saudi Arabia's Arab Light, Dubai's Fateh, Venezuela's Tia Juana Light, and Mexico's Isthmus. They are all different from each other; some are best for gasoline production, others best for aromatics. Crude oil is composed of thousands of different compounds that boil at different temperatures. A mixture boils at a range of temperatures (100°F to 600°F) displayed in Table 9-2. If the crude oil mixture in Table 9-2 is heated in a crude distillation tower, we can predict the order that the cuts (fractions) will leave the tower in side draws and we can predict the percentage of each cut.

Fractional distillation is the separation of different fractions of a mixture by vaporization and separate recovery of the vapors and residue. The mixture in Table 9-3 contains three different compounds that can be separated by fractional distillation since the boiling points of the components of the mixture are so far apart. This mixture can be sent to a distillation tower because a distillation tower separate mixtures by boiling point differences. Pentane, since it has the lowest boiling point, will boil off first. As pentane vaporizes and leaves the liquid mixture it can be collected as relatively pure pentane by condensing the vapor. This is what happens in a depentanizer tower. Hexane and heptane remain as tower bottoms. Now the remaining mixture contains predominantly hexane and heptane. We can send this mixture to the dehexanizer tower and hexane, which has the next lowest boiling point, will boil, vaporize, and be cooled in a condenser. The liquid remaining is almost all heptane in the tower bottoms. Boiling, condensing, and collecting have separated a mixture

Table 9-2   Boiling Range of Crude Oil Fractions

Crude Oil	
Component	**Boiling range**
Gasoline	100–400°F
Kerosene	350–550°F
Gas oil	500–900°F
Diesel fuel	400–600°F

Table 9-3   Three Component Mixture

Component	**Boiling Point**
Pentane	96.9°F
Hexane	155.7°F
Heptane	209.2°F

of three different hydrocarbons. No chemical reaction has occurred. The point being made is that large molecules boil at higher temperatures than smaller molecules.

ASTM is the acronym for American Society for Testing and Materials. This is a society that creates standards and procedures for testing various materials. Most processing industries use ASTM approved procedures that are recognized, accepted, and used around the world.

A distillation test is a physical test because the test does not change the structure of a compound. To determine if a kerosene sample meets specifications, a sample is collected and distilled in the laboratory. The ASTM Distillation Test simulates what goes on in a fractionating (distillation) tower, but it does it using only about 100 milliliters of sample and distills the sample on an automatic distillation apparatus that is programmed to heat up at a certain rate until a final temperature is reached. The temperature at which the first drop of condensate is collected is called the **initial boiling point (IBP)**. This first condensate contains the smallest molecules (lightest molecules) of the sample. The *endpoint* (EP) is the highest temperature the thermometer records. The compounds in the end point temperature (highest temperature) range are the large, heavy molecules. At the end of the test a print out will list the initial boiling point, the temperature of the various cuts, the end point, and what percent of the sample distilled in each temperature range.

Let's look at the scenario that follows to see how boiling point can reveal important information about our process. In this scenario the laboratory reported the following information about a unit's kerosene sample:

IBP Specification	**IBP Result from the Distillation Test**
340°F–355°F	365°F
EP Specification	**EP Result from the Distillation Test**
510°F–535°F	575°F

The IBP data reveals the unit's IBP is too high. What does this tell a tower operator? The tower is running too hot and driving too many large molecules overhead. The distillation results reveal that we were still collecting distillate at a temperature of 575°F, thus the kerosene's EP is too high. This tells an operator that his unit's kerosene contains more large molecules (called **heavies** or **heavy ends**) than specification allows and is not salable. It must go to an off-specification tank. The tower technicians must now verify that their tower temperature profile is within standard operating conditions, and if not, must troubleshoot the process to bring it back within SOCs. If the tower is indeed running too hot, that is the cause of off-specification product.

However, there is an alternate scenario to the off-specification kerosene. Suppose the tower temperature profile is within standard operating conditions. What could be the cause of off-specification product? The most logical conclusion is the tower feed has changed. Now the technicians must look at the laboratory results for the tower feed and see if they appear normal.

Naphtha boils in roughly the same range as motor gasoline. In general, the naphtha distillation range spans from about 100°F through 300°F–400°F, depending on the intentions and needs of the refiner. Refiners often produce two separate naphtha cuts, a light and a heavy fraction. They have rule-of-thumb boiling ranges of about 100°F through 175°F–200°F and 175°F–200°F through 300°F–400°F.

Look at Figure 9-1. It is a printout of an automated distillation of heavy cat naphtha (HCN) showing the data revealed to the processing unit. Analyses like this are entered into the laboratory's information management system that uploads it to the processing unit. Process technicians access this information on their consoles and review it for indications of unit problems. The IBP is 150.7°F, the EP is 433.1°F. When 50 percent of the sample has been collected the temperature is 296.5°F. This sample boils in a range between 150°F and 433°F. Remember, small molecules boil at lower temperatures than larger molecules. Naphtha with an IBP of 150°F has a larger amount of low-boiling material (lights) than naphtha with an IBP of 225°F.

```
DIST86-L, HCN
Sample        1788771
Operator      LAB

Start of distillation: 01/20/2006 03:41
End of distillation  : 01/20/2006 04:18

(Result with barometric correction)

      Obs/Corr   %/min        Evap.   Heat
  %      CF     or time    %    CF      W
IBP   150.7    7min18s    IBP  150.7
  1    183.0                1           80
  2    191.7                2   183.0   80
  3    197.5                3   191.7   74
  4    202.8                4   197.4   71
  5    207.2      63s       5   202.8   69
 10    221.3      5.4      10   219.2   51
 20    240.8      4.1      20   238.7   62
 30    260.0      4.4      30   258.3   67
 40    278.8      4.4      40   277.3   70
 50    296.5      4.4      50   294.9   74
 60    313.1      4.3      60   311.1   80
 70    331.9      5.3      70   330.5   83
 80    351.5      3.7      80   348.5  108
 90    376.5      3.8      90   372.9  127
 93    387.9    21min19s   93   383.0  149
 94    393.0                94  387.8  128
 95    400.5                95  393.1  128
 96    409.9                96  400.5  128
 97    421.9                97  409.7  128
 98                         98  421.7
FBP   433.1    3min 2s    97.5  433.1

Percent recovery       :     98.0%
Percent residue        :      1.0%
Total recovery         :     99.0%
```

Figure 9-1 Naphtha Automated Distillation Printout

Table 9-4   Distillation Results for Kerosene Sample

Result	Heater Oil °F	Kerosene °F
IBP °F	—	320
10% cut	406	—
95% cut	472	—
End point °F	553	508

Table 9-5   Specification Sheet for Kerosene

Test 8F	Heater Oil Specifications °F	Kerosene Specifications °F
IBP°F	—	330–365
10% cut	345–410	—
95% cut	NLT 470	—
End point °F	NLT 555	NMT 520

Note: NLT = not less than; NMT = not more than

ASTM Distillation tests can be used to indicate the volatility (ease of evaporation) of a product. Small molecules are more volatile than large molecules. A distillation curve showing significant amounts of low-boiling material represents a product of relatively high volatility. Now, let's look at some data and see how well we understand the importance of the boiling points of a mixture. Table 9-4 lists the results of an ASTM Distillation test. Compare the results with the product specifications listed in Table 9-5.

Compare the IBPs for kerosene. The unit is not meeting specifications. The kerosene is too light and more volatile than desired. Compare the 10 percent cut (10 percent collected) for the heater oil. The unit is okay on this sample. The EP for kerosene indicates the unit is also okay there. However, based on specifications the unit cannot sell this kerosene because it is off specification. The consequence of making kerosene that is off specification by just a few degrees is to sell it at a discount. The sales department will probably sell it at a loss, or if lucky, break even. Or the site might be able to blend the kerosene into some other product so it won't have to be sold at a loss. Or they might be able to blend the kerosene back to within specification. Whatever the unit does, management will look at the bottom line and ask, "Why did I pay the salaries of five operators, spend big bucks for electricity, steam, wear and tear on equipment, and end up with a product I can't sell?" Remember, *a processing unit is built to make a profit, not to reprocess material or sell material at cost or a loss.*

Reid Vapor Pressure Test

Vapor pressure is another common test of physical properties for hydrocarbon fuels, especially gasoline. Vapor pressure is the pressure exerted by a vapor in equilibrium with its liquid or solid phase. This means that as many molecules are leaving the liquid as a vapor as are condensing and returning to the liquid as a liquid. Vapor pressure is due to the speed of movement of the molecules in a vessel. When vessel pressure is high, the molecules are moving rapidly and striking the wall of the vessel more often. Molecules in a low-pressure vessel are moving slower and make fewer impacts. A common test used in refineries is the Reid vapor pressure test, an analytical method for the determination of the vapor pressure of fuels for spark-ignition engines.

If a mixture of small and large molecules is vaporized in a closed container the vapors exert pressure on the walls of the container and on the surface of the liquid. The greater the amount of vapor generated by the liquid, the higher the pressure exerted. Vapor pressure is important when determining safety procedures and the handling and storage of chemicals. Materials with high vapor pressures evaporate more easily and must be more carefully handled than materials with lower vapor pressures. Special precautions must be taken in transporting, handling, and storing substances with high vapor pressures. A high vapor pressure means high volatility and highly volatile substances tend to leak out of containers, piping, and vessels easier than less volatile substances. Vapor pressure is very important for trayed distillation tower control. Technicians responsible for the proper operation of trayed distillation towers should become aware of the importance of the vapor pressure differential between the trays.

A Reid vapor pressure test is conducted by putting some sample in a closed container and heating it in a water bath for a specified amount of time. Since vapor pressure changes as temperature changes, the vapor pressure is measured at a specific temperature of 100°F. The vapor pressure is the pressure produced by a substance at 100°F in a closed container.

Flash Point Test

The lowest temperature at which vapor is generated fast enough to flash when ignited is called the flash point. Flash point is important primarily from a fuel-handling standpoint. A very low flash point will cause fuel to be a fire hazard, subject to flashing, and possible continued ignition and explosion. In addition, a low flash point may indicate contamination by more volatile and explosive fuels, such as gasoline. A very important reason to maintain the flash point as high as possible is due to the electrostatic hazards in pumping distillate fuels. Never pump kerosene or diesel fuel into a tank or container that previously contained gasoline or other flammable material. These highly flammable residues may be ignited by static electricity generated from filling the tank.

There is an ASTM procedure for determining flash point and automated laboratory instruments that perform the test. Flash point tests are performed by heating a sample and exposing its vapor to a flame. As the sample increases in temperature, vapors leave the sample surface in greater amounts. The lowest liquid temperature at which the vapors flash is the sample's flash point. A flash is just a brief flame that burns all the available vapors and then goes out. Samples should be properly identified to warn laboratory technicians of possible flash point hazards. If a mixture containing small hydrocarbon molecules (lights) is blended with one containing larger ones (heavies) and then heated, the small molecules will vaporize first. The flash point of a blend is only slightly higher than the flash point of the lightest component. Remember this when submitting samples to the quality control laboratory.

The information contained in flash points can help technicians in several ways. One way is if the flash point of a product is normally quite high, but a test shows it to be low, the product may have been contaminated by a material with a lower flash point. Another way flash points help technicians is as safety information. Assume one of the petroleum naphthas below will be used as a cleaning solvent for equipment. Which would be the safest for a technician to use?

Naphtha A   Flash point 125°F   **Naphtha B**   Flash point 170°F

The answer is naphtha B. It is less volatile because it has a much higher flash point. As another example to emphasize this point, gasoline has a lower flash point than kerosene. Higher flash points provide increased safety. A person is safer around kerosene than gasoline.

Test for pH

The acidic or basic quality of a substance is a measurement of its *pH*. The pH scale ranges from a value of 0 (very acidic) to 14 (very basic), with 7 being neutral. Why is it important to know the pH of various substances and streams on a process unit? The pH of natural water is usually between 6.5 and 8.2. Most aquatic organisms have adapted to a specific pH level and may die if the pH of the water changes even slightly, which is why the pH of water discharged into the environment is carefully monitored for environmental compliance. No petrochemical site wants to make the local television news because they are responsible for a large fish kill. Also, certain substances are corrosive. They eat tissues and destroy flesh. They corrode piping and vessels, cause clotting of process materials, stop up drains and valves, and may initiate unwanted or dangerous chemical reactions. When a process unit sends a sample in for a pH test there is an important reason for that test. The technician should know and understand the reason.

Viscosity Test

Viscosity is the tendency of a fluid to resist flow. The property of a liquid being thick or thin can help distinguish one liquid from another. A thick, heavy liquid flows less easily than a thin liquid. Viscous petroleum liquids are usually composed of larger hydrocarbon molecules and boil at higher temperatures. Viscosity changes as temperature changes. As a liquid's temperature increases, the molecules move apart and also move more easily. A liquid is less viscous and flows more easily at higher temperatures.

Viscosity tests are conducted by measuring the time it takes for a specified amount of liquid to flow through a restriction or an orifice in a calibrated viscometer tube. Since viscosity changes as temperature changes, liquid viscosity is measured at a specified temperature. Viscosity tests are usually run in temperature controlled baths.

There are several systems for measuring viscosity. One common one is the *Saybolt viscosity* system that uses a Saybolt viscometer to measure viscosity as the number of seconds required to drain a fixed volume of liquid from the instrument. If 100 seconds is required to drain the instrument, the Saybolt viscosity is 100 seconds. Another way to measure viscosity is by *kinematic viscosity*. This is the ratio of the absolute viscosity of a fluid to the density of the fluid. Fluid is drawn into a kinematic viscometer, the viscometer is inserted into a liquid bath set at a certain temperature, and the time it takes for the fluid to move from a starting point to an end point is measured in hundredths of second. The time is multiplied by a conversion factor for the specific tube used.

Heavy, viscous liquids require more time than less viscous liquids to drain from the viscometer and have higher readings than low viscosity liquids. The fact that viscosity changes when temperature changes is an important consideration for lubricants. A lubricant that thins out very much as temperature increases is a poor lubricant for internal combustion engines. Technicians submit samples for viscosity tests for (1) product specification purposes, and (2) troubleshooting contamination of a liquid by another liquid. Technicians should also realize that their equipment requires lubricants with certain viscosity characteristics and using the wrong lubricant can damage their equipment. This is discussed in more detail in Chapter 11.

Specific and API Gravity Tests

Specific gravity (ASTM D792) is a measure of the ratio of mass of a given volume of material at 23°C to the same volume of deionized water. Specific gravity (SG) is especially relevant because plastic is sold on a cost per pound basis and a higher specific gravity means more material per pound. Specific gravity is a relative relationship and has no labels, no grams per milliliter, or pounds per cubic foot. By this, we mean a substance's specific gravity is related (relative) to another substance which acts as the standard. In the case of liquids, water at 60°F is the standard. For gases, air is the standard. Water has a specific gravity of 1.0 at 60°F. Specific gravity tests are very simple and many operators perform the test on their units for troubleshooting purposes and verifying tank truck and rail car contents of certain liquids.

A specific gravity test is conducted by pouring the liquid to be tested into a hydrometer cylinder and inserting a thermometer to determine the solution's temperature. Next, the appropriate hydrometer is lowered into the solution so that at least 25 percent of the stem is covered by solution. A hydrometer is a device for measuring the gravity of liquid substances (see Figure 9-2). The hydrometer is read where the surface of the liquid cuts across the hydrometer stem. A temperature correction chart is used to correct the hydrometer reading. The specific gravity is converted to a concentration, usually weight percent, using a curve or table.

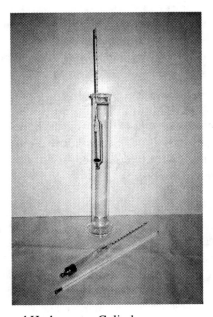

Figure 9-2 Hydrometer and Hydrometer Cylinder

API gravity is an arbitrary scale developed by the American Petroleum Institute in the early years of the petroleum industry. The API gravity scale was constructed so that API gravity increases inversely to density, thus the heavier (denser) the fluid the lower the API gravity. Higher value petroleum products (benzene, naphtha, gasoline) have higher API gravities. While the densities of most petroleum products are less than one, the API gravity scale was constructed so that most petroleum product values are between 10 and 70. The specific gravity of water is equal to 10 API.

The density of petroleum products is usually measured in units of API gravity with a hydrometer and hydrometer cylinder. The procedure is almost identical to that for specific gravity. If the hydrometer is placed in a low-density product, it will sink. The lower the API gravity value of a sample, the denser the sample. The higher the API value, the lighter the sample. Thus, if API gravity increases, the density of the sample decreases. API gravity samples should be analyzed at 60°F but since this is not always practical, the sample temperature is taken and the gravity is corrected to what it would be at 60°F using an API gravity temperature correction table.

TESTING FOR IMPURITIES

There are many tests for impurities in products. Impurities in finished products cause problems in whatever process or equipment they are used. The tests we will discuss are bottom sediments and water, cloud point, pour point, sulfur, and copper strip corrosion.

Bottom Sediment and Water (BS&W) Test

Water in crude oil is a harmful impurity and must be removed before it causes equipment problems. Water in fuel oil decreases the burning quality of the fuel. Solid particles in fuel (sediment) can clog valves, filters, carburetors, and injectors and decrease engine efficiency. Most products should be free of water and solids. ASTM requires gasoline to be visually free of undissolved water, sediment, and suspended matter. In the *bottom sediment and water (BS&W) test*, the sample is centrifuged, separating out the water and sediment to the bottom of the centrifuge tube. The tube has calibration marks so that the amount of water and sediment can be estimated and reported as a volume percent of the sample. In a similar manner, process equipment removes bottom sediment and water with *knockout pots* that remove water and filters that trap solids, such as rust and scale.

Cloud Point and Pour Point Tests

Cloud point and pour point are measures of winter temperature behavioral properties of distillate fuels. Many hydrocarbon products contain dissolved wax, which at low temperatures, may solidify and cause serious problems in fuel and lubrication systems. Both the cloud and pour point tests for the presence of wax.

Cloud point is the temperature at which paraffin (wax) first forms in fuel. Above certain temperatures the wax remains in solution, but as the hydrocarbon chills the wax accumulates in small particles, which makes the sample cloudy. Hence, the name *cloud point*. In practice, cloud point helps to determine the temperature at which paraffin crystals will begin to block fuel filters and lines and cause starting and stalling problems for diesel engines.

The *pour point* is the temperature at which the sample will no longer flow (pour) when the sample container is tilted at 90 degrees. The sample has solidified, or more correctly,

124

the oil has been boxed in by wax and can't flow. In practice, the pour point indicates the lowest temperature at which fuel can be pumped.

To determine the cloud point (and pour point), the sample is placed in a cold bath (usually made up of water and ice or alcohol and hot ice). The sample is contained in a long narrow jar that is corked and with a thermometer inserted through the cork. The sample is removed from the bath at specific temperature intervals (i.e., every 5°F decrease) and observed for the appearance of clouding. At the first appearance of clouding, that temperature is the cloud point of the sample. The pour point test is performed similarly to the cloud point test in that it uses the same bath, sample container, and thermometer.

Test for Sulfur

Sulfur is a chemical element that is normally found in varying amounts in all petroleum products. It is important that finished products have very little sulfur because sulfur is corrosive and some sulfur compounds have a very disagreeable odor. When a fuel containing sulfur is burned sulfur dioxide gas is formed. This gas combines with water and forms sulfurous acid. If this reaction occurs in an engine, the reaction can corrode the engine parts. A low sulfur value is very important for fuels and is also important for many chemical reactions. Sulfur can add color, odor, or undesirable compounds during chemical reactions.

Most sulfur tests are performed on gas chromatographs (discussed at the end of the chapter) or specific sulfur detection instruments that destroy the sample when it is injected into the instrument. Sulfur and some sulfur compounds react with hydrocarbons and become part of the hydrocarbon. They can't be removed by physical means, thus they must be detected by reacting the sample or destroying the sample in a test. Most sulfur detection tests are chemical tests.

Copper Strip Corrosion Test

The **copper strip corrosion test** is used to detect the presence of sulfur or other compounds that are corrosive to copper. If the sample corrodes copper, it will also corrode the metals that make up the piping and vessels within a processing unit. The purpose of the copper corrosion test is to ensure that hydrocarbon substances will not create excessive corrosion in vehicle fuel systems or equipment.

The test is performed by placing a brightly polished flat strip of copper into a small jar and covering the strip with sample. The jar is placed in a temperature controlled bath (usually 100°F) and left for a fixed amount of time. After the time has expired the copper strip is removed and examined for tarnish. A tarnished copper strip indicates enough sulfur to cause some degree of corrosion. The tarnish may be slight or severe. The tarnished strip is compared to a set of copper strip standards of varying degrees of corrosion. Each standard has an assigned numerical value (1, 2, etc.). Customers buy hydrocarbon products with a very low copper strip corrosion value to minimize corrosion of their piping and equipment.

APPEARANCE TESTS

Appearance tests are based on the appearance of a sample. Two of the most common appearance tests are *color* and *haze*.

Test for Color of Petroleum Products

Color specifications apply to many products, but in general, color provides an indication of contamination. The development of color in a product can usually be associated with low or poor product quality. Petroleum products may be divided into two general classes of color, *white* and *darker*. White colors vary from colorless (water-white) to a pale straw color and are usually made up of the smaller hydrocarbon molecules.

Saybolt color test is a physical measure of the clarity of a hydrocarbon and is used as an indication of the overall purity of the hydrocarbon. It is a useful parameter that assures freedom from trace contamination with heavier molecules, which may render the product unsuitable for designated critical applications. Such applications include home heating use, where conformance to minimum specifications is essential to ensure the proper burning of the fuel and safety.

Saybolt color is not only useful in determining heavier contaminants but is also helpful in determining the degradation of a product over long periods of storage. Since kerosene has a tendency to degrade faster than most other commercially available petroleum products, such as gasoline and diesel fuel, a higher kerosene inventory turnover is desirable. Some of the most frequent problems indicated by the Saybolt color test are improper storage of materials, careless handling, and using the wrong dispenser hoses (the hydrocarbon reacts with the hose material).

A Saybolt chronometer is the test instrument that measures the color of white products by comparing the depth of sample necessary to match the color of a glass standard. A liquid sample is poured into a long slender glass tube. A companion tube has no sample but a series of glass standards that can be rotated until a standard is found that matches the sample color. The technician peers through a binocular eyepiece to match the samples and releases sample from the sample tube until the sample color and glass standard match. Saybolt colors range from a +30 (lightest color) to a −16 (darkest color). The lightest color is often described as *water white*.

The Saybolt color test is for samples that are white or nearly white. Darker colors are measured with an ASTM colorimeter. A sample is poured into a colorimeter tube, inserted in the instrument and the technician peers through binocular eyepieces and rotates a wheel of colored glass standards until one matches or closely matches the color of the sample. Sample colors range from dark yellow to red, orange, and darker colors. Standards have color values of 1, 2, 3, and so on, with 1 being the lightest color. If the sample color falls in between two standards, the color may be designated as 1+, meaning greater than one but less than 2.

Test for Haze

Water and some chemicals can cause products to look hazy or cloudy. The haze test is conducted to determine if samples are relatively free of entrained water or substances that cause haze. The water that causes haze is not *free water*, which is water separate from the sample and usually located at the bottom of the container and easy to see. Water that causes haze is suspended in the sample (entrained) and since water and hydrocarbons don't mix, the sample appears hazy instead of being clear. The haze test is performed on a test instrument (spectrophotometer) that compares the amount of light passing through the sample to the amount of light passing through a standard (usually distilled water). Haze testing is done on product samples that must be relatively free of entrained water. Depending on the product, water can cause adverse chemical reactions, or in the case of fuels, pose a risk to icing up and blocking fuel lines.

GAS CHROMATOGRAPHY

Gas chromatography (GC) has been put in a category all by itself for two reasons: (1) it is a very complex subject that, due to time constraints, will be treated lightly and (2) it is one of the most common of on-line analyzers on processing units. GCs are such good analytical tools that quality control laboratories have dozens of them.

Gas chromatography is similar to distillation in that it separates a mixture but it does it much better. The separation allows identification and quantification of individual components of a mixture. The basic components of a complete gas chromatographic system include (1) a carrier gas supply, (2) a syringe for sample introduction, (3) an injection port, (4) a column and oven, (5) a detector, and (6) a data collection system.

The sample (a mixture) is injected with a syringe into the chromatograph through the injection port. The high temperature of the injection port vaporizes the sample components and keeps them in the gaseous state by the temperature of the column which is kept in an oven heated to several hundred degrees. The vaporized components are swept onto the column by the mobile phase, an inert carrier gas (usually helium or nitrogen). The components move through the column and separation occurs based on the affinity (attraction) of each component for the stationary phase. The GC column is composed of a coiled, tubular column and the stationary phase within the tube. The stationary phase is a dense, organic liquid compound coated on the column walls or the internal packing. Components of the mixture with a high degree of affinity for the stationary phase are slowed down while components with low affinity for the stationary phase move rapidly through the column. As a consequence of the differences in mobility due to affinities for the stationary phase, sample components separate into discrete bands that can be qualitatively and quantitatively analyzed. In other words, the GS will identify the different compounds in the sample and determine how much of each compound is in the sample.

Individual components of the mixture leave the chromatographic column and are swept by the carrier gas to the detector. The detector generates a measurable electrical signal, referred to as peaks, that is proportional to the amount of a component present. The position of the peaks on the time axis serve to identify the components and the area under the peaks provide a quantitative measure of the amount of each component.

AUTOMATED ONLINE ANALYZERS

As chemical processes become more complex, the rapid feedback of analytical results to the control room becomes more critical. The quality control laboratory analyzes samples for all the units in the plant, often on a first-come, first-served basis with samples stacked in line waiting for analysis equipment to be available. This results in long waits for analytical results (1 to 2 hours). The development of online analyzers helped resolve this long wait. Online analyzers automatically collect a sample, measure the required properties, and send the data to the control room without any help from an operations or laboratory technician. Analyzers can report data approximately every 15 to 30 minutes, if needed. Several analytical techniques are available for online analytical instrumentation, such as gas chromatography (GC), infrared analysis (IR), ultraviolet analysis (UV), and pH.

Online analyzers have some advantages. They analyze process material much faster than the laboratory and data is automatically uploaded into the unit database. The most important advantage is rapid sample analysis. Online analyzers also have some disadvantages: (1) they need attention to keep them calibrated and operating properly and (2) they are expensive, ranging from $25,000 to $100,000 or more per unit. Process units that need composition data as fast as possible rely on analyzers to help control their unit. Critical analyzers will be spared, but as online analyzers keep coming down on price more and more will be added to processing units. Instrument technicians are responsible for the maintenance and calibration of the analyzers. Process technicians are responsible for reviewing the analyzer data and notifying the instrument group when they suspect the analyzer to be malfunctioning.

SUMMARY

One of the most important jobs of an operator is to catch valid samples and submit them to the quality control laboratory for analytical testing. Analytical testing of unit samples verifies the unit is on specification, alerts the operator to a developing problem, and helps troubleshoot unit problems. Samples are collected routinely and analytical tests run on a regular schedule to assist a processing unit in remaining within standard operating conditions and producing products that consistently meet specifications. Technicians should understand the basic chemistry of their unit and what analytical tests are revealing about their process. They do not have to be a chemist to understand the information revealed by analytical test results.

Test results are no better than the sample submitted. A unit that submits bad samples (contaminated, dead-leg, fractionated) gets bad information. A valid (representative) sample is required to get useful data. The results are useless for controlling the unit or verifying product for shipment if the sample is contaminated or in any way non-representative.

Petroleum products have distinguishing physical properties, some of which are vapor pressure, API gravity, color, viscosity, and so on. Physical properties are measured by physical tests. Physical tests do not change the chemistry of the original sample. When kerosene is tested for its physical properties it still remains kerosene after the tests. It is unchanged. A chemical test involves a reaction that changes the characteristics of the original sample. Analytical tests can be separated into several categories, some of which are physical tests, tests for impurities, and appearance tests.

As chemical processes become more complex, the rapid feedback of analytical results to the control room becomes more critical. The quality control laboratory analyzes samples for all the units in the plant, often on a first-come, first-served basis with samples stacked in line waiting for analysis equipment to be available. This results in long waits for analytical results (1 to 2 hours). The development of online analyzers helped resolve this long wait. Online analyzers automatically collect a sample, measure the required properties, and send the data to the control room without any help from an operations or laboratory technician. Analyzers can report data approximately every 15 to 30 minutes, if needed.

REVIEW QUESTIONS

1. Describe three ways analytical tests support process units.

2. Explain the hazard(s) created when a sample is mislabeled.

3. What is a *check sample?*

4. Define *physical test* and *chemical test.*

5. List five physical tests.

6. Distillation towers separate mixtures based on _____.

7. Define *end point* and *initial boiling point.*

8. Discuss the importance of collecting valid samples.

9. Explain how a distillation test helps technicians troubleshoot a tower.

10. Explain how vapor pressure can inform a technician of the hazardous nature of a chemical.

11. Explain the information a flash point test yields.

12. Explain why a technician should understand the importance of pH results to his area of responsibility.

13. List two reasons why technicians submit samples for a viscosity test.

14. How are specific gravity tests helpful to a technician?

15. What does the cloud point of a substance determine?

16. Give two reasons why sulfur or sulfur compounds in finished product is a problem.

17. Explain why tests for color are important.

18. What does a *haze test* determine?

CHAPTER 10

Routine Unit Duties

Learning Objectives

After completing this chapter, you should be able to

- *Explain the purpose of a processing unit.*

- *Describe the duties of the outside technician (field technician).*

- *Define standard operating conditions.*

- *Explain why area rounds are important.*

- *List the things that a technician should address during area rounds.*

- *Describe how data is collected during area rounds and what is done with the data.*

- *Describe the duties of the inside technician (board person or control room technician).*

- *Describe the duties of the lead technician (chief technician).*

INTRODUCTION

A process unit is manned by crews, and members of these crews are given specific duties that encompass a job. Their jobs are described by jargon, such as the "inside man" and "outside man." Outside technicians have duties that keep them outside most of the time monitoring and maintaining the equipment of their area. On most process units there is more than one outside technician and they will have names referring to the outside section they are working, such as the "truck loading technician" or "reactor section technician." Inside technicians spend their time inside in front of a control board monitoring the unit and, where necessary, making minor adjustments. Some production sites have a technician called the "lead technician," a person who represents management and is responsible for the operation of the process unit. This chapter discusses the responsibilities of process unit technicians when their unit is operating normally and their duties are routine.

THE PROCESS UNIT AND ITS CREW

A process unit is built to make money and to show a profit. The process technician is hired to run the unit profitably. How long a unit remains in production depends on its ability to make a profit. That's the bottom line. The technician aids in making their unit profitable by knowing their job well and having the skills and knowledge to perform their job as required. A process technician is responsible for turning raw materials into salable products in a way that makes the operation profitable. The technician does this by operating a complex process efficiently and by a set of standard operating conditions (SOCs). A process unit that does not make money will be shut down or sold off and with it goes the jobs.

Operating units are composed of crews and typically have four crews on 12-hour shifts. Two crews are on duty while two are off. One crew is working the night shift while the other is working the day shift. Crews are usually named alphabetically, such as the *A crew* and *B crew*. Depending on the size and complexity of the operating unit, the crew size may be as few as two or as many as six or more. Each crew functions as a team and technicians are cross-trained to perform all jobs. Thus a technician having problems with equipment in their area of responsibility can rely on other technicians to assist them in controlling the area and keeping product on specification.

All process technicians should realize they were hired with the intent that they would learn their duties, perform them correctly, and make salable products safely and efficiently while seeking to continuously improve their process. Any technician not applying that attitude to their job is not seeking long-term employment.

DUTIES OF THE OUTSIDE TECHNICIAN (FIELD TECHNICIAN)
Outside technicians, or **field technicians**, are assigned to one of the outside areas of an operating unit, or they may be assigned to the docks or tank farm. They are responsible for the ongoing operation, maintenance, and security and safety of that area. Outside technicians spend a significant amount of time making *area rounds* to assess the status of their area. They should be highly qualified in their area and capable of training new hires or other technicians who are to be cross-trained in the outside technicians' areas. Because their crew functions as a team, the outside technicians (and inside technicians) will know the personality, moods, and skills of their fellow crew members and be able to work smoothly with them. They know that they are part of a team that can only function as well as its weakest member, and they work to ensure that they are doing their part for the team. Outside technicians know that their job will require some manual labor in all kinds of weather (rain, snow, heat, etc.) but that manual labor is the least important aspect of their job. More and more, process technicians are becoming knowledge workers and information and knowledge are critical to the success of their work.

Operate Unit Equipment

The proper operation of their area equipment is one of the most important duties of outside technicians. Damaged pumps are a tremendous expense to the processing industry. A moderate-sized unit, such as a styrene unit, may have from 100 to 200 pumps. Valves are expensive and valve seats are easily damaged. There are thousands of valves on a unit. Each process unit has an operating budget and when that budget is exceeded because of damaged equipment, unpleasant consequences can result for both management and labor.

As a rule, critical equipment to the operation of the process unit will be installed in pairs. For example, there will be two feed pumps to the reactor, a primary and a spare. If the primary feed pump fails, the technician must bring the spare on line to keep making product. Then the technician will lockout, tagout, purge, and write a priority work order for immediate repair of the spare. Keep in mind a technician's responsibility is *their area*. They are expected to have the skills and knowledge to keep their area within SOCs. No one will be standing at their side telling them what to do, when to do it, and how to do it.

When something happens to critical equipment, such as pumps that are spared, the damaged pump should be prepared immediately for maintenance personnel and the work order completed. The operating pump that is keeping that area of the unit running can fail and without a spare the unit goes down. There go the profit margins. A unit grossing a million dollars a day is down because of a $25,000 pump! The sooner the damaged pump is repaired and returned to duty as standby the better. Repair work should not be delayed. Spare equipment is useless if it isn't available as a spare.

When process equipment is taken down for maintenance, the outside technician is responsible for clearing the equipment to ensure no harmful material remains in it to pose a hazard to the maintenance technicians who will repair the equipment. Clear up of the equipment is done by purging with an inert material that will remove any hazardous material. The materials most often used for purging are nitrogen and steam, inert materials that will not mix or react with other chemicals to form harmful materials or create dangerous situations. Nitrogen, especially if it is heated, will clear most hydrocarbons and other contaminates found in chemical plant equipment. If the technician is dealing with a heavy or viscous material, such as heavy oil or molten polymer, steam may be used instead of nitrogen for purging.

Once the equipment is cleared and safe for maintenance personnel, the outside technician must ensure that it can not be started up if it is being repaired on site. This is done using the site lockout and tagout procedure. When maintenance declares the equipment to be repaired, the outside technician must verify the equipment is repaired. They should not just accept the word of the maintenance crew. They should briefly put the repaired equipment in operation and verify the equipment operates normally. It is a failure of responsibility not to do so. If the equipment does not operate normally and this is not detected before it is placed back in service as a spare, there could be severe economic consequences when the repaired spare is discovered to still be defective. The technician may be disciplined.

Monitor and Adjust Equipment

Occasionally outside technicians will have to make adjustments to their assigned equipment. By experience, they know the standard operating conditions (SOCs) of the process variables of temperature, flow, pressure, and level when their equipment is working properly. They also know the upper and lower limits (range) of each equipment variable.

Standard operating conditions (SOCs) are defined as an upper and lower limit that a variable must remain between to be acceptable. For example, if the technician is monitoring the discharge pressure of a pump and the acceptable limits are between 30 and 40 pounds per square inch gauge (psig), this range is considered the standard operating condition for that variable. Anything above 50 psig or below 30 psig is outside SOCs and must be corrected. Somewhere in the mid-range of the SOCs is the desired pressure. This desired pressure is called the **setpoint**. It is the responsibility of the technician to ensure that all their assigned equipment runs within the desired operating ranges.

Sampling and Sample Analysis

Shift technicians are required to collect unit samples several times a shift. At most sites this is three times during a 12-hour shift. Samples are collected to verify that all unit streams are within specification. In most plants the majority of samples are analyzed by the quality control laboratory, however, at some processing sites technicians have small lab rooms on the unit where they perform some testing. As discussed in Chapter 8, the trend today is for units to go to more on-line analyses (analyzers) and submit fewer samples to the quality control laboratory.

Samples will be collected from all important unit streams, some of which are unit feed, intermediate streams, utility samples, and finished product. A sample schedule indicates what samples must be taken, the type and size of sample container to use, and when to collect each sample. Samples may be solids, liquids, or gases. Sample containers may be plastic or glass bottles, metal cylinders for high-pressure gases (called sample bombs), plastic bags, boxes, buckets, and so on (see Figure 10-1).

If the processing unit has its own small laboratory, the unit technician will be trained to analyze some of their samples. Most of the tests will be simple physical tests that do not require complex analytical skills or knowledge. The laboratory may have automated test instruments that only require the technician to measure out a certain quantity of sample into the instrument and the instrument does the analysis. The technician will be required to understand the chemistry of their unit, what the analytical test is revealing about their unit, and how to properly analyze their samples.

Many samples will go into the unit sample box to be picked up by the control laboratory, or in some cases, the processing unit delivers their samples to the control laboratory. The control laboratory has specially trained laboratory technicians and more complicated testing equipment for sample analysis.

Samples are collected and analyzed to reveal composition of the unit streams, which in turn reveals whether the streams are within specification. Collecting samples is a simple, routine duty but it is very easy for an impatient technician to collect *bad* samples. A bad sample is a sample that does not represent the true nature of the material being sampled.

Material Handling

Material handling is the loading and unloading of tank trucks and rail cars, loading finished product, unloading solid raw materials, and the movement of materials from storage to

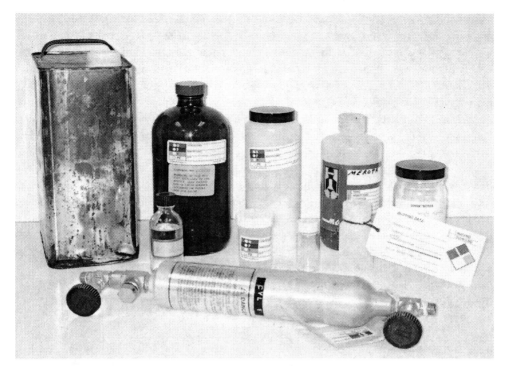

Figure 10-1   Sample Containers

processing units. This requires connecting and disconnecting hoses to vessels and manifolds, operating pumps, moving bulk materials in drums or large bags for discharge, etc. It may require that the technician operate forklifts. Material handling is covered in detail in Chapters 13, 14, and 15.

Housekeeping

Good housekeeping is mandated in site safety policies because good housekeeping is necessary to maintain a neat and safe work area. All employees have a responsibility to keep their work area clean. The work area is not just the outside area, it is also the bathrooms, lunchroom, changing room, etc. A technician can trip over a utility hose in their work area and injure their back or they can slip in a puddle of spilled coffee in the lunchroom and fracture their skull. Housekeeping is important. It requires that the technician ensure that all trash is picked up and all spills are washed down in their operating areas. They should also make sure that all hoses, ladders, and other equipment are in their assigned places when not in use.

Especially important in housekeeping is the lunchroom area. Technicians who have handled toxic materials should not be allowed to enter the lunchroom wearing their field gear. Toxic materials from their clothing, gloves, and hard hats may contaminate lunch tables, tabletops, and refrigerators. A contaminated lunchroom poisons food and drink.

Making Area Rounds

It is important for the outside technician to be aware of the status of equipment and process conditions in his area at all times. **Making rounds** is the one activity that keeps them aware of the equipment and conditions in their area of responsibility. A technician makes a round by going out into their area of responsibility and checking the equipment. Rounds are normally made every two hours, or more frequently, if needed. Usually, each outside technician has a **rounds list** that identifies what equipment they are to look at and what they are to look for. They must fill out and turn in (or upload) the list at the end of the shift. The rounds list is a long list of equipment to be checked, instrument readings to be recorded, inspections to be made, leaks to be detected, etc. Any problem detected that cannot be immediately corrected should be written up on the rounds list, reported to the lead technician or the shift supervisor, noted in the operations logbook, and then requested to be corrected or repaired by maintenance via a work order.

For a technician just coming on shift, the first round is very important. They have a fresh set of eyes and senses not jaded by the 12-hour shift and having already made several rounds. On the technician's first round they will be more alert and more critical of conditions. This is when their chances are best at detecting a developing problem. Later, after they have spent 10 hours on the unit and made three previous rounds, they may not be as attentive on their last round.

Many companies have done away with paper rounds lists and gone to programmed, hand-held mobile computers that allow data input with keystrokes (see Figure 10-2). The device may have jacks and leads that connect to inputs on electric motors or other equipment that upload data from the equipment, such as motor rpms, bearing temperature, etc. It also may have a bar code scanner that allows it to scan a bar code on equipment or an area and bring up the data page for the equipment or area. The advantage of such devices is that it can upload all the round data into a database onto a computer in the control room. Later, the technician or an engineer can pull up the data and view trends (temperature, vibration, rpm, etc.)

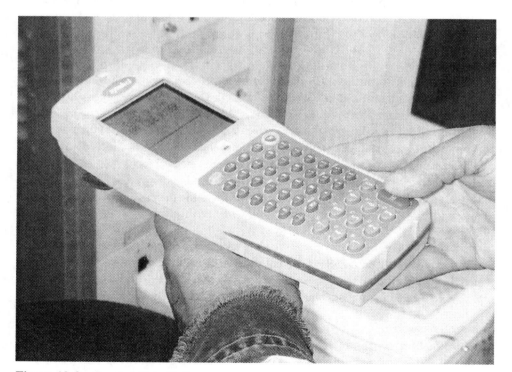

Figure 10-2 Symbol Data Gatherer

for the area equipment. This aids in predictive maintenance, troubleshooting, and turn-around planning.

The human senses are like fine-tuned instruments gathering data. Technicians should be alert to what their senses are telling them when making their rounds. They should know which process lines should be hot or cold and a prudent use of their sense of touch can verify line temperature. Normal pump and motor bearing temperatures range from 110°F to 120°F. The sense of touch is a useful indicator of a pump operating at normal temperature. Technicians should be able to hold their hand on the pump and motor casing above the bearings if they are less than 130°F. A motor bearing that needs lubrication or an overloaded motor sounds quite different from a properly functioning motor and the sense of hearing can detect these mechanical problems. Chemical leaks or overheated electrical equipment can be detected by olfactory senses. The sense of sight provides information on instrument settings, gauge readings, equipment alignment, leaks, etc.

Outside technicians should not be complacent when making rounds. The purpose of making rounds is surveillance, which is defined as *to watch over*. There are thousands of feet of pipe, thousands of valves, countless joints and flanges, and scores of pieces of equipment, all of which are capable of failure or leaks. Good technicians will not hurry through the rounds because they want to avoid the heat, mosquitoes or rain. Failure to be alert to a small leak of flammable material can lead to a dangerous situation and a big mess to clean up. More time will be spent outside cleaning up the mess than would have been spent making a proper round. While walking the area, technicians should be alert to low areas, such as sumps, sewers, and drains, and the high overhead areas, such as pipe racks, upper levels of towers, or other elevated equipment. They should be aware of any non-routine activity dictated by daily orders of the unit engineer such as a temporary process change or new equipment. Technicians should also be aware of activities in areas adjacent to their area (turnaround on the adjacent unit, street repairs, etc.), which may affect their operations.

When necessary, technicians will make adjustments to levels, flows, temperatures, and pressures to keep the process at SOCs. Usually, this will not be necessary since so many units are highly automated today, however, some manual adjustments may have to be made for waste streams, utilities, drains, etc. Good technicians should be aware of the name, location, function, and normal operating parameters of every piece of equipment in their area. They should know how these parameters are controlled and the consequences if they are not controlled properly. While making changes in the field, outside technicians should radio the board person of these changes. Good communication is important at all times.

Safety and Emergency Equipment

Outside technicians should be familiar with safety and environmental procedures and regulations that apply to their unit. They should know the hazards of the process materials, especially if these materials are toxic, flammable, or explosive. They should know the reportable quantities of each material and how to minimize environmental damage if a spill or release occurs. Outside technicians should remember to occasionally glance up at the windsocks at the top of the distillation towers that indicate the wind direction, which is primary safety knowledge. In the event of a release or fire the technicians should move upwind to avoid toxic or dangerous vapors, gases, or hazardous combusted products.

Technicians are protected by the safety and emergency equipment in their area of responsibility. Defective or outdated safety equipment offers no protection. Technicians should regularly inspect all safety equipment in their area of responsibility and report or replace defective equipment. Usually, a weekly inspection of area safety equipment is included in the rounds list. Safety equipment to be inspected might include

- Fire extinguishers
- Fire monitors
- Breathing air equipment
- Safety showers and eye wash stations
- Relief valves
- Fire blankets
- Unit fire hoses

All emergency equipment in a technician's area of responsibility should be routinely checked to ensure that it is ready for use in an emergency situation. Emergency equipment could be spare radios, medical supplies, breathing air cylinders, uninterruptible power supply (UPS) system, diesel generators, etc.

Equipment and Preventive Maintenance

The outside technician has extensive preventive maintenance duties. These duties include lubricating equipment, tightening valve packing nuts for packing leaks, changing and cleaning numerous in-line filters, replacing gaskets, changing small valves, servicing steam traps, cleaning cooling tower and blower suction screens, etc. These are simple mechanical duties similar to replacing a spark plug in a car or a water faucet outside of the house. More complicated maintenance items are left to plant maintenance or contract maintenance. Technician maintenance duties are discussed in greater detail in Chapters 11 and 12.

Technicians will monitor lubrication systems, rotating equipment vibration, and monitor pump glands and seals. Lubrication of rotating equipment is one the most important tasks of outside technicians. Good technicians will make sure that all rotating equipment in their area are lubricated properly. They will check the flow, level, pressure, and temperature of all lubricant streams and will follow the lubrication schedule described on their rounds list.

About 50 percent of all bearing failures on rotating equipment are caused by improper lubrication. Many pieces of rotating equipment have separate oiler systems that require monitoring the oil temperature. Critical pieces of equipment may have high temperature and low oil level alarms. Technicians must develop a feel for the normal sound and vibration of their rotating equipment. Large, critical equipment may have vibration monitors that alarm or shutdown the equipment before severe damage or a hazardous condition occurs. Rotating equipment should be checked frequently to ensure vibration does not become excessive and hazardous. Any pieces of rotating equipment with excessive vibration should be shut down as soon as possible and repaired.

Packed pumps are designed for a small leakage of the lubricating oil to the atmosphere. If this leakage becomes excessive, the packing gland should be adjusted to prevent wasting oil to the sewer. Many pumps are equipped with mechanical seals that are designed to keep the pumped material contained within the pump. Seals can develop leaks.

Other Duties

Technicians should become familiar with all process lineups in their area of responsibility. Improperly aligned valves can result in wasted product, cross contamination of materials, environmental incidents, or pump damage. They should learn the normal operating position of all the control valves in their area of responsibility and verify the valve positions when making their rounds. Technicians may have a large number of control valves in their area, but it only takes a second to note the position indicator on the control valve. A change in control valve position could be an indication of improper flow rates, leaking instrument air systems, or unusual wear of the valve trim. Technicians should also be aware of any control valves in their area that are operating with the bypass partially open.

Good technicians should be constantly looking for leaking piping and equipment. Leaks, no matter how large or small, have consequences. Leaks, depending on the nature of the material leaking, are a source of fugitive emissions and a hazard due to their toxins or flammable nature. They also represent lost raw materials or salable product. As soon as a leak is detected it should be logged and plans made to stop the leak. See Table 10-1 for a list of the duties for outside technicians.

DUTIES OF THE INSIDE TECHNICIAN (BOARD PERSON)

The **inside technician**, also called the **board person**, is responsible for the operation and control of the entire process unit from the board screen (console) in the control room and making adjustments as needed. When the unit is lined out and running smooth it is an easy, laid back, air-conditioned job, but when the unit is upset, it is a high stress job. In the past, only a few technicians were trained as inside technicians and these were considered the brightest and best. The trend today is that all technicians are trained on the control board.

Knowledge of the Process Unit

Outside technicians work with the individual pieces of equipment in their area. The outside technicians might think of their area as one large system or several small systems that make up one large system. The inside technicians look at the entire processing unit. If there are four outside areas, the inside technicians look at all four at the same time or individually, depending on which screen they pull up. They have it all right in front of them, four big systems or a dozen small systems that make up the four big systems. Their concern is making all these individual systems work together as one processing unit that makes an on-specification product. The inside technician must monitor and control pressures, temperatures, flow rates, and levels of the entire process. They must understand how the various systems that make up the process interact. When things go wrong management calls them and asks, "How did this happen and when are you going to fix it?"

In the past the rule has been that a technician should not be an inside technician, or a board person, until they have experience in all the outside areas of the processing unit. This is not the rule today. It stands to reason that the inside technician, viewing the entire unit, should be familiar with and understand the entire unit. Simply put, the inside technician should be an expert on the process. This is the best case scenario.

Inside technicians must understand their unit. They must understand what causes undesirable side reactions, produces-off specification materials, or has the potential to start a runaway reaction. Inside technicians should be familiar with the critical SOCs of the unit and

Table 10-1 Outside Technician Duties

Task	Skills and Knowledge Required
1. Prepare equipment for maintenance during normal operation and turnarounds	Safety regulations for lockout/tagout, clearing procedure, paperwork, and forms
2. Assist lead operator and other technicians as required	Know all job positions on the processing unit
3. Prepare for step-up as lead operator	Know all duties of the lead operator
4. Assist in training other operators	Know all duties; know how to instruct and communicate
5. Communicate effectively	Good writing skills for logbooks and work orders; good radio protocol; good oral communication for interacting with support groups and shift passdown
6. Perform material handling	Know how to load/unload rail cars, tank cars, barges, and ships; transfer material between storage tanks
7. Verify area and job documents	Read and verify logbooks, loading reports, certificates of analysis, chemical consumption, etc.
8. Comply with all DOT, OSHA, EPA, and state and local regulations	Know and understand the regulations and where to access them
9. Make rounds and collect samples	Equipment location and operating parameters, sample points, proper sampling technique, hazards
10. Contribute to quality improvement and cost savings	Use quality tools to improve the process in your area; membership in a quality team or cost reduction committee
11. Promote teamwork	Understand and apply teamworking skills and knowledge to strengthen your team
12. Respond to emergencies	Know the location of emergency equipment and the correct response to unit spills and releases

Note: The duties above are meant to be comprehensive but not all inclusive. The operator will be expected to perform duties not specifically listed but necessary to continue the safe and efficient operation of the unit.

understand how to control the unit's reactions if it becomes necessary to make corrections. When things go wrong and the distributed control system screen starts sounding process alarms and flashing out of SOC warnings, they must remain calm and not overreact. This is hard to do when what had been a peaceful, dull distributed control system screen is now flashing abnormal variable values and sounding alarms. The inside technicians are feverishly acknowledging the alarms and trying to determine what the heck happened. The worst thing they can do is panic and overreact. To do that would be to make a bad situation much worse.

The inside technicians should have a thorough knowledge of the major pieces of equipment on the unit. They should know the purpose and limitations of the equipment and its location and piping and valve arrangements so that they can work and coordinate with the outside technicians to start up, run, and shut down any piece of equipment or part of the unit. In non-routine situations radio contact and close coordination of the inside and outside technicians will be critical.

Process Control

Inside technicians should be familiar with the various types of process control systems. Local mounted control panels may control parts of the process or the control room may have a combination of control systems, such as pneumatic controllers, electronic controllers, and a distributed control system (DCS). With a pneumatic or electronic control system, a diagram of the process may be located on the wall of the control room. The diagram will make reference to the controllers, gauges, strip charts, and other instruments mounted on panel boards on the control room walls. The technicians should be familiar with the location and function of each controller.

A unit controlled by DCS will have a control room distinctly different from a control room using pneumatic or electronic controllers. A distributed control system (DCS) uses a computer screen to interface with the process. All controlling is done from a keyboard and bank of computer monitors. Each DCS will have a series of screens schematically displaying major equipment and piping. As an example, for a distillation tower there may be five screens. One screen would show the entire tower and associated equipment, the others would show the overhead section, the feed section, the tower bottoms section, and tower temperature and pressure profile. Each screen will provide a dynamic diagram of the section, using color-coding to indicate operating equipment, valves, and piping. Below each piece of equipment the operating parameters of temperature, pressure, flow rate, and level will be displayed. By use of a computer mouse or keyboard, the controller for the individual piece of equipment can be brought up and adjusted. An inside technician can make adjustments and then return to the area screen to view how the system responds to the change. Interlock alarms will also flash on the screen, which a technician will have to acknowledge and then seek the cause for the interlock activation. The DCS software keeps a record of all changes, alarms, alarm acknowledgments, and activated interlocks. Inside technicians can view this data back several days or weeks to search for trends and troubleshoot unit problems.

Coordinating Work with the Outside Technician

The DCS can't do everything. Sometimes controllers or instruments fail and those functions are taken over manually. The outside technicians and inside technicians must closely work together and stay in contact principally by two-way radios set to the unit's frequency. An inside technician is monitoring and controlling the entire process unit and may radio an outside technician to make an adjustment in their area or to check a piece of equipment that isn't responding correctly. Outside technicians inspect and troubleshoot pieces of equipment and keep the inside technician informed so that, if needed, they can plan operations around this piece of equipment.

Coordinate Maintenance Work

Inside technicians are the coordinators of maintenance work. The function of the maintenance division is to keep equipment running and fix equipment that is not running.

However, they cannot accomplish their mission independently. Maintenance cannot go into an operating area and start working without coordinating and scheduling their work with the operations group. This requires coordination with an inside technician or lead technician, who in turn may delegate the coordination to the outside technician with the damaged equipment. Inside technicians know when they can schedule a piece of equipment for maintenance work. Inside technicians make the equipment available for maintenance by arranging for the outside technicians to bring the spare equipment online.

When critical equipment fails and must be pulled and repaired immediately, the inside technicians, maintenance personnel, and unit engineer must plan to get the job done with a minimum or, better yet, no amount of downtime. Can the unit or parts of the unit be put in circulation mode? Can they bring down just one section? Should they call out extra personnel on overtime? Unit operations personnel will need to get the process stabilized, get the faulty equipment cleared, locked out, and ready for maintenance. With an inside technician as coordinator, maintenance personnel will work closely with operations personnel during the repair work.

Quality Control

Inside technicians should be familiar with the specifications of all unit streams, and especially unit final products. The control laboratory data and analyzer data will be uploaded into a database, and then accessed from the unit DCS. Inside technicians can pull up the data screen and view the data, which they should do on a regular basis. Out of range data will be flagged and catch their attention. Inside technicians should understand what the analytical data received from the control laboratory is telling them about their process. Understanding the data guides them in making minor adjustments to keep the unit producing on-specification product.

When laboratory or analyzer data reveals out of range variables the inside technicians should not jump to conclusions and decide that either the unit or a section on the unit is in trouble. The proper action to take is to view that section of the unit on their screen and check the SOCs of the equipment and streams. If they look normal, they should radio an outside technician and ask if they are aware of any problems in this area. Next, they would question if the outside technician noticed anything unusual about the sample that yielded the bad results. Lastly, to be on the safe side and to satisfy the International Organization for Standardization requirements, ISO 9002-2000 (if the unit is ISO certified), the inside technicians will request the outside technician to catch another sample and submit it for analysis. Often, bad analytical results are due to a bad sample.

Inside technicians may also tabulate data necessary for the final product lot records and certificates of analysis. They also may work closely with the shipping coordinator, verifying there is enough material in the product tanks for the quantity of rail cars or tank trucks to be loaded the next day.

Emergency Situations

The inside technicians, or board person, must be ready to respond to a variety of different emergency situations, such as

- Loss of electrical power
- Loss of cooling water
- Loss of instrument air or steam
- A major fire or release on the unit

The Occupational Safety and Health Administration's (OSHA's) Process Safety Management Standard, 29 CFR 1910.119, mandates that procedures for non-routine situations be in place. Assume there is a loss of steam to the unit. An inside technician can access the procedure for this emergency and begin implementing it. While the inside technician is doing this, engineering controls built into the unit during the design phase will have caused certain valves to fail open or fail close in just such an emergency. Long term loss of steam is a rare situation because most sites have more than one boiler. It may be possible to keep part of the unit up and on circulation while a lineup is completed to route in steam from another boiler. In severe situations the inside technicians, plant manager, and unit engineer must make a decision:

- Bring the entire unit down (very expensive).
- Keep sections of the unit on circulation (unit is not down but is not making product).
- Limp along as best as possible (making some product but not enough to pay expenses).

If the unit must be shut down, it must be come down as safely and efficiently as possible. The inside technicians will be aware of emergency back-up systems, such as emergency power generators and the rerouting of steam. Even as they are reacting to the emergency situation and bringing the unit down or into the circulation mode, they are already thinking of bringing the unit back up and making the unit profitable again. Inside technicians (and every technician) should think like businessmen. They should be aware if their unit is making money or going in the red. Employment is dependent on profits. Inside technicians and outside technicians will be planning the steps to bring the unit back up after the emergency situation has been corrected. During shutdown they will flush or purge certain vessels and lines to remove viscous or polymerizing materials that could plug lines and equipment.

In the event of a catastrophic situation, such as a major fire or toxic release within the site, the board person becomes responsible for communications as well as process control. If the emergency occurs on an adjacent unit, they will keep personnel on their unit informed about the situation and plan how their unit is to respond if the situation begins to affect the unit.

HEALTH, SAFETY, AND ENVIRONMENTAL ISSUES

Inside technicians, like outside technicians, should be familiar with safety and environmental procedures and regulations. They should know the hazards of the materials they are monitoring and controlling, especially if these materials are toxic, flammable, or explosive. They must also know the reportable quantities of each material and how to minimize environmental damage if a spill or release occurs. Inside technicians must know how to report a release or spill to federal and state agencies, and management. See Table 10-2 for a list of inside technician duties.

DUTIES OF THE LEAD TECHNICIAN

The position of **lead technician**, or **chief technician**, used to be more common 30 or 40 years ago but was done away with at many sites when shift supervisors assumed the same duties. Shift supervisors were part of management and usually were technicians who left an hourly position and accepted a salaried position with management. In unionized plants, shift supervisors could not do technician work so their duties were restricted to

Table 10-2 Inside Technician Duties

Job	Knowledge and Skills Required
Answer control room phone	Phone etiquette, communication skills
Monitor/use control room radio	Radio protocol
Monitor the physics and chemistry of the process unit	Basic chemistry and physics
Operate the DCS	Place instruments in auto or manual mode, monitor variables, respond to alarms, look for trends, etc.
Coordinate unit emergencies	Use radio and DCS to assist outside operators; effective communication
Remain within regulatory compliance	Assist in controlling spills and releases
Monitor control laboratory analytical data for the unit	Respond to suspicious data, request new samples
Fill and schedule overtime	Understand the overtime guidelines and fill out overtime log
Maintain the unit's ISO 9001-2000 system	Understand and follow policies and procedures developed to maintain unit ISO registration
Create shift schedule	Understand crew rotation sequence and who is qualified to work which areas

Note: These duties are meant to be comprehensive but not all inclusive. The technician will be expected to perform duties not specifically listed but necessary to continue the safe and efficient operation of the unit.

payroll, planning, scheduling, preparing bills of laden or certificates of analysis, etc. Today, because of the fiercely competitive environment, corporate management realized that the shift supervisor was a perfectly good technician who was no longer doing technician duties. Shift supervisors are going away and lead technicians are coming back because lead technicians can help perform technician duties plus do payroll and scheduling. Today's lead technicians may or may not be in a union, and if they are a member of the union, they still will perform what in the past had been management duties.

A lead technician is not a management position though they do what was once considered to be management work. The incentive for the position is an increased pay rate. A lead technician assigns unit jobs to technicians, schedules maintenance and product loading activities, holds shift meetings to relay information about unit operating changes and discuss accidents and incidents within the plant, and schedules overtime.

SUMMARY

A process unit is built to make a profit. The process technician is hired to run the unit profitably. How long a unit remains in production depends on its ability to make a profit. That's the bottom line.

Process technicians fall into three categories: outside technician, inside technician, and lead technician. Each has different duties and responsibilities. Outside technicians, also called field technicians, are assigned to one of the outside areas of an operating unit or to the docks or tank farm. They are responsible for the ongoing operation of that area, for the maintenance of the area equipment, and the security and safety of the area. Outside technicians will spend a significant amount of time making area rounds to assess the status of their area. Inside technicians, also called board persons, are responsible for the operation and control of the entire process unit from their console in the control room. Their primary goal is to monitor the board screen and make adjustments as needed. Lead technicians assign unit jobs to technicians, schedule maintenance and product loading activities, hold shift meetings to relay unit information and discuss accidents and incidents within the plant, and schedule overtime.

REVIEW QUESTIONS

1. List the three technician positions on a process unit.

2. Briefly describe the duties of the outside technician.

3. Briefly describe the duties of the inside technician.

4. Briefly describe the duties of the lead technician.

5. Explain what the outside technician should do when maintenance informs them that their equipment is repaired.

6. List three maintenance duties of the outside technician.

7. Explain why technicians collect samples.

8. Define a *bad sample*.

9. Explain why technicians should not wear their field gear in lunch rooms.

10. What is the purpose of a technician *making rounds*?

11. For a technician coming on shift, why is the first round so important?

12. Describe some of the knowledge an outside technician should have regarding his safety and health duties.

13. List five pieces of safety equipment the outside technician should inspect.

14. What would a noticeable change in control valve position indicate to a technician?

15. List several consequences of leaking piping and valves.

16. Describe the course of action a technician should take when laboratory or analyzer data reveals out of range data.

CHAPTER 11

Maintenance Duties: I

Learning Objectives

After completing this chapter, you should be able to

- *Define the acronym MTBF and explain why it is an important concept.*

- *State the factor responsible for the majority of equipment failure.*

- *Discuss why equipment breakdowns occur.*

- *List six benefits of operator maintenance.*

- *List eight maintenance duties of technicians.*

- *Discuss the importance of lubricating equipment properly and with the correct lubricant.*

- *Explain how lubricants work.*

- *Explain how water gets into lubricants.*

- *List five lubrication systems.*

INTRODUCTION

The process technician's chief concern is efficient process operations. Any process that is not properly cared for will not operate efficiently just as any automobile not properly cared for will not operate efficiently. Proper equipment maintenance is critical to efficient process operations. More and more plants are assigning preventative maintenance duties and responsibilities to the process technician. They are in the best position to do the routine maintenance and preventative maintenance required to keep their equipment running properly because they work with their equipment everyday. In addition, the process technician knows the equipment well enough to sense when it is not working correctly.

EQUIPMENT RELIABILITY

There are usually only three basic equipment maintenance requirements: (1) cleaning, (2) lubrication, and (3) bolting (tightening). The role of a process technician is critical in achieving equipment reliability. Just as the most reliable and well-designed automobile will fail in the hands of a thoughtless or inexperienced owner, the best and most reliable equipment will not perform optimally if the process technician lacks training, resources, or motivation. As the driver of an automobile accepts responsibility to monitor the car's dashboard instruments, an operator must accept equipment surveillance as their prime responsibility.

When process technicians make their rounds they are surveying their area of responsibility. The purpose of surveillance is to spot deviations from normal operations. Some surveillance (electronic) can come from the control room console, but in most cases, only the technicians' eyes and ears will do the job. An example of this is the fact that some refineries in the United States have centrifugal pumps with a *mean-time-between-failures (MTBF)* of eight years while others report MTBFs of only two years. Each refinery purchased the same pumps from the same manufacturer, have both unionized and non-unionized employees, and new versus old sites. The difference in equipment reliability and the millions of dollars saved is due to the training and knowledge of site process technicians in preventative maintenance.

Some processing sites have the philosophy that accidents do not just happen as some random event but are caused. The same philosophy can be applied to equipment breakdowns. They are due to human errors, such as negligence and ignorance. Most of the time there is a human factor behind equipment failure. Since breakdowns rarely happen due to random events, they are preventable. Failure investigation may point to various reasons for failures but the root cause of most failures is human, as revealed in Table 11-1. One survey found that approximately 45 percent of rotating equipment failures were due to operator control (or lack of control). Equipment shock (rapid temperature changes, allowing equipment to run dry) accounted for 25 percent of failures and contamination of the lubricant and/or lubrication systems accounted for the other 20 percent. The other ten percent could not be explained.

While ignorance and lack of skill can be overcome by proper training, the human attitude and mindset toward equipment failure is more difficult to change. Failure investigations tend to find the root cause in things such as components, electronic circuits, loads, etc., but often ignore the human aspect. Breakdowns will occur unless all people associated with the equipment change their way of thinking and behavior from an "all equipment fails" attitude to one of "how do we prevent equipment failure?"

Equipment is said to have broken down if it fails to perform its standard function, which means there is disruption of normal equipment operation. This means that not only function

Table 11-1   The Most Common Causes of Equipment Failures

Failure	Fault
Faulty design	Manufacturer
Faulty selection	Technical division
Faulty installation	Contractor or maintenance
Misalignment	Maintenance
Excessive load	Operations
Excessive heat	Operations
Excessive lube oil particles or moisture	Operations
Repetitive failure	Operations or maintenance

loss but also function reduction is regarded as equipment failure/breakdown. Equipment stoppage is catastrophic breakdown, not just breakdown. In the function reduction condition, continued equipment operation results in defective product, reduced output, frequent stoppages, noisy operation, reduced speed, or unsafe conditions. Performance deterioration due to wear and tear, loss of fit between parts, low-voltage, poor insulation, or leakage is termed **function deteriorating breakdown**. For instance, a dim or flickering fluorescent light is a function reduction breakdown. It is normal to assign importance to catastrophic breakdowns where there is total stoppage of equipment and/or production loss. However, function deterioration/reduction conditions that are not given due attention can lead to catastrophic breakdown.

Why Breakdowns Occur

Most breakdowns occur during startups and shutdowns, however, equipment failure of recently overhauled equipment could also be due to poor maintenance. Causes that evade notice are termed **hidden defects**. The key to achieving zero breakdowns is to uncover and correct these hidden defects before a breakdown actually occurs. This is the function of preventive maintenance.

Breakdowns are only the small visible tip of the iceberg. Hidden defects such as dust, dirt, sticking, abrasion, looseness, leakage, corrosion, erosion, deformation, scratches, cracks, temperature, vibration, and noise are the collection of abnormalities that lie beneath the iceberg's surface. The tendency to overlook minor defects results in minor defects becoming major defects. Various justifications are given for not attending to detected minor defects, such as the perception that such defects will not contribute to a major breakdown, the maintenance cost is not worth the cost of eliminating defects, and the inclination to obtain the maximum life of the component before it fails (i.e., run-to-failure). The decision to run a piece of equipment known to be in a function deterioration mode until it fails is usually based on economics. Although the run-to-failure approach can lead to emergency outages and very costly failures, there may be a good reason for not dismantling running equipment for checks or overhauling, such as when advanced condition monitoring systems are available to check the equipment. Depending on the piece of equipment and production schedules, the run to failure philosophy can be a common practice.

Preventive maintenance jobs such as lubrication, cleaning, bolting, and minor adjustments that do not require any major equipment disassembly cannot be neglected. Preventive maintenance has its own justification and reward. When equipment breaks down catastrophically, defects multiply and more components need to be replaced. Where timely minor cleaning or adjustment would have required the replacement of only bearings, catastrophic breakdown may require the replacement of bearings, shaft, races, etc. Breakdowns rarely occur due to a single major defect but are usually due to the combined effects of several minor defects. The interaction of minor defects influences each other and magnifies the damage. Minor defects can have profound effects on component life. If minor defects are not removed as soon as they are noticed, the initial failure damage may be obscured by subsequent damage and the root cause may not be known and corrected. And because the subsequent damage obscures the initial damage repetitive or chronic failures may occur.

It is very important when technicians are doing a failure analysis not to overlook minor defects. Failure analysis reports should include a "lessons learned" section that is human-oriented. As with accidents as well as equipment breakdowns, every failure has a lesson to teach. Unless we learn from the failure and apply the lessons, failures will repeat themselves.

Adhering to Operating Conditions

All machines are designed for certain operating conditions such as voltage, rpm, pressure, temperatures, utilities, etc. Operations and maintenance staffs should be thoroughly familiar with the equipment operating/rated conditions and should not run it under off-specification conditions. Running the equipment off-specification can lead to deterioration. There are two types of deterioration: (1) natural deterioration due to aging, wear, heat, etc., and (2) forced deterioration due to improper lubrication, choking, cavitation, dust, abrasives, and so on.

The probability of failure in a well-oiled and well-maintained machine is negligible. Lubrication retards wear and tear. Cleanliness helps to prevent corrosion due to leaks and moisture. Hence, maintaining basic conditions such as lubrication, cleaning, and decoking are essential for reliable and trouble-free equipment. Initially, preventative maintenance should be based on vendor recommendations and later updated by actual operating experience.

Operator Maintenance

When the equipment owner does maintenance, it is known as *operator maintenance*. Operators are considered to be the owner of its equipment. Operator maintenance has several benefits (see Table 11-2) and is a major step toward achieving zero breakdowns. A good system does not allow equipment to wait for maintenance. Operators are available around the clock near the equipment so they can continuously watch for problems while the maintenance staff is usually available only during the general shift in their maintenance workshops. In operator maintenance, the operator provides first aid to the equipment as soon as symptoms are noticed.

The process technician should look after the cleaning, oiling, bolting, servicing, minor adjustments, and minor repairs, such as attending to leaks, daily checks and preventive maintenance. The maintenance technician can then concentrate on specialized repairs, overhauling, alignments, reliability improvement measures, and solving chronic failures. If the maintenance staff has to do daily routine jobs that operations can do, and also do overhauling and failure analysis it will be too busy putting equipment back into service to have

Table 11-2 Benefits of Operator Maintenance

Benefits of Operator Maintenance Programs
Significant drop in unplanned maintenance, emergencies, and chronic failures.
Uninterrupted production, good quality output, less rejections.
Reduced catastrophic breakdowns, eventually tending toward zero breakdown.
Reliable and safe operation.
Reduced downtime during repairs because of timely maintenance.
Low maintenance, inventory, and operating costs.
Reduced manpower requirement.
Cleaner plant, more pleasant atmosphere to work in.
Enhanced knowledge and understanding of the equipment for operator and maintenance staffs.
Reduced employee stress or tension for operations and maintenance staff.

time for a thorough failure analysis. Thus, failures will continue because the mechanic has to skip certain critical measurements because of the backlog of work requests.

Daily equipment checking by look, listen, and feel should be a duty of the operator. Table 11-3 shows some of the maintenance duties of an operator. The precise maintenance duties of operators will vary from plant to plant. Some plants do not allow any operator maintenance. Some plants that practice operator maintenance have reported a 99 percent drop in unplanned maintenance. Vigilant process technicians have saved costly shutdowns and process trips that would have shutdown a unit. Operator maintenance is not a magical wand to rid equipment of all problems but the results will be lower failures and increased machine availability and reliability.

LUBRICATION

Lubrication training is as important as any other training program for plant personnel because the increased knowledge and skills of personnel responsible for applying the lubricants can substantially increase equipment reliability. In addition, trained operating personnel will be more alert to equipment malfunction and report conditions before the equipment actually fails. Lubrication training of plant personnel includes instruction in basic principles of lubrication, scheduling, and lubrication procedures.

Good Lubrication Practices

All equipment with moving parts needs lubrication. This includes everything from the high-speed turbine rotating at 10,000 to 15,000 rpm down to the hinge on the control room door. A lack of proper lubrication is the quickest way to shorten the useful life of any piece of moving equipment. The net result of improper lubrication is major equipment damaged, possible costly unit downtime, and expensive equipment repair. Field technicians are responsible for the proper lubrication of the equipment in their area. Lubrication tasks include changing the oil in equipment that uses oil and greasing bearings. Technicians should also understand when lubrication oil is too contaminated to continue using.

Table 11-3 Maintenance Duties of Process Technicians

Maintenance Duties of Process Technicians
Lubrication—oil top up and replacement, valve greasing, plummer block bearings, gear couplings, open gears, etc.
Routine maintenance—adjustments, opening, cleaning, and assembly of filters; opening, cleaning, and reclosing of strainers up to 4 inch size
Decouple small equipment for maintenance
Open and reinstall drain plugs, caps, etc.
Attend/arrest leaks of ferrule joints, threaded connections, plugs, flanges up to 2 inch size
Adjust tension and replace V-belts
Arrest flange joint leaks by tightening
Observe equipment repair/overhauling
Measure overall vibration using data collectors
Measure temperature using noncontact thermometer
Adjustment off-center conveyor belts
Fix coupling and belt guards
Tighten/adjust gland of reciprocating pumps and isolation valves
Maintain steam traps
Assist specialized maintenance staff during major repairs
During shutdowns/turnarounds, assist maintenance in maintenance work.

Contaminated oil spells rapid death for hydraulic machinery and lubricated equipment. Fine tolerance equipment can have clearances between parts of 5 to 10 microns (0.005 to 0.01mm, 0.0002 to 0.0004 of an inch). Solid particles (contaminants) larger than the clearance gap will jam into the space and be broken-up and mangled while ripping out more material from the surfaces. Solids suspended in oil are like grinding paste. They scour and gouge surfaces, block oil passages, and make the oil more viscous. The longer the oil is left dirty, the faster the rate of failure. Even expensive synthetic oil is of no use if it is contaminated by solid particles.

Processing units will have lubricants available for rotating equipment. How lubricants are stored, labeled, and the containers used to apply them can affect contamination of the lubricants. Poor storage practices can result in contamination of lubricants. As an example, if a lube oil is stored in a metal drum outside and exposed to the elements, then contamination is inevitable. If rainwater collects on top of a drum, changes in ambient temperature can cause water on the top of the drum to be drawn into the drum via capillary action through the drum metal. The drum may be closed and sealed, but some water will enter the drum and will accumulate over time to a level that will require the remaining oil in the drum to be discarded. Lubricant storage containers should be kept undercover and out of the elements.

Large lubrication containers, such as 55-gallon drums, will normally be placed on a drum rack with a spigot and vent cap inserted into the drum for drawing out the lubricant. Some vents are just a vent that is used to prevent vacuums that would form while the lubricant was being drawn from the drum. Allowing air into the drum breaks the vacuum but also intro-duces water vapor and dust into the drum. Dust, though very fine, is an abrasive. The water vapor condenses into water. Both dust and water contaminate the lubricant. Plants focused on equipment reliability should use vents that contain a filter that removes particles (dust) and a desiccant that absorbs moisture.

Technicians should be careful not to use wrong or degraded lubricants. The right lube viscosity at the operating temperature for the load and shaft speed is critical. Only clean lube with the correct specification for the bearing at operating conditions is acceptable. When the bearing lube deteriorates, dispose and replace it with fresh lubricant. How do you know the lubricant's condition, contaminants, and quality at any point in time? If you are not sure whether it should continue in service get rid of it. The cost is a lot less than a breakdown.

High temperatures on rotating equipment can reduce lubricant viscosity and thin the lubri-cant film. External sources of heat may also cause equipment temperature to rise. Is the nearby process radiating heat onto the equipment or is it in direct sunshine? Is the absorbed heat preventing the bearing from dissipating (shedding) its own heat? Is the process heat being transferred down the shaft and into the bearing? Is the bearing housing covered in dust, dirt, rags, or rubbish that act like insulation and retain the heat? Does the equipment need an extra flow of cooling air to remove the heat? If oil lubrication is used, should an oil cooler be installed? Does the equipment housing need more surface area for radiation and convection cooling? For a lubricant to maintain its viscosity and do effective lubrication it must be kept cooled to the temperature of the required viscosity.

As a final note, certain lube oil additives frequently attack galvanized steel. Metal lube oil dispensing containers should be replaced with plastic dispensing containers.

Lubricants and the Necessity of Lubrication

Any substance that can serve as a cushion or protective film between moving mechanical parts can be considered a lubricant. Most people are familiar with the oil used in an auto-mobile or the general-purpose utility oil bought at hardware stores. A lubricant can be a solid, semi-solid, liquid, or gas. Graphite powder is an example of a solid lubricant and is used in lubricating lock mechanisms. Thick bearing grease is a good example of a semi-solid lubricant. Automotive motor oil is an example of a liquid lubricant. Last, high-pressure nitrogen used to separate the moving parts of special bearings is an example of gas used as a lubricant.

Lubricants can be characterized by their properties of viscosity, temperature stability, chemical resistance, and chemical composition. For example, lower viscosity oils are use-ful in lubricating machinery at low temperatures since they will flow more easily between the moving parts when the equipment is cold. They are also used for lubricating small precision parts since it is easier for the low viscosity lubricants to work their way into small clearances. The other extreme is high viscosity oils and semi-solid greases used where heavy-duty lubricants are needed to withstand high temperatures. If the temperature is extremely high, heat resistant lubricants such as silicone-based greases may be used since hydrocarbon-based greases breakdown.

When moving parts in machinery or equipment rub against one another, they produce friction. Friction erodes surfaces, wastes energy, and produces heat that must be dissipated to prevent equipment damage. The wear caused by friction can quickly damage or destroy equipment. The heat generated by friction can cause rotating parts to expand and seize, and the extra energy needed to operate improperly lubricated equipment can increase operating costs or even overload and damage motors. In theory, lubrication is based on the physical separation of the metal-to-metal surface by a film of fluid. Under the most ideal conditions of correct grade, adequate supply, and cleanliness, lubrication keeps friction and wear held to a minimum. By adding a slippery film between moving parts, it reduces friction, which in turn reduces equipment temperature and energy demand.

The reduction of friction is a major reason for proper lubrication, but it is not the only reason. Proper lubrication can also

- Reduce wear
- Cool moving parts
- Dampen shock
- Prevent corrosion
- Flush out contaminants
- Seal systems

These are discussed in more detail in the following paragraphs.

Reduce Wear. Surfaces, though they may look and feel smooth, have microscopic ridges and valleys. As non-lubricated parts move against one another the ridges on the contacting surfaces are broken off, causing the surface to gradually wear away. With time, this gradual erosion becomes noticeable. If a thick enough film of lubricating material is introduced between the moving surfaces, it prevents the two surfaces from touching and little or no wear will occur.

Cool Moving Parts. Heat generated by friction can cause moving parts to expand and seize (lock together). Lubricants reduce heat by reducing friction that generates heat, plus the lubricating oil also carries heat away from the moving parts. The specific heat of lubricating oil is about one-half that of water, so oil will not remove heat as efficiently as water-cooling would. However, a pressurized oil system which pumps oil to and from the lubricated parts will carry significant heat away. If the lubricating system also has an oil cooler that transfers the heat from the oil to some other heat sink, it will remove even greater amounts of heat.

Dampen Shock. Machines will run more quietly and last longer if the metal-to-metal impact or shock that occurs as one moving part meets another can be dampened (reduced). Lubrication does this to a degree. When a force acts on a liquid volume, mechanical impact is converted to fluid motion, and the force is distributed over the surface area that comes in contact with the liquid. In this way, liquid lubricants can dampen the concentrated forces that occur as moving parts meet one another, such as when gear teeth mesh. Spreading the concentrated force over a greater surface area reduces the impact stresses and increases the life of the machinery.

Prevent Corrosion. Corrosion is caused by the reaction of oxygen with metal parts to form metal oxides (rust). Lubricants form protective films on metal parts that prevent air (oxygen) from reacting with the coated metal surface and greatly reduce or eliminate normal corrosion.

Flush Out Contaminants. Dirt or other contaminants that work their way between moving parts act as an abrasive and cause additional wear on moving parts. Lubricants can flush such materials out of the system in several different ways. The flowing oil in pressurized lubricating systems flushes contaminants from the bearing and other moving parts as the oil is circulated. An in-line filter removes particles from the oil before it is re-circulated back to the lubricated parts. Manually greased parts (grease guns) can also flush out contaminants when new grease is added and the old grease containing contaminants is forced out of the lubricated area.

Seal Systems. When moving parts are properly greased, the lubricant will fill the space between the parts, keeping them separated. A lubricant with the right viscosity will remain in place, sealing the system as the parts move. If there are any openings that expose the grease to air, the lubricant will oxidize forming a hard scab that keeps the system sealed. Sealing in this way keeps the lubricant in place and prevents dirt form entering the system.

Factors Affecting Lubrication
Factors that determine the effectiveness of lubrication include

- Pressure at the bearing surface
- Temperature
- Viscosity of the lubricant
- Operating speed
- Alignment
- Condition of the bearing surface
- Starting torque
- Purity of the lubricant

Some of these factors are interdependent. For example, the viscosity of any oil is dependent upon temperature, and temperature is affected by operating speed and ambient temperature. Also, viscosity is dependent on operating speed because the lubricant must be able to cling to the bearing surfaces and support the load at all operating speeds and operating conditions. A more adhesive lubricant is required at high operating speeds than at moderate or low. At low operating speeds a more cohesive (ability of particles to stick together) lubricant is required to keep the lubricant from being squeezed out from between the bearing surfaces.

Technicians have control over three very important lubrication factors. They have control over (1) viscosity, (2) purity, and (3) the proper amount of lubricant. They control viscosity by picking the proper lubricant, control the purity of lubricant by preventing contamination of it, and should know the proper levels and amounts of lubricant to add.

Process technicians should never guess which lubricant to use or assume one lubricant is as good as another. The proper lubricant for each piece of equipment will be identified on their rounds lists, maintenance procedures, and even on oil reference charts posted where the lubricants are stored. If the unit is out of the proper lubricant, *don't substitute*; borrow from a neighboring unit. They will likely have identical equipment and lubricants. Also, if a technician notes that a piece of equipment is low on lubricant, they should not wait to "oil up" at the end of the shift. Bearing damage could occur before then.

MOISTURE—A DESTRUCTIVE LUBRICANT CONTAMINANT

Moisture is often considered to be the second most destructive lubricant contaminant; particulates (dust, dirt, rust, etc.) are considered to be the most destructive. Moisture can enter lubricated bearing systems in several different ways, resulting in dissolved, suspended, or free water. Both dissolved and suspended water can promote rapid oxidation of the lubricant's additives and base stock and diminish lubricant performance. Rolling element bearings may experience reduced life due to hydrogen embrittlement caused by water-penetrated bearing surfaces. Many other moisture-induced wear and corrosion processes are common in both rolling elements and journal bearings.

Moisture is considered a chemical contaminant when suspended in lubricating oils. Its destructive effects in bearing applications can reach or exceed that of dust or dirt contamination, depending on various conditions. Once water enters the casing of a machine where bearings are used, such as an engine, turbine, or gearbox, it may move through several chemical and physical states. Generally, water will enter an oil in one of the four following ways:

1. Absorption—Oil can absorb some moisture directly from the air. The relative humidity of the air and the saturation point of water solubility in the oil influence the amount of moisture that can be absorbed. Depending on temperature and pressure, this solubility limit will vary from about 100 parts per million (PPM) for low additive oil to several thousand PPM for high additive and certain synthetic oils. Absorbed water is always dissolved in the oil at first, but later may be condensed out to a free or emulsified state due to temperature/pressure changes.
2. Condensation—Humid air entering oil compartments will often cause moisture condensation on the walls and ceilings above the oil level. Frequent temperature change cycles may greatly increase the rate of condensation. Eventually the condensation will coalesce and run down the casing walls to the bottom, forming a layer of free water.
3. Heat Exchangers—Corroded or leaky heat exchangers (lube oil coolers) are common sources of water contamination in lubricating fluids.
4. Free Water Entry—During oil changes or the addition of makeup lubricant, water can be introduced to the oil compartment. Condensation of water in storage containers is the most common origin of this water.

Once water enters an oil, it is in constant search of a stable existence. Unlike the oil, the water molecule is polar, which greatly limits its ability to dissolve. If water molecules are unable to find polar compounds to attach themselves to, the oil is termed *saturated*. Any additional amount of water will result in a supersaturated condition causing free water to be suspended or settle in puddles at the bottom of the sump. This supersaturation can also occur as a result of lower oil temperature. When free water is suspended, a colloidal suspension or emulsion exists that causes a visible cloud or haze in the oil.

Effects of Water on Additives and Base Stock Lubricants

The chemical and physical stability of lubricants are threatened by minute amounts of suspended water. Water promotes several harmful chemical reactions that attack rust inhibitors, viscosity improvers, and the oil's base stock. The effects are undesirable by-products such as varnish, sludge, organic and inorganic acids, surface deposits, and lubricant thickening (polymerization). Large amounts of emulsified water can lower viscosity and reduce a lubricant's load-carrying ability.

Effects of Water on Bearing Surfaces and Bearing Life

Water etching is a common type of corrosion occurring on bearing surfaces and their raceways that is caused primarily by the generation of hydrogen sulfide and sulfuric acid from water-induced lubricant degradation. This occurs as a result of the liberation of free sulfur during hydrolysis reactions between the lubricant and suspended water.

The elastohydrodynamic lubrication associated with rolling element bearings demands consistent oil viscosity. When water invades the lubricant, this important property is compromised. High local area pressures under bearing contacts can reach 100,000 to 500,000 pounds per square inch gauge (psig), depending on dynamic loading and bearing size. At such pressures the lubricant film thickness is reduced to 0.1 to 0.3 microns (μm) and forms a momentary solid. When moisture is present, this thin oil film can fail allowing the bearing and its raceway to come in contact with each other. If sustained, the result will be a marked reduction in bearing fatigue life.

For journal bearings, the hydrodynamic pressures between the shaft and bearing surfaces may not exceed 1000 psig. And, depending on such factors as speed, load, viscosity, and bearing size, film thickness can range from as low as 0.5 μm to as high as 100 μm. Moisture can reduce lubricant load-carrying ability in journal bearings causing shaft and bearing contact. Water also contributes to various forms of corrosive and cavitation damage to journal bearing surfaces. Babbitt bearings, consisting mostly of lead and tin, are easily oxidized in the presence of water and oxygen.

Water Control

The universal presence of water limits the success of totally preventing water from combining with lubricating oils. However, its entry can be greatly minimized and its effect on lubricant life and machine damage greatly reduced. The first step in any proactive maintenance effort is to set limits beyond which a particular contaminant level must not exceed. With moisture, the target is a level of moisture in the lubricant that must not be exceeded. To a great extent this target will vary by application (steam turbines, diesel engines, screw compressors, and industrial gearboxes). As a general rule, moisture at a level of 100 parts per million (PPM) is a reliable limit for many applications in terms of lubricant and bearing life. A moisture contamination control program should include routine, on-site monitoring of lubricant moisture levels to insure these levels are within target limits.

Table 11-4   Effects of Moisture Content on Machine Life

Effect of Mositure Content on Machine Life **Life Extension Factor (LEF)**						
		2X	**3X**	**4X**	**5X**	**6X**
Current Moisture Level (PPM)	1000	250	130	90	63	50
	500	125	65	45	31	25
	250	63	33	23	16	13
	100	25	13	9	6	5

Source: Data from H.P. Block, "Quantifying effects of moisture-contaminated lube oil on machine life." *Hydrocarbon Processing*, October 2002, p. 13.

Most of the time lubrication oil samples are submitted to the site quality control laboratory for moisture analysis. A new technology has been introduced for user-level (technician) moisture detection in the form of a hand-held probe and microprocessor that can collect and analyze data. At the tip of the probe, which is submerged in an oil sample, is a miniature heating element. During a test, this heating element glows at constant temperature causing suspended moisture to vigorously vaporize emitting a distinctive acoustic signal known as crackling. A microphone mounted adjacent to the heating element picks up this signal and electronically passes it to the data collector for analysis. The unit is able to detect suspended moisture to as low as 25 PPM and as high as 10,000 PPM. A typical test takes less than 30 seconds.

The plant quality control laboratory can also analyze samples for water at the PPM level. Technicians will collect lube oil samples from systems that recirculate the lube oil and submit the samples to quality control laboratory for a *water in lube oil* test. Another name for the test is the *Karl Fischer Test for Water*. Most of the lube oil recirculating systems have a water jacket that cools the oil before it is returned to the operating equipment. Water can get into the lube oil from the atmosphere through worn bearing seals and labyrinth seals that have an air gap. Atmospheric water vapor will condense out and contaminate the oil. Also, the oil's water cooling system may leak water into the oil reservoir. Some moisture (water) in oil can be tolerated but not excessive moisture. Most process units that submit lube oil samples for moisture have limits, such as 50 PPM, and when they are exceeded, the oil is drained and replaced. Table 11-4 graphically illustrates the effect of moisture content on machine life. Consulting the table, we see that reducing the water content of a lube oil from 500 PPM to 45 PPM would increase the machine life by a factor of 4 or 400 percent.

LUBRICATING SYSTEMS

A lubricating system may be as simple as a squirt type oilcan or as complicated as a pressurized, circulating oil system complete with pump, reservoir, cooler, filter, and pressure instrumentation. Regardless of the design, it must supply the required amount of the proper lubricant in a trouble-free and economical way. The major lubricating system designs include

- Manual lubrication
- Gravity lubrication

- Splash lubrication
- Constant level oilers
- Pressure lubrication
- Oil mist lubrication

Descriptions of how each of these function and guidelines for proper use are discussed in the following paragraphs.

Manual Lubrication

Manual lubrication is the simplest and oldest form of lubrication. It includes oil supplied by squirt-type or plunger-pump oilcans and grease supplied by manual grease guns. It can also include grease supplied in a screw-type grease cup and oil or grease applied with a brush or stick. Operation of this type of equipment is straightforward. A good preventative maintenance program will have a lubrication schedule for all manually lubricated equipment that indicates where to lubricate, how often to lubricate, and what type of lubricant to use.

Gravity Lubrication

Gravity lubrication uses drip-feeders, vibrating pin feeders, or wick oilers to supply lubrication to the needed areas. All of these systems have a small reservoir of oil elevated above the lubrication point and a pipe or channel to direct lubricant flow to that point. Lubricant flow from a drip-feeder is regulated by an adjustable needle valve, or by turning a small knurled nut on top of the oiler or by using a screwdriver to adjust a screw within the housing. The vibrating pin feeder has a small rod or pin that extends through the oil supply channel and rides on a rotating shaft. The shaft movement causes the pin to vibrate, which regulates oil flow down the pin to the rotating shaft. The wick oiler has a fibrous wick immersed in the oil. Gravity and capillary action help the wick transfer oil to the lubrication point. Regardless of the type of oiler used, the technician should check equipment oilers while making area rounds. There should always be a supply of oil in the reservoir and the feed rate in drops per minute should be checked against the required feed rate for proper lubrication.

Splash Lubrication

Splash lubrication systems have an oil sump or oil reservoir built into the equipment housing. This sump holds a specific level of oil that allows some of the moving parts to be partially immersed in oil. The moving parts splash oil onto the parts that need lubrication. The oil level in the sump is checked using a dipstick. A gear reducer is an example of a system that uses splash lubrication. Portions of the larger gears are immersed in the oil and their rotation splashes oil onto other gear teeth, lubricating them.

A *ring oiler* is a special type of splash lubrication system. Ring oilers are frequently used to lubricate the bearings on low-speed rotating shafts. A ring of brass or steel, with a diameter about one and one-half times that of the rotating shaft is placed on the shaft so that the two will rotate together. The ring picks up oil out of the sump as the ring and shaft rotate and splashes oil onto the shaft and bearings to provide lubrication. This works well for low-speed applications but not for high speeds.

Constant Level Oilers

Constant level oilers maintain a constant oil level in the equipment they lubricate. They consist of an oil bowl mounted on the equipment by a connecting pipe. Oil flow is driven by gravity and initiated by oil level in the equipment sump. The oil bowl is clear, usually

Figure 11-1 Conventional Constant Level Oiler

glass or plastic, so that the oil level in the bowl (reservoir) can be determined by a walk-by inspection. Oil in the equipment being lubricated must be checked for oxidation and contamination. If either condition exists, oil is drained from the sump and fresh oil from the oil bowl flows into the sump to the appropriate level. See Figure 11-1 for an image of a conventional constant level oiler.

Pressure Lubrication

A *pressure lubrication* system uses a pump to circulate the lubricant to some high-speed equipment. The simplest system uses a reservoir below the points of lubrication to catch and hold oil to be circulated back to the equipment. The oil is pumped from this reservoir back to the lubrication points. The oil that drains from the lubrication points returns to the reservoir to be reused. More complex systems may also have an oil cooler to cool returning oil and strainers and filters to remove foreign materials from the recirculating oil.

Technicians monitor the oil level in the reservoir and the oil pressure generated by the pump. If the system has an oil cooler, they also note the oil temperature. If the system is large or complex, additional instrumentation will be provided. If oil pressure and/or oil temperature is critical to the operation, alarms (or interlocks) may be provided to alert technicians that the equipment is about to deviate from safe operating conditions.

Oil Mist Lubrication

Oil mist lubrication is the lube application of choice. Plant wide piping (headers) distributes the mist to a wide variety of equipment. Oil mist is easily produced and its flow to bearing easy to control. Flow is a function of reclassifier size and piping pressure. Unless plugged by an unsuitable lubricant, reclassifiers have a fixed flow area. Header pressures range from 20 to 35 inches of water. Modern units are provided with controls and instrumentation that maintain these settings.

Oil mist systems are available as open systems and closed systems. Open systems (dry-sump) will lubricate, preserve, and protect both operating and stand-by rolling element bearings. At all times only clean, fresh oil will reach the bearings, then vent to the atmosphere. The open systems use oil formulations that are neither toxic nor carcinogenic and supply a constant stream of fresh oil. Closed oil mist systems (wet-sump) recover the oil for reuse and the oil may accumulate contaminants. The wet-sump oil level is maintained by an externally mounted constant level lubricator.

A 1999 survey on oil mist usage in U.S. refineries found that almost 50 percent of the refineries use oil mist lubrication extensively. The survey was based on large-scale systems with a fully monitored oil mist generator (OMG), which serves multiple pumps and drivers. The average pump population per system is 30 to 50 pumps and drivers, usually using pure mist or dry sump applications. Oil mist is an aerosol that is a mixture of one part oil to 200,000 equal parts of air. It is a lean mixture that will not support combustion and will not explode. The appearance of oil mist resembles cigarette smoke or steam drifting from a pump or motor through a vent line. Using oil mist lubrication, a typical application in the hydrocarbon processing industry will reduce oil consumption up to 40 percent over the traditional oil sump method of lubrication.

Oil mist is generated by passing high velocity air over or through an orifice that pulls oil into the air stream. The high velocity air shatters the oil into particle sizes of one to three μm, thus the resemblance to cigarette smoke. Airflow transports these small oil particles through a piping system to the equipment to be lubricated. Prior to being applied to a bearing, the small particles of oil are passed through an orifice, reclassifier, or mist fitting, causing the small particles to impinge on each other and grow in size. The heavier particles are then large enough to wet the surfaces and provide adequate lubrication for most rolling element bearings. (See Figure 11-2 for a schematic of the oil mist lubrication process.) It is excellent lubrication for bearings operating at 1,800 to 3,600 rpm and it is often the preferred method of lubrication for bearings operating in the 10,000 to 15,000 rpm range where splash lubrication is ineffective. Because an oil mist system has no moving parts, there is little chance of failure. The only requirements for generating oil mist are clean air and clean oil. Electric monitors control and help maintain a constant mist density.

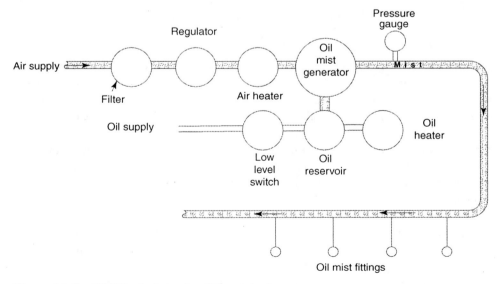

Figure 11-2   Oil Mist Lubrication Schematic

Companies using oil mist lubrication see the mean-time-between-repair (MTBR) increase from 24 to 36 months up to 48 to 60 months. With the average repair cost per pump between $5,000 and $10,000, the reduced repair costs attributed to oil mist lubrication can have a dramatic impact on a refinery's bottom line. Oil mist lubrication provides a number of important benefits, including

- Reduced bearing failures
- Decreased manpower for lubrication tasks
- Elimination of oil changes
- Improved mechanical seal life

One of the most favorable benefits of pure oil mist lubrication is that it reduces the operating temperature of the bearing it lubricates, typically 20°F to 35°F. Heat is generated by friction created from rolling the bearing through an oil sump at a high speed. When the oil sump is drained away, there is no oil to generate heat and no sump in which to retain it, so the temperature drops. With every 10°F reduction in temperature, the bearing fatigue life is increased by 11 percent.

Oil mist lubrication, while often superior to the traditional oil sump lubrication, not only lubricates equipment during operation but it also protects and preserves internal components (bearings and gears) when equipment is idle. Idle or spare equipment is subject to surface corrosion on internal components due to the ingress of atmospheric moisture, and the possibility of fretting wear caused by ground vibration. Idle equipment will also experience thermal cycling, or breathing in of contaminated air as temperature changes, many times throughout a 24-hour period. This causes the surrounding atmosphere to enter the bearing or gear cavity. The atmosphere may be laden with moisture, dust, or even acid fumes. The atmospheric contaminants accumulate on machine surfaces, causing corrosion and reducing equipment life. In contrast, oil mist builds a slightly positive pressure, approximately 0.25 inches of water, in the cavity. This pressure prevents the intrusion of the airborne contaminants, and internal surfaces and components are coated with a protective film of oil. The oil film on internal components guards against corrosion and protects against fretting wear by creating a thin film of separation.

SUMMARY

Any process that is not properly cared for will not operate efficiently just as any automobile not properly cared for will not operate efficiently. Proper equipment maintenance is critical to efficient process operations. More and more plants are assigning preventative maintenance duties and responsibilities to process technicians. They are in the best position to do the routine maintenance and preventative maintenance required to keep their equipment running properly because process technicians work with their equipment every day. In addition, process technicians know the equipment well enough to sense when it is not working right. Most equipment breakdowns are due to human errors, such as negligence or ignorance.

Lubrication training is as important as any other training program for plant personnel because the increased knowledge and skills of personnel responsible for the applying lubricants can substantially increase equipment reliability. In addition, trained operating personnel will be more alert to equipment malfunction and report conditions before the equipment actually fails. Lubrication training of plant personnel includes instruction in basic principles of lubrication, scheduling, and lubrication procedures.

When moving parts in machinery or equipment rub against one another, they produce friction. Friction erodes surfaces, wastes energy, and produces heat that must be dissipated to prevent equipment damage. The wear caused by friction can quickly damage or destroy equipment. The heat generated by friction can cause rotating parts to expand and seize, and the extra energy needed to operate improperly lubricated equipment can increase operating costs or even overload and damage motors. Lubrication, by adding a slippery film between moving parts, reduces friction which in turn reduces equipment temperature and energy demand. The major lubricating system designs include manual lubrication, gravity lubrication, splash lubrication, constant level oilers, pressure lubrication, and oil mist lubrication.

REVIEW QUESTIONS

1. Explain the purpose of a technician's rounds.

2. Most of the time there is a _____ behind equipment failure.

3. Equipment stoppage is termed a _____ breakdown.

4. List five causes of equipment failure.

5. List five examples of equipment function reduction.

6. List six hidden defects that lead to equipment failure.

7. Explain what is meant by the phrase "every failure has a lesson to teach."

8. Explain the two reasons why equipment deteriorates.

9. List six benefits of an operator maintenance program.

10. List eight maintenance duties of process technicians.

11. A _____ of lubrication is the quickest way to shorten the life of moving equipment.

12. Explain how leaving a steel drum of lubricant outside in the weather can result in the lubricant becoming contaminated.

13. List four characteristics of lubricants.

14. List five major reasons for the proper lubrication of moving equipment.

15. The first most destructive lubricant contaminant is _____; the second is _____.

16. Explain how water gets into lubricants.

17. Explain the harmful effects of water on lubricants.

18. Describe how a process unit would detect the level of water in its lubricants.

19. List five lubrication systems.

20. Describe one lubrication system.

CHAPTER 12

Maintenance Duties: II

Learning Objectives

After completing this chapter, you should be able to

- *Explain why process technicians are critical to MTBR.*

- *Explain how to prevent bearing and wear ring problems in centrifugal pumps.*

- *List three checks to be made on electric motors.*

- *Describe the importance of steam traps to energy conservation.*

- *List four types of steam traps.*

- *Discuss what causes steam traps to fail.*

- *Describe three methods for troubleshooting steam traps.*

- *List three ways to unplug process lines or equipment.*

- *List several problems caused by piping vibrations.*

INTRODUCTION

This chapter discusses more maintenance duties of process technicians and how these maintenance duties contribute to the bottom line. Many plants are in the process of changing their philosophy from that of shutting down on a time-scheduled basis to that of running a unit until the condition of the unit machinery indicates it is time to shutdown. This shutdown condition is determined by instrumentation available to technicians and this type of maintenance philosophy is called **predictive maintenance**. Predictive maintenance is very important where the rotating equipment is the limiting factor in a plant maintenance schedule.

Process technicians may become involved with predictive maintenance by making rounds with a portable data acquisition device, a hand-held microprocessor that includes special programs for rotating equipment. They will collect data from rotating equipment and upload it to a control room computer where unit engineers will access it and determine the running condition of the equipment. In plants using predictive maintenance, machine turnarounds are based on machine condition rather than elapsed time in use.

PUMPS AND PUMP MAINTENANCE

Chemical processing plants face enormous pressure to improve profitability by increasing internal operating efficiencies. Improved profitability increases based on how well a plant is able to decrease its operating costs. One way to do this is by a systemic plant-wide effort to improve the performance and reliability of rotating equipment. Centrifugal pumps are prime candidates for such attention because they are so numerous and widespread in the industry. The ability to improve equipment reliability has a direct impact on bottom-line costs in several ways. It increases mean-time-between-repair (MTBR) and reduces maintenance costs, equipment acquisition costs, unit down time, purchasing and inventory activities, and the production of off-specification product.

Most medium-sized plants have from 500 to 2,500 pumps. It is common for plants without a systematic equipment reliability and performance program to have a MTBR of 18 months; those with established programs average 60 months MTBR. A medium-sized facility can save millions of dollars a year by increasing MTBR from 18 months to 60 months. Process technicians are critical to the MTBR because they operate and monitor the pumps 24 hours a day, 365 days a year.

Bearings and Wear Rings

For our purposes, the term *bearing* may be applied to anything that supports the motion of sliding and rolling surfaces on rotating shafts in process equipment. The rotating shaft of any machine (centrifugal and reciprocating pumps, steam turbines, compressors, etc.) has the main purpose of maintaining radial (side-to-side, up-and-down) clearance, and axial (back and forth) clearance between stationary and rotating parts. These clearances must be small and precise for efficient and long-term operation. There are many variations of bearing types but they are generally classified as sliding surfaces (friction) or rolling contact (anti-friction) bearings. Friction bearings are called *sleeve*, *journal*, or *thrust* bearings. Anti-friction bearings are called *ball* or *roller* bearings. The terms *friction* and *anti-friction* bearings can be misleading because both types are highly efficient in removing friction but neither is entirely frictionless. All bearings require proper lubrication to attain trouble-free operation and a normal wear life.

Although **wear rings** are not normally classified as bearings, under certain conditions of operation they act as bearings. In centrifugal pumps there must be a running or rotating joint formed by a portion of the impeller and a portion of the casing. Wear rings are placed between the impeller and casing suction head of the pump for the purpose of sealing without wearing the two rings out. This rotating joint made of a flexible man-made material separates the suction and discharge chambers. The leakage of the pumped liquid through the clearance of this running joint, from the discharge side back into the suction side, would result in a gradual erosion of the impeller and casing if not for wear rings. The rate of erosion would depend on the amount of grit and foreign particles in the pumped fluid. Wear rings can be removed and replaced at a fraction of the cost of impeller or casing replacement. Clearance around the wear rings is uniform and quite small (i.e., 0.012 inch) in a small pump.

Pump Trouble Spots—Bearings and Wear Rings

The two main trouble spots in the centrifugal pump are the bearings and wear rings (see Figure 12-1). Other failures are packing leakage, mechanical seals, shaft breakage, worn shaft sleeves, or loose impellers. Mechanical seals and bearings represent 75 percent of centrifugal pump failures.

The term *wear ring* is misleading because these rings should not wear. When a pump is running properly the rings don't touch each other and a film of liquid separates them. The clearance between them must be very close to prevent back flow of liquid. The liquid keeps the rings separated and lubricated. Sometimes liquids evaporate at the eye (suction) of a pump because of the low pressure. The liquid evaporates because the suction pressure is less than the vapor pressure of the liquid at pumping temperature. When a pump begins to lose suction the rings begin to touch because the fluid pressure is no longer steady. This condition overheats the rings and causes expansion. The rings will seize each other and bring the pump to a halt.

The most important aspect of bearings to an operator is lubrication. Many of the newer but smaller pumps have sealed bearings and no lubrication is required. Ball bearings seldom wear out if treated properly. Bearings are designed to operate with a very fine film of

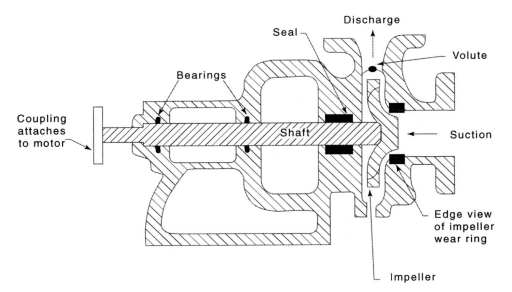

Figure 12-1 Simple Schematic of a Centrifugal Pump

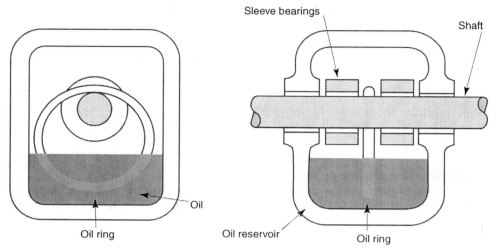

Figure 12-2   Oil Rings

lubricant surrounding each ball, and as a result, the balls could roll for years without wear. This is proven with bicycle ball bearings. The close fit of bearings means smooth operation but it also means trouble if dirt enters the bearing housing. A tiny particle of dirt can do enough damage to destroy the entire bearing. Bearings are made of steel and may become rusty in a wet atmosphere. Moisture must be kept out of bearing housings. Technicians should use only clean oil or grease and keep the lubricant in a clean, covered container.

Some of the older pumps are lubricated from an oil reservoir at the bottom of the bearing housing. Either part of the bearing itself dips into the oil reservoir or an oil ring carries oil up to the shaft (see Figure 12-2). Lubrication by oil rings is usable on relatively low-speed pumps. For high-speed pumps where lots of heat is generated in the bearings, lubricating oil must be force-fed by a separate lube oil pump.

ELECTRIC MOTORS
Electrical motors account for two-thirds of U.S. industrial consumption of electricity. Pumping systems account for 25 percent of electric motor usage. The large electric motors used in the processing industry are expensive and it is an important technician job to keep them under surveillance and maintain them in normal operating condition.

Starting the Motor
If possible, before a motor is started, the technician should verify the motor shaft is free to rotate. If the motor shaft is not free to rotate, it should be freed before any attempt is made to start the motor. If nothing happens when the starter button is pushed, the main switch should be checked followed by the fuses and circuit breakers to verify that power is being supplied to the motor. The starter button should be pressed and released at once. If there is an overload, jiggling the button or holding it down would pit and damage the contact points of the starter switch. A slow start can be caused by an overload. Where possible, a pump coupled to a motor should be checked to see if the pump shaft is free to rotate before the motor is started. If a motor does not start, a careless or inexperienced operator may *bump* the motor by working the start and stop buttons several times or try to hold or wedge the switch contacts or relays closed. Such efforts are more likely to damage the motor than start

it. If a motor kicks out repeatedly during attempts to start it, no more than two or three attempts to start it should be made.

Experienced technicians know that there is a problem with an electric motor after they have pressed the starter button three times and the motor does not start. After they have checked the power lines and verified the motor has power, they know is time to report the motor as down and write a work order.

Motor Vibration, Temperature, and Lubrication

Each motor must be checked for excessive vibration. Certain motors, based on their criticality, have electronic vibration monitors that alarm and shutdown the motor if vibration becomes excessive. A simple and rough estimate of vibration can be determined by the technician feeling the motor housing. Motor bearings must be checked for excessively high temperatures. Overheated bearings may be found by feeling the motor housing. (**Caution:** First touch the motor lightly to determine if it is hot enough to burn you.) A bearing is too hot if it feels much hotter than another bearing on the same piece of machinery. This is a very subjective determination and comes with years of experience. If a bearing is too hot, the lubricating action and the oil level should be checked. Vibration and temperature can also be checked with data acquisition devices that have attached wands for sensing temperature and vibration. On a routine basis technicians should check the oil level of oil-lubricated bearings. If oil cups need refilling more than once a day, bearings may be worn or oil may be leaking from seals. Oil seepage or leaks may damage motor windings.

Cleaning and Ventilation

Electric motors have fans inside to force air through the windings. The winding is the wire wound around the armature in the motor. Motors should be kept clean because oil and grease on motor windings can collect dirt that will restrict circulation of air through the motor. If cooling air is blocked, the windings may get too hot. Oil leakage from motor-driven pumps should not be allowed to enter the motor and damage the windings. If water is used to wash the motor area, care should be taken to keep it out of the motor. The internal fan in a motor may draw water or other moisture into the motor and short it out. A motor with a cooling-air fan has a guard screen before the fan. The screen keeps out dirt and insects. Dirt may build up on the screen and restrict airflow to the motor, which makes it run hotter. Technicians should inspect and clean the screens.

Overloaded Motors

Whenever possible, a motor that stops should be relieved of the full load before restarting. Full-load starts are likely to draw too high a starting current. Overloading the motor with excessively high starting current is likely to trip its overload protection device. All proper adjustments to the driven equipment should be made before putting power back to the motor. If a motor smokes, it is probably overloaded and should be shut down unless the load can be reduced quickly. Quick action prevents serious damage to the motor.

Bearings and Lubrication

In an electric motor the rotor rotates on its shaft. Bearings support the shaft and center the rotor in the stator. The bearings may also limit the **thrust** (end-to-end-movement) of the rotor shaft. Movement of the rotor shaft along its length is called thrust. A bearing's *thrust capacity* is its ability to resist end-to-end movement of the rotor shaft. Thrust capacity of the bearings prevents thrust from knocking the shaft against the end covers of the motor.

The rotor is supported by sleeve bearings. The sleeve bearings usually found in motors have a very limited capacity to resist thrust. Under continuous thrust, sleeve bearings quickly become hot and need to be protected against wear or damage. Oil lubrication is used with sleeve bearings in horizontal-shaft motors. A loose-riding oil ring on the shaft carries oil from the reservoir to the top of the shaft. The oil ring can carry oil to the shaft only after the motor has started to run because the oil ring is rotated by the motor shaft.

To provide greater thrust capacity, grease-lubricated ball bearings may be used instead of oil-lubricated sleeve bearings. Small- and medium-sized motors are usually fitted with grease-lubricated ball bearings. Most bearing housings have two openings, one to introduce fresh grease or oil and the other to drain oil or excess grease. To add fresh grease, the pipe plug that closes the pressure-relief and drain opening is removed. Before grease is added, grit or other foreign matter on the tip of the grease gun or on the grease fitting must be removed. Dirt ruins the bearings in a motor in a very short time. With the pipe plug removed, old grease is forced out as fresh grease is pushed in. Forcing too much grease into a bearing harms the bearing and the excess grease may drip onto the motor windings. Excessive grease in a bearing can cause the bearing to overheat, which makes the grease thinner. Thin grease may creep or be thrown out of the bearing onto the motor windings. Grease on motor windings attracts and holds dirt and dust. *Most bearing failures result from greasing too much or too often.* Some electric motors may have *bearing buddies*, a special fitting that delivers a consistent grease supply pressure over an extended period of time.

STEAM TRAPS

Too much steam is wasted and complacency and ignorance about steam traps is costing steam users much more than they realize. A plant that maintains its boiler but neglects the rest of its steam system can experience large losses. Losses can include not only wasted energy but replacement of damaged equipment and misuse of man-hours. It is not uncommon to discover system losses in the hundreds of thousands of dollars. In order to create savings by producing steam system efficiencies, it is important to understand the basics of a steam system.

Steam loss can occur in both the supply and return side. Steam traps are automatic valves that release condensed steam (condensate) from a steam space while preventing the loss of live steam. They also remove air and non-condensables from the steam space. Steam traps are designed to maintain steam energy efficiency by performing specific tasks such as maintaining heat for a process. Once steam has transferred heat and becomes hot water, it is removed by the trap from the steam side as condensate and either returned to the boiler via condensate return lines or discharged into the atmosphere (a wasteful and expensive practice). If condensed steam is allowed to remain as water in piping it can cause water hammer that can damage equipment and piping. Steam traps located strategically in the steam line remove this condensate. In addition, when steam is used as a heat source, much of the heat energy in the steam is present as latent heat. When steam exits a heat exchanger, as much of the steam as possible should be captured to recover the maximum latent heat energy. Placing a trap in the steam line exit of the heat exchanger prevents live steam from blowing through the exchanger and wasting energy while allowing condensed steam to flow out of the system and be recovered as steam condensate.

Types of Steam Traps

There are many types of steam traps because there are many different types of applications. Each type of trap has a range of applications for which it is best suited. Steam traps are best understood if described in terms of their generic operation modes, such as "continuous flow" and "intermittent flow." Continuous flow traps will, to one degree or another, continuously discharge condensate. These are float, fixed orifice, and thermostatic traps. The thermostatic trap is a hybrid. It can be considered either a continuous flow or an intermittent flow, depending on the condensate load. Under heavy condensate load or at start-up, it will tend to have a continuous discharge. Intermittent traps will cycle open and closed. They have a pattern of hold-discharge-hold. These traps are the thermodynamic, inverted bucket, and bimetallic.

Trap Characteristics

Float traps consist of a ball float and a thermostatic bellows element. As condensate flows through the body, the float rises or falls, opening the valve according to the flow rate. The thermostatic element discharges air from the steam lines. They are good in heavy and light loads and with high and low pressure, but are not recommended where water hammer is a possibility. When these traps fail, they usually fail closed. However, the ball float may become damaged and sink down, failing the trap in the open position.

Fixed orifice traps contain a set orifice in the trap body and continually discharge condensate. They are said to be self-regulating. As the rate of condensation decreases, the condensate temperature will increase, causing a throttling in the orifice and reducing capacity due to steam flashing on the downstream side. An increased load will decrease flashing and the orifice capacity will become greater There is the possibility that on light loads these traps will pass live steam. There is also a tendency to waterlog under wide load variations. Fixed orifice traps can become clogged due to particulate buildup in the orifice and at times impurities can cause erosion and damage the orifice size, causing a blow-by of steam.

Thermostatic traps have, as the main operating element, a metallic corrugated bellows that is filled with an alcohol mixture that has a boiling point lower than that of water. The bellows will contract when in contact with condensate and expand when steam is present. Should a heavy condensate load occur, such as in a unit start-up, the bellows will remain in a contracted state, allowing condensate to flow continuously. As steam builds up, the bellows will close. Therefore, there will be moments when this trap will act as a "continuous flow" type while at other times it will act intermittently as it opens and closes to condensate and steam, or it may remain totally closed. These traps adjust automatically to variations of steam pressure but may be damaged in the presence of water hammer. They can fail to open should the bellows become damaged or due to particulates in the valve hole that prevent closing. At times the tray can also become plugged and the trap can't close.

Inverted bucket traps have a "bucket" that rises or falls as steam and/or condensate enters the trap body. When steam is in the body, the bucket rises and closes a valve. As condensate enters, the bucket sinks down and opens a valve that allows the condensate to drain (see Figure 12-3). Inverted bucket traps are ideally suited for water hammer conditions but may be subject to freezing in low-temperature climates if not insulated. Usually, when this trap fails, it fails open. Either the bucket loses its prime and sinks, or impurities in the system may prevent the valve from closing.

Thermodynamic traps have a disc that rises and falls depending on the variations in pressure between steam and condensate. Steam will tend to keep the disc down or closed. As condensate builds up it reduces the pressure in the upper chamber and allows the disc to

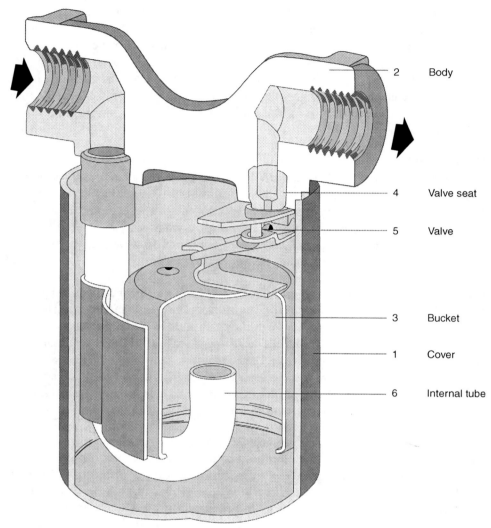

Figure 12-3 Bucket Trap

move up for condensate discharge (see Figure 12-4). This trap is a good general type trap where steam pressures remain constant. It can handle superheat and water hammer but is not recommended for process, since it has a tendency to air-bind and not handle pressure fluctuations well.

When Traps Fail

Many traps fail in the open mode. When this occurs boilers may work harder to produce the necessary energy for process equipment to perform their task, which in turn can create high back-pressure to the condensate system. This inhibits the discharge capacities of some traps, which may be beyond their rating, and cause system inefficiency. While most traps operate with back-pressure, they'll do so only at a percentage of their rating, affecting everything down the line of the failed trap. Steam quality and product is affected. A closed trap produces condensate back up into the steam space and the equipment will not produce the intended heat.

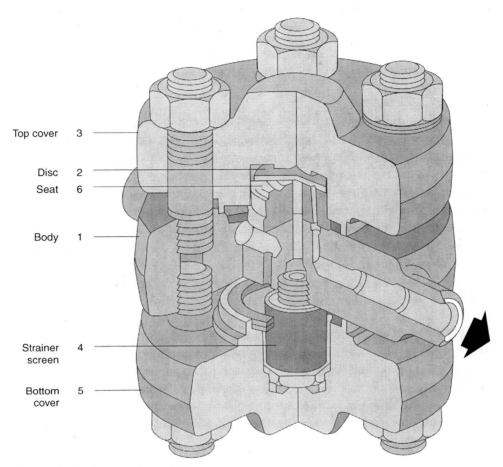

Top cover   3

Disc   2
Seat   6

Body   1

Strainer   4
screen

Bottom   5
cover

Figure 12-4   Thermodynamic Trap

Dirt is another cause of trap failure. Dirt (scale and rust) is always being created in a steam system. Excessive build-up can cause plugging or prevent a valve from closing. Dirt is generally produced from pipe scale or from over-treating of chemicals in a boiler.

When steam traps cause a back up of condensate in a steam main, the condensate is carried along with the steam. It lowers steam quality and increases the potential for water hammer. Not only will energy be wasted, but equipment can be damaged. Water hammer occurs as slugs of water are picked up at high speed. In some systems, the flow may be at 120 feet per second (about 82 mph). When the slug of condensate reaches an obstruction, such as a bend or a valve, it is suddenly stopped, creating a loud noise similar to a hammer striking the piping. The damaging effect of water hammer is due to steam velocity, not steam pressure. Water hammer can be as damaging in low-pressure systems as it can in high and can produce a safety hazard if it ruptures a valve or strainer.

Free-flowing condensate in a steam system is destructive. It can cause valves to become wire-drawn (etched) and unable to hold temperatures as required. Small beads of water in a steam line can eventually cut many small orifices that steam can pass through. Wire-drawing will eventually cut enough of the metal in a valve seat that it prevents adequate closure, producing leakage in the system.

Troubleshooting Steam Traps

Technicians can use three simple methods for checking steam trap performance:observation, sound, and temperature measurements. The best results are achieved by using a combination of methods or by using one method to check the results obtained with one of the other methods.

Observation. Observation of condensate discharging from the trap is the easiest way to verify the trap's performance. The technician should know the difference between flash steam and live steam. **Flash steam** is the lazy vapor that forms when hot condensate is discharged from a steam trap to the atmosphere. The presence of flash steam is normal. If a mixture of condensate and flash steam is being discharged several times a minute as the trap cycles, the trap is operating properly. **Live steam** is steam discharged from the trap with a high velocity. It is not a lazy vapor. Generally, live steam indicates the trap has failed and must be repaired or replaced. If the trap discharges into a closed condensate return system, observation is not possible unless a test valve is installed downstream of the trap. Then the trap can be checked by closing the isolation valve on the return line to the steam system and opening the test valve to grade for observation of exhaust steam.

Sound. The sound method checks steam traps by listening to them as they operate. Variations of the sound method range in precision from an ultrasonic testing device down through an industrial stethoscope and to homemade listening tools such as a two-foot length of a 3/16 inch steel rod or a screwdriver. The operation of the trap internals can be heard with any of these homemade devices merely by placing one end of the tool against the trap bonnet and the other end to the ear. See Table 12-1 for examples of proper and improper operating sounds of various types of traps.

Temperature Measurement. Temperature measurement is the third way of checking the operation of steam traps. A steam trap is essentially an automatic valve designed to pass condensate and hold back steam. Since the steam side will be hotter than the condensate side of the trap, measuring pipeline temperature immediately upstream and downstream of

Table 12-1   Typical Operating Sounds of Various Types of Traps

Type of Trap	Proper Operation	Failure
Disc (impulse of thermodynamic)	Opening and snap closing of the disc	Normally fails open and chatters rapidly as steam blows through
Mechanical (bucket)	Cycling sound of bucket as it opens and closes	If fails open, sound of steam blowing through. If fails closed, no sound.
Thermostatic (bellows)	Sound of periodic discharge if load is medium to high; possibly no sound if load is light	If fails open, sound of steam blowing through. If fails closed, no sound.

Source: *Yarway Steam Traps: Simple Techniques for Surveying Steam Traps* (Miami, FL: Tyco Flow Control): 2005, Table 3.

Table 12-2   Pipe Surface Temperature Versus Steam Pressure

Steam Pressure (psig)	Steam Temperature (°F)	Pipe Surface Temperature Range (°F)
15	250	238–225
50	298	283–268
100	338	321–304
150	366	348–329
200	388	369–349
450	460	437–414

Source: *Yarway Steam Traps: Simple Techniques for Surveying Steam Traps* (Miami, FL: Tyco Flow Control): 2005, Table 2.

the trap can indicate performance. The two requirements for this method are a simple contact pyrometer (an instrument that measures temperature) to measure surface temperature of the pipe and knowledge of line pressure upstream and downstream of the trap. For each steam pressure there is a corresponding steam temperature. Table 12-2 shows typical pipe surface temperature readings corresponding to several operating pressures. Steam tables are available that provide such information for all pressures. Pipe temperatures should be about 90 to 95 percent of the saturated steam temperature shown in the steam tables because of heat losses around the steam trap.

As an example, assume the upstream pressure in the piping system is 150 pounds per square inch gauge (psig) and the pressure downstream of the trap is 15 psig. An upstream temperature measurement with the pyrometer is 338°F and a downstream reading is 229°F. The table shows that for an upstream pressure of 150 psig a pyrometer reading between 329°F and 348°F should be obtained. And for a downstream pressure of 15 psig, a pyrometer reading of between 225°F and 238°F is desirable. Thus, the trap is functioning properly. Now, assume the same pressures but a pyrometer reading of 338°F upstream and 300°F downstream from the trap. The insufficient spread between the two temperatures indicates that live steam is passing into the condensate return line. The trap has failed open. Steam trap troubleshooting charts are available at the end of this chapter in Exhibits 12-1, 12-2, and 12-3.

PACKING GLAND ADJUSTMENT

Often, when a leak develops around the packing of a valve or similar equipment, adjusting the packing gland can stop it. This is a simple job, but although the job may seem simple, safety considerations may make the job more involved and complex. It is important to consider the condition of the valve and the packing adjustment nuts or bolts. Corroded nuts or bolts may be frozen and will snap off when adjustment is attempted. The task may have to be assigned to maintenance personnel and may require re-routing flow through that line.

The first step in stopping a valve packing leak is to plan the job and prepare a work permit to ensure that the job will be done safely. Process conditions and materials should also be

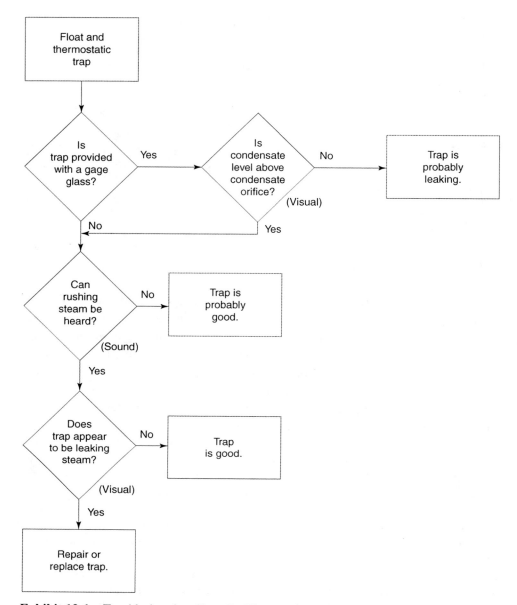

Exhibit 12-1   Troubleshooting Chart for Float and Thermostat Steam Trap

considered when planning the job. Leaks on high-pressure systems (above 150 psig) or systems containing highly toxic materials should not be tightened under pressure. The equipment should be depressurized to a slight positive pressure before attempting adjustment. Plant policy may require that a standby person be available while the repair is in progress. The **standby** is responsible for the safety of the individual or crew working on the valve. The primary function of the standby person is to serve as rescue for the person performing the work, not to help with the work.

UNPLUGGING PROCESS LINES

Some process materials are prone to plugging lines. These include materials that (1) contain solids, such as slurries, (2) have relatively high freezing points, such as molten

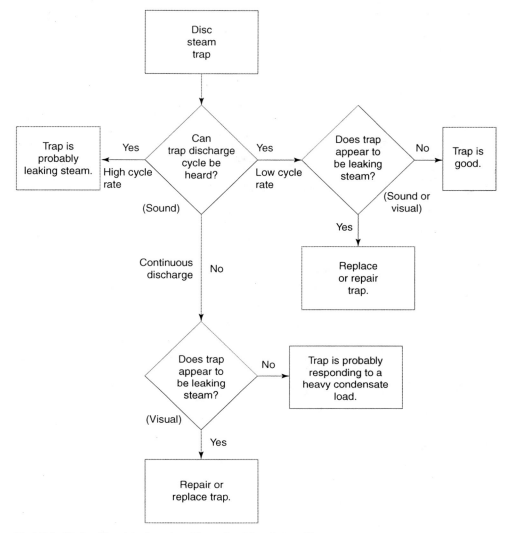

Exhibit 12-2   Troubleshooting Chart for Disc Steam Trap

polymers, and (3) tend to cake when wet, such as dry solids. These materials can plug process or utility lines. Line unplugging is usually a job for operations and maintenance technicians working together. Careful thought should be given to unplugging lines safely. Material safety data sheets and site safety procedures should be checked to determine possible material hazards and the required personal protective equipment (PPE) to use for the job.

Four basic techniques are used to deal with most unplugging and cleaning jobs involving process piping and equipment:

1. Removing the plug by using pressure
2. Mechanical methods (drilling, rodding, or rattling)
3. Applying external heat
4. Cleaning with a solvent, acid, or caustic solution

177

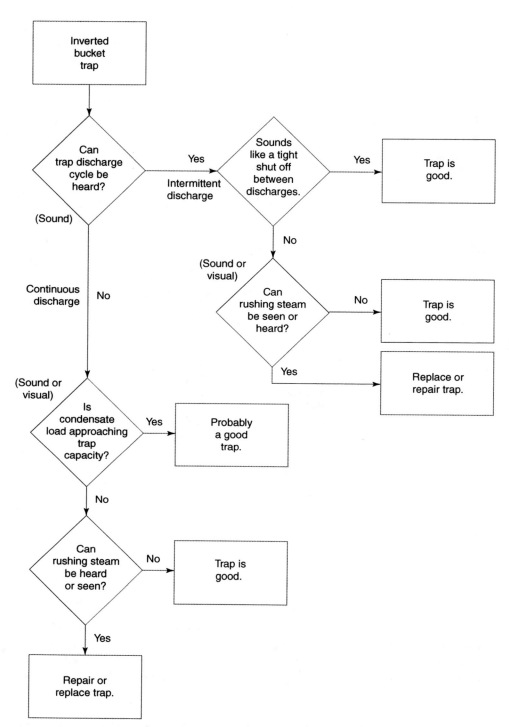

Exhibit 12-3 Troubleshooting Chart for Inverted Bucket Steam Trap

Pressure

Pressure is used for clearing plugged lines in a variety of situations. The source of the pressure may range from a manually operated hydraulic pump or a hand operated grease gun to a high-pressure positive displacement pump or compressed gas cylinder. The pressurized material may include water, air, nitrogen, steam, solvent, process materials, or high viscosity greases. Safety should be a factor in the choice of pressurized materials. Compressed gases should be avoided if a liquid or semi-solid material can be used because the stored energy in a compressed gas makes it more hazardous to use. If the plugged system contains flammable materials, air should not be considered as a pressure source because of the possibility of creating an explosive mixture. Process water, not potable water, should not be used because of potential drinking water contamination if process materials back into the potable water system.

The line used to connect the pressure source to a plugged system should be made from solid piping or tubing rather than hoses. The design should include a pressure bleed-off valve, a pressure-indicating gauge, and a pressure-flow control valve located in close proximity to the gauge. Gauge range should be twice that of the pressure source. The drain valve will also be useful in alternately pressuring and depressuring the line in an effort to move the plug.

If the plug in a line is being pressured towards an open end or open flange in the line, the end of the line must be securely anchored to withstand the reactive force of the sudden discharge of pressure. The reactive force could cause the end of the pipe to leap upwards or sideways. The open end of the line should also be covered or shielded to minimize spray and the area at the open end of the line should be barricaded.

Mechanical Methods

Mechanical methods for clearing plugged lines or equipment include drilling, rodding, vibrating, rattling, and so on. Drilling techniques are frequently used for unplugging heat exchanger tubes, injection nozzles, or other small diameter openings that plug with solid materials. Vibrating and rattling techniques are used most often with powdered or granulated solid materials that tend to cake or bridge. Rodding is used whenever safe access to the plug is possible.

Drilling. *Drilling* is frequently used to unplug heat exchanger tubes. Long hydraulically (hydroblasting) or pneumatically driven drills are used for this work. Heat exchangers that use steam or some other heated medium on the shell side should be completely vented and cooled before drilling starts. If the shell side remains hot and the drilling operator uses water to cool and lubricate the drill rod, the water could flash to steam and blow back on the operator or other personnel in the area. The area at the end of the pipe or equipment opposite the drill end should be barricaded and a backstop set up to deflect any material discharged when the drill breaks through. Additional precautions should be taken if there is a potential for release of toxic or flammable materials during the drilling process. Diking or containment materials should be available, plus fire fighting equipment.

Rattling and Rodding. *Rattling* and *rodding* are other mechanical ways to vibrate or prod the plug material to free it up so that flow will continue. This may be something as simple as a hammer to vibrate the sides of piping and vessels or a metal rod to probe a plugged nozzle. Or it may involve sophisticated pneumatically driven rattlers permanently installed on the walls and cone of silos and feed hoppers to encourage the flow of powered or granulated solids.

Drilling, rodding, or rattling a section of pipe or equipment where pressure or liquid may be trapped can be hazardous. A large amount of material can be released from a tank or vessel when a plug breaks loose while rodding out a nozzle. High-pressure piping can have a large quantity of gas trapped between two plugs. Verify proper valving, pressure relief devices, and wear PPE that protects against the material in the line.

Heat

Heat can unplug process lines if it can melt the materials that formed the plugs or if it can increase the solubility of the plug materials so that they can be flushed from the system with a solvent. Heat can also reduce the viscosity of liquid materials and improve their flow characteristics. Heating process equipment or piping to remove a plug must be done carefully if process material is trapped in a line between two plugs. Heating can increase internal pressure between the two plugs and overpressure the system. Whenever possible, the system to be heated should be opened and heating should start at the open end and work gradually back toward the plug.

Cleaning with a Solvent

Solvent cleaning uses a liquid that dissolves the plug material and acts as a flushing medium that removes plugs and fouling materials in the process. Solid piping should be used whenever possible for connecting solvent circulating equipment to the equipment being cleaned. If a hose is used its material should be compatible with the solvent and the hose should be secured to prevent excessive strain on couplings and whipping in the event of a failure. If the solvent is reactive with the material being removed and evolves gases, venting will be necessary, especially if the gas evolved is flammable. Arrangements should be made for the safe disposal of the materials produced during cleaning or unplugging. Blinds should be installed or physical disconnects made in all lines where contamination is possible and wherever disconnects are made.

PIPING VIBRATION

Vibration is the motion of a machine or a machine part back and forth, or relative from one part to another. Piping vibration is measured in *displacement*, which is the distance the pipe moves. Vibration problems in piping occur in both new and existing systems with abnormal operating conditions. While making rounds, process technicians should be alert to piping vibrations that could lead to failure. Surging, two-phase flow, unbalanced machinery, fluid pulsation, rapid valve closures, local resonance, or acoustic problems can cause excessive vibration. A large processing unit may have miles of piping, all of which must be supported correctly to prevent failure from vibration. If lines are suspected of excessive vibration, technical support personnel should be requested to measure the vibration. Lines vibrating enough to cause failure should be marked and a work order written to modify the piping support with gussets, hangers, or snubbers (See Figure 12-5).

Excessive piping vibration can cause serious problems. Threaded connections can loosen, flanges can start leaking, pipes can be knocked off their supports, and in extreme cases, pipe fatigue failure can occur. There are two major categories of piping vibration and several

Figure 12-5 Piping Hangers

vibration types within each category. An engineer must know the type of piping vibration being dealt with before the vibration can be treated effectively. The two major vibration categories are *steady-state* and *transient*.

Steady-state vibration is forced, repetitive, and occurs over a relatively long period of time. The force causing the vibration can be generated by rotating or reciprocating equipment (especially compressors), pressure pulsations, or fluid flow (liquid or gaseous). Excessive steady-state vibration can cause a fatigue failure in the pipe due to a large number of high-stress cycles. This failure will probably occur at a stress concentration point, such as a branch connection, threaded connection, or elbow. Steady-state vibration can also cause failures at small diameter connections and tubing (e.g., instrument lead lines) or cause flange leakage due to the loosening of the studs. Steady-state vibration can be either low frequency (<300 Hz) or high frequency (≥300 Hz). Low-frequency vibration will typically cause lateral movement of the pipe, while high-frequency vibration can cause flexural vibration of the pipe wall itself.

Transient vibrations occur for a relatively short period of time and are usually caused by larger, exciting forces. Pressure pulses traveling through the fluid are the most common cause of transient vibration in piping systems. These pressure pulses exert unbalanced dynamic forces on the pipe that are proportional to the straight length of pipe between bends. A common transient piping vibration is water hammer, which may be caused by rapid pump starts or stops or by quick valve closing or opening.

Piping vibration can result in two types of problems—*technical and psychological*. A technical vibration problem is one that has the potential to cause a failure (or has already caused one) and must be dealt with. In many situations, a pipe can vibrate forever and never cause a failure. However, it might be a psychological problem to the personnel in an operating unit who see it every day and are concerned about it. If it is easy and inexpensive to fix a psychological vibration problem, it is usually better to do that rather than try to convince operating personnel that everything is really fine with that pipe shaking. Process technicians that suspect potential problems from vibrating piping or equipment should write work orders requesting technical support to inspect the vibrating piping.

SUMMARY

Chemical processing plants face enormous pressure to improve profitability by increasing internal operating efficiencies. Improved profitability increases based on how well a plant is able to decrease its operating costs. One way to do this is by a systemic plant-wide effort to improve the performance and reliability of rotating equipment and other equipment.

The two main trouble spots in the centrifugal pump are the bearings and wear rings. Other failures are packing leakage, shaft breakage, worn shaft sleeves, or loose impellers. Mechanical seals and bearings represent 75% of centrifugal pump failures.

The large electric motors used in the processing industry are expensive and it is an important technician job to keep them under surveillance and in normal operating condition. Technicians should know what to do when a motor will not start, how to monitor vibration and temperature, and lubrication procedures.

Steam traps are design to maintain steam energy efficiency by performing specific tasks such as maintaining heat for a process. There are a variety of steam traps that operate on different principles. The types of traps are float, fixed orifice, thermostatic, inverted bucket, and thermodynamic.

Four basic techniques are used to deal with most unplugging and cleaning jobs involving process piping and equipment. These are removing the plug by using pressure, mechanical methods (drilling, rodding, or rattling), applying external heat, and cleaning with a solvent, acid, or caustic solution.

REVIEW QUESTIONS

1. Explain how technicians are more involved with predictive maintenance.

2. Explain why the technician is critical to the MTBR for pumps.

3. List several ways the ability to improve equipment reliability has a direct impact on bottom-line costs.

4. List the two main trouble spots in centrifugal pumps.

5. Describe how wear rings get damaged.

6. Before an electric motor is started technicians should verify the _____ is free to rotate.

7. Explain what is meant by the term *bumping* a motor.

8. How does bumping a motor harm the motor?

9. List two important predictive maintenance checks to make on electric motors.

10. Explain why cleaning and ventilation is important to electric motors.

11. Explain the function of steam traps.

12. What is wrong with dumping condensate to grade?

13. Leaving condensate in steam lines can cause _____.

14. List four types of steam traps.

15. Describe the operation of one type of steam trap.

16. List two reasons for steam trap failure.

17. List the three simple methods for troubleshooting steam traps.

18. List the four basic techniques for unplugging and cleaning piping and equipment.

19. List three problems caused by excessive piping vibration.

CHAPTER 13

Material Handling I: Bulk Liquids

Learning Objectives

After completing this chapter, you should be able to

- *Describe the different ways that chemical materials are packaged in the process industry.*

- *Describe the different ways chemical materials are transported in the process industry.*

- *List various railroad tank cars and tanker trailers used to transport chemicals.*

- *Describe the limitations of the various trailer and car designs.*

- *Discuss the hazards involved with the material handling of bulk liquids.*

- *Explain the steps in loading and unloading tank trucks and tank cars.*

INTRODUCTION

This is the first of three chapters that discusses material handling. This chapter is directed at large bulk liquid carriers, such as tank trucks and rail tank cars. Chapter 14 will discuss the different type of containers used for all material processes and the material handling of solids, and Chapter 15 will discuss material handling with the emphasis on tank farms and storage tanks. This chapter describes the various containers for packaging product materials and discusses the various cargo trailers used to transport materials in bulk. The remaining pages focus on the material handling of bulk liquids.

Process technicians work tank farms, loading racks, and bagging and drumming operations that require them to load containers and tanks, transfer materials, or unload materials. This movement of chemical materials is called **material handling**. The process of material handling requires the knowledge and skills to

- Obtain hazard information about the material being transferred (contained in the material safety data sheet [MSDS])
- Do a valve lineup
- Operate pumps
- Inspect cargo vessels for mechanical integrity
- Verify the correct placarding and/or placard a vessel
- Read and fill out a bill of laden
- Know spill containment and cleanup procedures
- Calculate equipment parameters that transfer the exact amount of material
- Know the characteristics and limitations of different types of shipping containers

Process materials may be stored and/or transported in a wide variety of containers or tanks. Some of the containers can be used for solids, liquids or gases. Some are specific for a particular state of matter, such as a gas. The different types of containers consist of

- Bags and boxes
- Drums and barrels
- High- or low-pressure gas cylinders
- Portable tanks and tote bins
- Sack containers
- Intermodal containers
- Tank trailers and hopper trailers
- Railroad tank cars, hopper cars, flatcars, and box cars
- Pipelines
- Stationary storage tanks

Each type of container is unique and technicians must become familiar with all of those used on their unit and know how to safely transfer material into and out of them. To fill smaller containers and move them to the warehouse may require the technician to operate fork trucks or small cranes for loading heavy containers onto flatbed trucks. They may also have to operate material handling equipment such as conveyors, elevators, and bagging and boxing equipment.

SMALL BULK LIQUID MATERIAL HANDLING

This section briefly describes drumming operations. We all have probably noticed the blue polypropylene 55-gallon drums and black metal drums. The filling, placarding, and preparing of shipment of liquids in drums is called **drumming**. A facility that loads and ships drums of materials will have a drumming room, drumming shed, or a drumming area within a building. The drum room will have scales with a digital readout, a manifold with hoses for loading materials from different tanks, a short roller belt for moving the loaded drum from the scale to a pallet, and Department of Transportation (DOT) hazard labels. (See Figure 13-1 for a visual of drums on a conveyor waiting to be filled.) The drumming

Figure 13-1   Drums on Conveyor Waiting to Be Filled

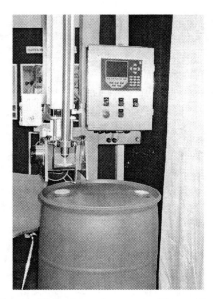

Figure 13-2 Drum Lances and Bung Tool

technician will place an empty drum on the scale, tare it (weigh it), valve the manifold for the correct product, load the drum to the correct weight through a drum lance or hose, then seal the drum with a bung tool and label it. (See Figure 13-2 for a visual of a drum lance and bung tool.) The technician will then fill out the bill of laden, shrink-wrap the drums to the pallet or band the drums together on the pallet, and move the drum(s) with a forklift to the product warehouse for shipping. Attaching the drums to a pallet is called *palletizing*.

Tote bins done in a later chapter will be filled in the same manner except they may require larger scales because of their greater size and greater weight when fully loaded. Generally, facilities that do much drumming make small quantities of a large variety of special chemicals. This is usually done by blending or in batch reactors.

LARGE BULK MATERIAL CARRIERS

Frequently, process technicians will be required to load or unload liquids, gases, and solids from railroad tank cars or truck trailers. Several different types of tank cars and tank trailers carry the array of processed chemicals. Some are designed for low-pressure application only, others for liquefied gases under significant pressure, corrosive acids or bases, compressed gases, cryogenic liquids, or dry bulk materials. Technicians should recognize each type of vessel and their limitations. Tank cars and tank trucks typically have at least four placards, one at each end and one on each side of the tank. The Interstate Commerce Commission and the DOT have established regulations for the safe transportation of hazardous substances. These regulations are binding on all shippers. Individuals in charge of loading bulk carriers should be familiar with these regulations. The plant shipping department is responsible for scheduling the movement of tank cars and tank trucks into and out of the plant. They also coordinate the arrival and departure of barges and ships at the plant docks if the product is transported by barge or ship.

Tank Trailers

Atmospheric Pressure Tank Trailers (MC-306/DOT-406). Atmospheric pressure tank trailers are used extensively to transport liquids that can be maintained at a vapor pressures less than 3 pounds per square inch gauge (psig). This includes flammable, combustible, and non-flammable liquids; toxic liquids; and liquid food products. Examples of transported liquids are orange juice, milk, gasoline and kerosene. Since atmospheric trailers are not pressure rated, pressure transfer techniques cannot be used for unloading. Generally, they have a 9,000 gallon maximum capacity.

Atmospheric tank trailers are usually constructed of aluminum and the cross-section profile is elliptical in shape. Internal stiffening rings provide additional support. Outlet piping, valving, and controls are located under the trailers. The trailers are divided into compartments by internal bulkheads with manhole assemblies located on top of the trailers for each compartment. Longitudinal rails along the top of the trailers provide rollover protection. (See Figure 13-3.)

Low-Pressure Trailers (MC-307/DOT-407). Low-pressure trailers are designed to carry chemicals with vapor pressures up to 40 psig at ambient temperature (70°F). They may carry flammable, combustible, non-flammable, and toxic liquids. Some typical cargos are styrene, paraxylene, gasoline, and jet fuel. Construction is usually steel or stainless steel with a double shell and insulation being common. Generally, low-pressure trailers have a maximum capacity of about 6,000 gallons. The cross-section profile of the tank is circular in shape. Rollover protection includes circumferential supports at each end and flashing boxes around the manhole(s). Discharge outlet, valving, and unloading controls are located either under the trailers or at the rear. Vents, pressure relief devices, and vacuum breakers are located on top of the trailers. External stiffening rings or a double shell with covered stiffening rings provide additional support. (See Figure 13-4.)

Figure 13-3   Atmospheric Tank Trailer

Figure 13-4   Low-Pressure Tank Trailer

High-Pressure Trailers (MC-331). High-pressure trailers are designed for carrying liquefied gases, such as anhydrous ammonia, propane, and butane, and are designed for pressures above 100 psig. They have a heavy, bulky appearance and both ends of the trailers have large hemispherical heads. Loading/unloading piping and controls are either under the trailers or in the rear. If located under the trailers, a sturdy guard cage provides protection. A bolted manway is frequently located at the rear of the trailers. Vents and pressure relief devices are located on top of the trailers. An excess flow valve protects outlet piping to prevent a major spill if a leak develops in the unloading system. The upper two-thirds is painted white or a highly reflective color.

Corrosive Liquid Trailers (MC-312/DOT-412). Corrosive liquid trailers are designed for carrying acidic and caustic materials and have a capacity of 5,000 to 6,000 gallons. Examples of chemicals transported in these type of trailers are sulfuric and hydrochloric acid, sodium hydroxide and potassium hydroxide. Since these materials are denser than hydrocarbons, the corrosive liquid trailers have a smaller volume than most other bulk liquid trailers. These are designed for relatively high pressures (up to 75 psig) and are made of steel- or corrosion-resistant alloys. They are similar in appearance to the low-pressure trailer, but are smaller in diameter and have a cigar shaped profile. They frequently have a top loading/unloading station with exterior piping extending the discharge connection to the lower rear of the trailers. (See Figure 13-5.)

Cryogenic Liquid Trailers (MC-338). Cryogenic liquid trailers transport gases liquefied at **cryogenic** temperatures (below −300°F, −150°C). Examples include liquid oxygen, nitrogen, carbon dioxide, and hydrogen. The trailers are double walled with the space between the walls maintained under vacuum. The evacuated space plus heavy insulation keeps cryogenic materials in liquid form while they are in transport. They are large, bulky trailers with fairly flat ends. Loading and unloading piping and controls are usually located at the rear of the trailers or just in front of the dual rear wheels.

Tube or Cylinder Trailers. Cylinder trailers carry compressed gases, such as air, hydrogen, nitrogen, helium, and refrigerants. They do not transport liquefied gases. The trailers consist of several large, horizontal gas cylinders that are connected by a manifold of piping. They are designed for pressures from 3,000 to 5,000 psig. (See Figure 13-6.)

Dry Bulk Trailers. Dry bulk trailers carry dry bulk materials, such as polymer pellets and granulated fertilizers. They are characterized by large, sloping, V-shaped bottom unloading compartments and frequently have an auxiliary air compressor used to pneumatically

Figure 13-5   Corrosive Liquid Trailer

Figure 13-6 Tube or Cylinder Tank Trailers

Figure 13-7 Dry Bulk Trailer

convey product during loading or unloading. These trailers will have top manholes, and/or exterior piping for loading and bottom unloading piping. (See Figure 13-7.)

Intermodal Containers. *Intermodal containers*, also referred to as ISO containers, are cylindrical tanks built inside a rectangular framework that permits lifting and stacking for transport on flatbed trailers or in the cargo hulls of ships. The cylindrical tanks are sized to hold about 40,000 pounds of material and fit on a truck trailer. Two or three such units can be loaded on a railroad flat car, and multiple units can be loaded in the cargo bay of an ocean freighter. (See Figure 13-8.)

Figure 13-8   Intermodal Containers

Railroad Cars

Railroad cars are similar to trailers but larger. Trailers may contain 6,000 to 9,000 gallons of material while some railcars hold over 20,000 gallons. Like trailers, railroad cars come in many different designs, including low-pressure and high-pressure tank cars, liquefied gas cars, high-pressure cylinder cars, cryogenic cars, hopper cars for granular solids, and gondola cars for larger solid materials. General-purpose tank cars have a working pressure of 100 psi or less. Pressurized tank cars have a working pressure greater than 100 psi.

General-Purpose Tank Cars. The non-pressurized or general-purpose tank cars transport most liquid materials, such as flammable and toxic liquids, molten solids (sulfur), liquid farm products, fruit juice, and many other products. The cars may be lined, insulated, double walled, or single shelled. They may have as many as six separate compartments and each compartment has the capability of carrying different cargoes since they each have separate loading and unloading fittings (see Figure 13-9).

Pressurized Tank Cars. *Pressurized tank cars* carry hazardous and non-hazardous liquids and liquefied gases. They are designed for service pressures up to 400 psig or more and equipped with a safety relief valve set to match the vapor pressure of the transported material at 105°F. They may be insulated or have double walls for greater protection. An off-white paint indicates sprayed on insulation. The cars have the loading and unloading valves on the top and a dome cover provides protection for the valves.

Dedicated Cars. Some railcars are treated as dedicated cars, which is usually done if (1) the product is unusually hazardous, such as hydrocyanic acid, (2) if the product can react with other materials, or (3) if the manufacturer wants to be sure that some other material left in

192

Figure 13-9   General-Purpose Tank Car

a car, called a **heel**, does not contaminate their product. In certain cases, the name of the material is painted on the side of the car, or if the product is especially hazardous, the tank car will be painted with an easily identified color and/or pattern. Tank cars carrying hydrogen cyanide, for example, will be painted white with two vertical red bands and a horizontal red stripe on the car.

MATERIAL HANDLING TECHNIQUES FOR BULK LIQUIDS

The hazards of loading petroleum products can be minimized by closely following company procedures. The greatest potential danger in loading volatile products into tank cars and trucks is from a fire resulting from the flammable mixture of organic vapors and air. However, fire is neither the only danger nor the most common type of accident. The loader, who works high off the ground, is subject to the dangers of slipping, falling, and overstraining.

Loading and Unloading Railroad Tank Cars and Tank Trucks

The procedure for transferring bulk liquids typically consists of three steps:

1. Weighing and spotting
2. Making connections and transferring the liquid
3. Disconnecting and releasing

Railcars and tank trucks are weighed before and after transfers (loading or unloading) to determine how much liquid material has been loaded. Railcars are weighed on track scales that can weigh 25,000 pounds or more. A scale house is located adjacent to the track scales that will have electronic equipment that tares (weighs) the railcar, calculates the difference between the loaded and unloaded railcar, and prints out a shipping paper or sends the information to the shipping department electronically. The same arrangement is set up for tank trucks but the scales are much smaller and usually located near a guard gate. Barges and ships are not weighed but loaded by volume. After weighing the railcar or trailer it is spotted. **Spotting** refers to placing a railcar, tank truck, or barge in the correct position at a loading rack or dock and inspecting it before the transfer has begun.

Safety Factors During Loading and Unloading

Technicians will receive on-the-job training that will make them aware of several safety factors during unloading and loading cargo vessels. The following sections discuss these safety factors.

Pulling a Vacuum and Collapsing a Cargo Tank. A vacuum can be created by pumping a liquid from a cargo tank without replacing the empty space left behind. Unless a tank is constructed to withstand a vacuum, it will collapse like a squeezed aluminum can when submitted to a partial vacuum. For example, assume the tank of a railcar is 50 feet long and 10 feet high. If a partial vacuum is pulled on the tank so that the tank has an internal pressure of 13 psig, then the total pressure exerted over the surface of the tank will be approximately 430,000 pounds. This cumulative pressure pushing in from all sides can cause the tank to collapse. To avoid pulling a vacuum, nitrogen or the liquid's vapors must be routed into the tank via a vent valve.

Buildup of Static Electricity. As hydrocarbon products are transferred, a difference in electrical potential can build up between the fill line or piping that the liquid flows through and the tank being loaded or unloaded. This buildup of electrical potential increases the chance of electrostatic sparking (creating static electricity). If the spark occurred near a tank's open manway where a mix of air and volatile vapor exists, it could cause a fire or explosion. A grounding cable used to drain electricity to a ground eliminates the static electricity hazard. The grounding cable should be attached before the tank manway is opened and remain in place until it is closed. One end of the cable should be attached to the metal frame of the tank and the other to the metal loading rack.

Containing Spills. Drain pads are shallow depressions in the concrete that drain spills to a collection point for recovery and disposal. When a hazardous material is being transferred drain pads may be required below the loading rack. Where drain pads are not feasible or installed, large drain pans catch spills and drips. If liquid spills onto the tank during a transfer, it should be cleaned off immediately to minimize the risk of individuals coming into contact with the tank later.

Emergency Shutdown Systems. Emergency shutdown systems will remove all control power from the loading bay, stop all pumps, and close all valves. This can occur through

- A fire eye detecting a flame
- Power failure in the 110VAC control power circuit
- The pressing of any emergency shutdown button on the loading platform
- Fire protection systems for the loading bay or platform that may include a gas detection system for emergency shutdown and/or a stationary foam system triggered by an ultra-violet detector.

Tank Truck Features

Tank trucks are designed and built to DOT standards. All tank trucks have valves, many of which are located in a manway cover on top of the truck. The valves may be bleed valves, vent valves, product outlet valves, and safety valves. Vent valves may be pressure vents and vacuum vents. Pressure vents release excess pressure that builds up in the tank; vacuum vents (breakers) admit air into the tank to avoid pulling a vacuum on the tank and collapsing it. Product outlet valves may be located both on top of the tank for top loading and unloading and under the tank for bottom unloading (see Figure 13-10). A spill dam surrounds the manway to contain any spilled product. Spills are removed for safe disposal through a drain hose in the dam. Crash protection devices called *rollovers* are built onto the back and front sections of the spill dam to protect the vents and valves in the manway from being damaged should the tank flip over. See Figure 13-11 for a top view of a tank truck.

Figure 13-10   Bottom Valve Arrangement on Tank Truck

Figure 13-11   Top View of a Tank Truck

OVERVIEW OF LOADING A TANK TRUCK

The first step in loading a tank truck is to weigh the truck, which is usually done on a scale located close to the guardhouse. The guards may inspect shipping papers, driver identification, and record the weight of the entire tractor-trailer rig. The truck is then directed to the appropriate loading rack (see Figure 13-12). At the loading rack the truck is spotted, the ignition turned off, and the parking brake set. The driver and process technician check the paperwork for the truck, then the technician makes an inspection of the tank truck, which is detailed in the section of this chapter titled "Pre-Transfer and Post-Transfer Inspections." After inspecting the truck, a grounding cable is attached to the tank and wheel chocks are placed under the truck tires. Then the traffic barrier is lowered to warn plant personnel that loading is taking place and prevent running vehicles, which are a source of ignition, from entering the loading area. The loader will don the required personal protective equipment (PPE) and a safety harness. The safety harness is required because the loader will be working high up on top of the truck's cargo tank to make the inspection and to load, if there is top loading (see Figure 13-13). The loader will be tied-off to the loading platform.

If the product is to be top-loaded through an open manway, the vent valve should be opened slowly and checked for unsafe pressure before the manway is opened. The manway cover is then unbolted after pressure has been relieved and the tank inspected to ensure it is clean and has no residue (*heel*) from the last shipment in the tank. Next, to prevent problems with static electricity, the loading arm is lowered straight down into the tank and stopped just off the bottom. The loading arm is secured in place and the valve on the loading arm opened. Before starting the transfer, the technician presets a loading meter to the desired amount of product to be loaded. The meter will shutoff flow automatically when the desired amount is reached. Next, the technician checks the valve lineup to ensure the product is being loaded from the correct tank in the tank farm, then opens the valve to the product storage

Figure 13-12   Tank Truck Loading Rack

Figure 13-13   Loader at Truck Loading Rack

tank. The transfer pump is started, and while loading proceeds, the technician remains alert for leaks or other abnormal conditions.

After the tank truck is loaded, it must be disconnected and released. First the pump is shut-down, then the loading valves are closed in the reverse order from the way they were opened. After the valve on the loading line is closed, the loading arm is removed from the tank. The technician may be required to take a sample of the truck product and submit it to the quality control laboratory. Then the manway cover can be closed and bolted back in place. In some cases, a safety seal and product tag are attached to the manway cover. The safety seal prevents the valves on the truck and the material in the truck from being tampered with. The product tag provides information about the material in the tank. The technician also attaches a safety seal at the outlet valve on the truck. Lastly, the grounding cable and wheel chocks are removed. Now that all safety precautions have been met and the transfer is complete, the technician signs the necessary paperwork. The truck returns to the weigh scales to be weighed again before being released. The difference between the empty tare of the truck and the filled truck is the weight of product being purchased.

OVERVIEW OF UNLOADING A TANK TRUCK USING PRESSURIZED NITROGEN
Depending on the product in a tank truck (flammable or non-flammable) and if the tank is designed to withstand the pressure, the tank can be unloaded with pressurized air or nitrogen. After being weighed, the tank truck is spotted beside the storage tank that will receive the tank's contents. Many tank farms have one or two manifolds for unloading tank trucks. A **manifold** is a metal rack with hoses, connections, valves, and unloading pumps that route the tank truck's contents to one of several tanks. The tank truck is inspected for safety and mechanical integrity, a sample may be collected and sent to the quality control laboratory for analysis, and then unloading begins.

A product line is attached to the product valve at the top of the tank. The product valve is attached to a dip tube that extends to within inches of the bottom of the tank. An unloading hose is attached to the tank's product valve and to the manifold that has already been lined up for transfer to the appropriate tank. Next, a nitrogen hose is connected to the vapor valve on top of the tank and the technician sets a pressure regulator to the proper pressure. The vapor valve is opened to allow nitrogen into the tank and the product valve is opened to allow the nitrogen to push the tank contents into the unloading line and down to the pump.

During the transfer, the technician checks for product flow by feeling the unloading hose for vibration caused by moving product or watching a flow meter. In most plants, the technician remains with the tank truck until unloading is complete; however, some plants allow the technician to leave the tank truck and do other tasks, but they must be able to keep the tank truck in their line of vision. When the transfer is completed, the vapor valve is closed and excess pressure is bled off the tank. The technician opens the manway and visually inspects the inside of the tank to ensure no significant amount of material is left. The nitrogen supply is then turned off, the pressure is bled off the nitrogen hose, and the hose is disconnected. The product valve is closed, the unloading hose disconnected, and the shipping papers signed. The tank truck will then return to the weigh scale to be weighed and the technician will clean and/or purge the unloading hose according to procedure.

It is important to note that a significant amount of residue called a heel may remain in a tank truck after unloading, thus placards identifying the material must remain on the truck. The placards are not removed until the tank is cleaned. The carrier (trucking company) is responsible for the cargo tank cleaning.

OVERVIEW OF LOADING A RAILCAR

Railroad tank cars can be roughly divided into two categories, general-purpose tank cars and pressurized tank cars. General-purpose tank cars have a working pressure of 100 psi or less. Pressurized tank cars have a working pressure greater than 100 psi. A typical tank car has on the top of the tank a safety relief valve, a dome covering the valve assembly, and a manway. Valves and vents (vacuum and pressure) are located under the dome cover (see Figure 13-14). Some of the valves are for pressure relief, others for product transfer or vapor recovery. On pressurized tank cars the liquid product valves are positioned parallel to the length of the tank and vapor valves are positioned perpendicular to the length of the tank. Tank cars that carry flammable compressed gases must have check valves (also called excess flow valves) in the product and vapor lines. Their purpose is to stop flow of product when flow rates exceed limits, such as when a hose ruptures.

Tank cars have loading racks similar to tank truck loading racks. In fact, it is not uncommon for a loading rack to be located adjacent to a railroad track and load railcars on one side of the rack and tank trucks on the other (see Figure 13-15). The basic tasks involved in loading a tank car are the same as those loading a tank truck: weighing and spotting, making connections and transferring the product, and disconnecting and releasing.

Switch personnel are workers who operate locomotives or railcar movers and move railcars about within the plant. They move the empty tank car to the track scale, tare it, and then move it to the loading rack. Placards on the spotted tank car are reversed, indicating it is empty. Derailers are locked in place to isolate the tank car. A *derailer* is a device that fits over the rails and prevents oncoming tank cars from running into the spotted tank car.

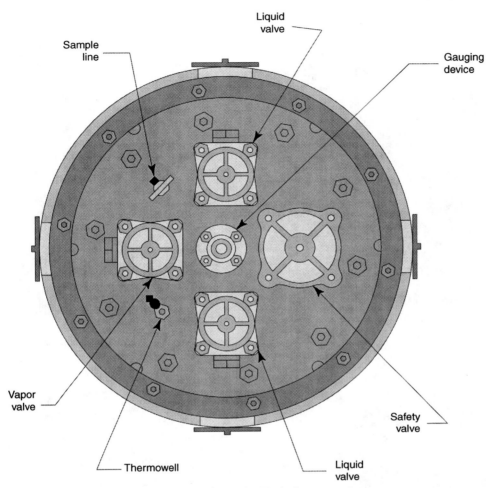

Figure 13-14   Typical Valve Arrangement for Tank Cars

Warning signs are also set on the tracks to isolate the tank car (see Figure 13-16). The tank car wheels are chocked to keep the car from moving and a grounding cable is attached to the tank car (see Figure 13-17). Generally, the loading pump is instrumented and controlled so that the pump will not start unless it senses the tank car is grounded. When the tank car is spotted and grounded, the technician inspects the tank car for damage or defects (discussed in greater detail later in the chapter in the section titled "Pre-transfer and Post-transfer Inspections").

After the ground level inspection, the technician presets a meter at the loading rack for the desired amount of product. The technician then checks the manway, the dome cover, and the safety valve on top of the tank. Next, the dome cover is lifted and the technician checks the condition of the safety valves, safety vents, and rupture disks. Pressure on the car must be relieved before the technician can open the manway to check the interior of the car to ensure it is clean and ready to receive the product. Heels left in the tank can contaminate the freshly loaded product. Hinges and bolts on the manway are inspected for damage or wear, and the manway gasket is inspected. Then the manway is closed and sealed.

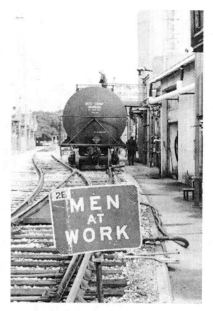

Figure 13-15   Railcar Loading Rack   **Figure 13-16**   Derailer and Warning Sign

Figure 13-17   Chocked Railcar

If the railcar is being top-loaded with a liquid, the technician closes any bleeder valves on the loading lines. **Bleeder valves** are installed on the loading lines for sampling and for bleeding air out of the lines. Next, the loading line is connected to the product valve on the tank car and the loading meter is preset to deliver a specific volume of product. Then a vent line is connected to the vent valve and the vent valve opened. Once everything is lined up and securely connected, the technician opens the product valve on the loading arm, starts the loading pump, and loading begins. Product flows into the tank and air and/or vapors in the tank are vented to the vent system. A sample may be collected and sent to the quality control laboratory for analysis and the creation of certificate of analysis.

After the tank is loaded, the technician follows the required procedures for disconnecting and releasing it. Valves are closed in the reverse order that they were opened and the railcar is then pressure tested with nitrogen to check for leaks around the manway and valves. The technician may leave a few pounds of nitrogen in the car to serve as a blanket over the product. Finally, the technician attaches a product tag to the sampling device at the dome, lowers and locks the dome cover, and puts safety seals in place on the dome cover and manway to prevent tampering with the valves or material in the tank.

To prepare the tank car for release from the loading rack, the technician reverses the placards to show it is loaded and removes the wheel chocks, the grounding cable, the derailers, and the warning signs. The technician performs their final inspection and notifies switch personnel the tank car is ready to be moved to the scales.

PRE-TRANSFER AND POST-TRANSFER INSPECTIONS

Before and after loading or unloading a cargo tank, a general inspection should be made of the tank (railcar tank, tank truck tank) and its associated equipment. The inspection may be made with an inspection check sheet that might include the following:

Railcar and Tank Truck Inspection

- Inspect the tank for cracks and obvious damage.
- Check the safety attachments on the tank (hand rails and ladders) and verify they are not bent or loose.
- Check the running gear (springs, bearings, brakes, axles) on tank cars for obvious signs of damage or problems.
- Check that there is no heel remaining in the tank.
- Ensure there are no leaks from valves, fittings, caps, plugs, and gaskets.
- Check that the bottom outlet on the tank is tightly sealed and a chain attaches the outlet cap to the tank.
- Inspect the dome cover on the tank to ensure it is securely fastened and there are no loose nuts or leaks. Leaks can be detected by smell, sound, and wetness.
- Check the placards and placard holders to ensure the proper placards are securely in place and that the data on the placards and shipping papers match.

Loading Platform Inspection

- Check the loading arms to ensure they move freely and have no leaks at the connections.
- Check hoses for kinks, cracks, and other signs of wear or damage.
- Ensure each hose is identified properly for the product being moved. Various products require hoses and hose lining of different materials.

- Check the O-rings and connection pieces at the end of the hoses to ensure they are not damaged or need replacement.
- Verify safety equipment (safety shower, eyewash, fire extinguisher) is in place and working.

LOADING AND UNLOADING TANK TRUCKS AND RAILCARS

Loading and unloading railroad tank cars and highway tank trucks follow very similar steps. This section presents a step-by-step procedure for unloading. Note: Exhibits 13-1 and 13-2 are schematics of unloading loading procedures. These illustrations use a railroad car, but a similar technique may be used for most tank trucks and other liquid containers. A similar arrangement may be used for a portable liquid container. An unloading pump is used in the first procedure. The second procedure uses pressure transfer, an alternate unloading technique that uses air to pressure the material from the cargo vehicle to a storage tank. For flammable materials or materials that might be degraded by air or contaminants in the air, nitrogen may be used to pressure the cargo vehicle.

Procedure for Unloading Sulfuric Acid Railcars Using an Unloading Pump

> It is plant policy that the technician wears the appropriate PPE for the chemical being handled. Failure to do so will result in disciplinary measures.

Check bill of laden (shipping papers) and verify vehicle identification and contents.

1. Spot railcar on railcar track scale ME-222, which has a weighing capacity up to 225 tons. The scale is interfaced with a keyboard in the scale house for the electronic transmission of net weight and tare of the railcar.
2. Move the railcar from the track scale and spot it in front of the unloading platform.
3. Block the railcar with standard wheel clamps to ensure it will not move during unloading.
4. Ground the railcar and set up barricades and signs to warn unloading is in progress. The ground detector XS-501 located on the control panel at the unloading platform indicates proper grounding.
5. Inspect the tank car for leaks or damage that could interfere with the safe unloading of the contents.
6. Verify the receiving tank has sufficient outtage to hold the contents of the tank car and still have 10 percent remaining unfilled (freeboard).
7. Open the dome cover and inspect unloading fittings and valves. Look for evidence of mechanical damage or fitting leaks.
8. Remove any cover or plug on the tank vent valve. **CAUTION:** Leaking valves may have caused a pressure build up under the cover or plug.
9. Carefully open the air connection valve and vent off pressure. If there is no air connection valve, the safety vent should be removed. **CAUTION:** Acid cars will frequently be under pressure due to thermal expansion or slow hydrogen generation. All pressure must be relieved to prevent acid spray.
10. If no venting occurs, verify the air connection and vent valve are not plugged.

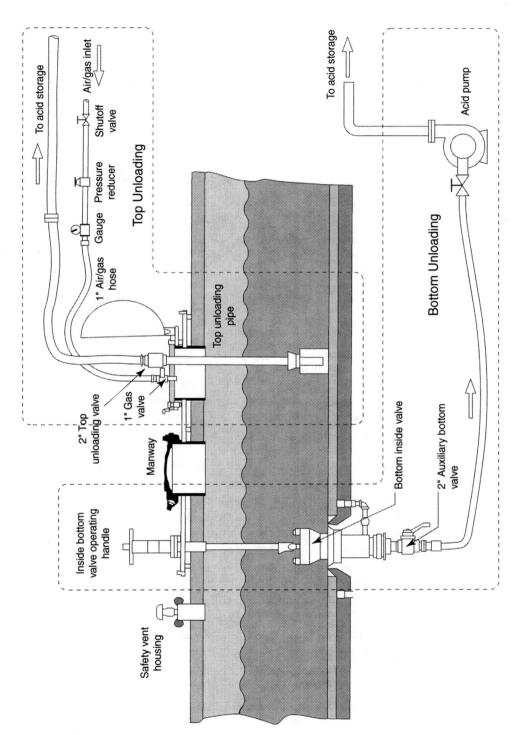

Exhibit 13-1   Railcar Unloading Schematic

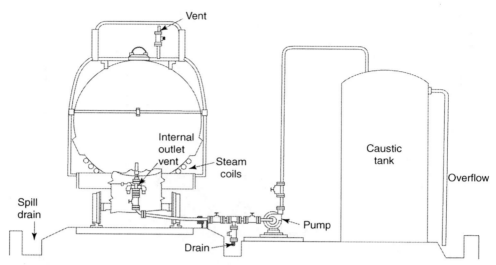

Exhibit 13-2   Layout for Bottom Unloading Caustic Solution by Pump

11. If a sample is required, open the filling hole and wait 15 minutes before lowering the sampling device into the tank car. **CAUTION:** This step allows any hydrogen gas to escape to the atmosphere. If the sampling device is lowered immediately after opening the filling hole, a stray spark could trigger an explosion by igniting generated hydrogen gas.

12. Remove the plug from the railcar unloading (outlet) valve at the bottom of the tank and inspect the condition of the threads, gaskets, and eduction (unloading) pipe. Pipe threads in acid service deteriorate rapidly when exposed to acid and atmospheric moisture.

13. Uncap the acid connection line to the storage tank and inspect for residual material in the hose from the previous transfer.

14. Connect the line to the unloading valve.

15. Line up for unloading the railcar.

16. Verify the storage tank vent is open.

17. Verify the fill hole cover is open. **CAUTION:** Adequate vacuum relief must be provided to prevent the collapse of the tank car.

18. Verify all vent and bleed valves on the unloading hose are closed.

19. Open the suction valve to the unloading pump.

20. Open the unloading valve on the bottom of the tank car. Wait several minutes and check for leaks.

21. Open the inlet valve to the receiving tank.

22. Start the pump from outside the probable spray area. Check the pump discharge pressure to insure that the pump is working properly. Gradually open the pump discharge valve until it is fully open.

23. Monitor the pump operation, receiving tank level, and trailer or car pressure frequently as unloading proceeds.

24. When product transfer is complete, shutdown the pump and close the pump discharge and suction valves.

25. Close the inlet valve on the receiving tank.
26. Bleed residue in the hose to the process drain. Break hose connections and replace the flexible hose cap to keep foreign material out of the hose. Purge or flush hoses, if required.
27. Replace plugs or caps on tank car valves.
28. Close and secure the car dome cover.
29. Complete the paper work. Obtain tare of the railcar for accounting and submit shipping papers and certificate of analysis to the shipping clerk.

Procedure for Unloading Caustic Soda Railcars Using Air Pressure

It is plant policy that the technician wears the appropriate PPE for the chemical being handled. Failure to do so will result in disciplinary measures.

Check bill of laden (shipping papers) and verify vehicle identification and contents.

1. Spot the railcar on the railcar track scale ME-222 and record the tare of the railcar.
2. Move the railcar from the track scale and spot it in front of the unloading platform.
3. Block the railcar with standard wheel clamps to ensure that it will not move during unloading.
4. Ground the railcar and set up barricades and signs to warn unloading is in progress. The ground detector XS-501 located on the control panel at the unloading platform indicates proper grounding.
5. Inspect the tank car for leaks or damages that could interfere with the safe unloading of the contents.
6. Verify the receiving tank has sufficient outage to hold the contents of the tank car and still have at least 10 percent unfilled °Freeboard).
7. Open the dome cover only if a caustic sample must be collected. After the sample is collected, close the cover and ensure the gasket is securely in place and the dome latches fastened.
8. Remove the cap on the unloading hose and inspect the hose for cleanliness and integrity.
9. Remove the plug to the eduction (unloading) pipe and connect the unloading hose to the eduction pipe.
10. Remove the cap on the air supply hose and inspect the hose for cleanliness and integrity. **NOTE:** If utility air is not available, use nitrogen.
11. Inspect the air supply hose relief valve and pressure gauge.
12. Remove the plug to the air inlet valve on top of the tank car and connect the air supply hose to the valve.
13. Open the tank car air inlet valve.
14. Open the air supply valve and slowly apply air pressure. Normal unloading pressure is 25 to 30 psig.
15. Verify the valve line up to the storage tank is correct.
16. Slowly open the eduction pipe valve and allow the caustic soda into the storage tank.

CARGO TANK LOADING REPORT

Date _____

Customer:_____ Product:_____ Storage Tank:_____ Trailer #_____
My signature confirms this information agrees with the reference document and the
Baytank service order   Signature_____

Note: Cargo tank is the proper DOT name for vessels also commonly called tank truck, tank trailer, and tank wagon.

PRIOR TO LOADING

1. Is the tank spotted at the correct loading area? — 1. Yes____ No____
2. Does the cargo tank have proof either by stamp or documentation
   that it has passed inspection in accordance with 49 CFR 396.17 through 396.23? — 2. Yes____ No____
3. What is the retest date?_____
4. Does the tank have proper DOT specification number, per the attached sheet? — 4. Yes____ No____
5. What is the proper DOT number?_____
6. Can all the outlines, including the domes of the tank, be sealed? — 6. Yes____ No____
7. Is the gasket on the dome in good condition? — 7. Yes____ No____
8. Does the dome have all bolts in good working condition? — 8. Yes____ No____
9. Is the outside of the tank in good condition? (No dents, etc.) — 9. Yes____ No____
10. Are ladders, safety railings, walkways in good condition? — 10. Yes____ No____
11. Is the tank clean, free of liquid, and odor free? — 11. Yes____ No____
12. If the answer to 11 is NO, is the heel compatible to the product to be loaded? — 12. Yes____ No____
    (Check with supervisor)
13. Are all the outlets and vents blind flanged tight or plugged? — 13. Yes____ No____
14. Are the belly lines free of liquid? — 14. Yes____ No____
15. Is the tank number the same as the number on the loading report? — 15. Yes____ No____

If the answer to any of the above questions is "NO" DO NOT LOAD THE CARGO TANK. Notify your shift supervisor.
If all answers are "YES", proceed with the following checks.

LOADING THE CARGO TANK

1. Is the ground cable connected properly? — 1. Yes____ No____
2. Is the trailer brake locked and the engine off? — 2. Yes____ No____
3. Is the internal valve operable and closed? — 3. Yes____ No____
4. Is the tank belly line valve closed with the cap off? — 4. Yes____ No____
   (If the valve leaks any time during loading, discontinue and notify your supervisor)
5. Are the steam heater coil caps removed? (On the tank with interiors) — 5. Yes____ No____
   (If the steam heater coil leaks any time during loading, discontinue and notify your supervisor)
6. Is the loading line connected safely? — 6. Yes____ No____
7. Is the proper safety equipment being used? — 7. Yes____ No____

AFTER LOADING CHECKS

1. Was the product loaded Methyl Acrylate? — 1. Yes____ No____
   (If yes, the backside of this report must be filled out.)
2. Has the product been loaded to the correct outage? — 2. Yes____ No____
3. Has a retain sample been taken and recorded? — 3. Yes____ No____
4. Has the loading line been disconnected? — 4. Yes____ No____
5. Has the dome cover been bolted tight? — 5. Yes____ No____
6. Has the ground cable been disconnected? — 6. Yes____ No____
7. Have all the valves been closed, capped, and no leaks? — 7. Yes____ No____
8. Have the proper placards been placed? — 8. Yes____ No____
   A. Record the hazard class: _____
   B. Record the four (4) digit number:_____
9. Have all the plant fittings been removed from the tank? — 9. Yes____ No____
10. Have all spills been cleaned? (side of tank, loading area) — 10. Yes____ No____
11. Have all the outlets and domes been sealed? — 11. Yes____ No____
12. Have all the seal numbers been recorded? — 12. Yes____ No____
    A. The seal numbers are:_____/_____ _____/_____

The answer to any of the above questions is "NO," DO NOT RELEASE THE CARGO TANK. Notify your shift supervisor. If all
the answers are "YES," release the cargo tank.

CERTIFICATION: THE CARGO TANK HAS BEEN PREPARED AND IS READY TO SHIP.

SIGNED:_____ DATE:_____

Exhibit 13-3 Cargo Tank Loading Report

17. If a leak develops
    a. Close the air supply valve and close the eduction pipe valve.
    b. Relieve air pressure on the tank car.
    c. Eliminate all air or solution leaks by tightening connections on the hoses or replacing the hoses.
18. Repeat steps 14 through 16.
19. Remain within sight of the tank car while unloading proceeds.
20. A drop in pressure or the sound of rushing air indicates the tank car is empty.
21. When the unloading line is drained, shut off the air supply line and allow the system to stand for 5 minutes to relieve internal pressure on the tank car.
22. Close the tank car air inlet valve and slowly unscrew the air hose, allowing the hose to depressurize.
23. Disconnect the air hose from the tank car air inlet valve and replace the valve plug.
24. Close the eduction pipe valve.
25. Open the tank car manway and verify the car is empty.
26. Close and tighten the manway.
27. Disconnect the unloading hose from the eduction pipe and replace the eduction pipe plug.
28. Close and secure the car dome cover.
29. Clean out the unloading hose with steam or water inside the containment area.
30. Complete paper work. Obtain tare weight of railcar for accounting and submit shipping papers and certificate of analysis to shipping clerk.

SUMMARY

Process technicians work tank farms, loading racks, and bagging operations preparing containers and tanks for loading, and then loading them. This movement of chemical materials is called *material handling*. Frequently, process technicians will be required to load or unload liquids, gases, and solids from railcars or truck trailers. Several different types of tank cars and tank trailers carry the array of processed chemicals. Some are designed for low-pressure application only, others for liquefied gases under significant pressure, corrosive acids or bases, compressed gases, cryogenic liquids, or dry bulk materials. Technicians should recognize each type of vessel and their limitations.

The procedure for transferring bulk liquids typically consists of the following three steps: (1) weighing and spotting, (2) making connections and transferring the liquid, and (3) disconnecting and releasing. Before and after loading or unloading a cargo tank, a general inspection should be made of the tank (railcar tank, tank truck tank) and its associated equipment. The inspection is usually made with an inspection check sheet.

After the tank is loaded, the technician follows the required procedures for disconnecting and releasing it. The loading valves are closed in the reverse order from the way they were opened and the car then is pressure tested with nitrogen to check for leaks around the manway and valves. The technician will attach a product tag to the sampling device at the dome, lock the dome cover, and put safety seals in place on the dome cover and manway. To prepare the tank car for release from the loading rack, the technician reverses the placards to show it is loaded, removes the wheel chocks, the grounding cable, the derailers, and the warning signs. A final inspection is performed and the switch personnel is notified that the tank car is ready to be moved to the scales.

REVIEW QUESTIONS

1. List six types of containers used for material handling of product.

2. Name the federal agency that regulates the transportation of materials on public highways and waterways.

3. List three types of trailers used for the transportation of bulk materials on public highways.

4. Explain why corrosive liquid trailers carry smaller volumes than most other trailers.

5. Most liquid materials transported by rail are carried in _____.

6. Explain why some cargo tanks are treated as dedicated tanks.

7. List the three steps for the transferring of bulk liquids.

8. Explain the purpose of a scale house.

9. List four safety factors technicians must consider when loading and unloading cargo tanks.

10. Explain how a loading technician will prevent hazards created by static electricity.

11. Explain why housekeeping is important to loading technicians.

12. List the types of valves found on a cargo trailer.

13. What is the purpose of a spill dam on a cargo trailer?

14. Briefly describe the pre-loading activities a loading technician will do when a tank truck drives up to his loading rack.

15. Why is the loading arm lowered just inches off the bottom of a cargo tank?

16. Explain why a safety seal is attached to a manway cover.

17. Who are switch personnel?

18. What is the purpose of a derailer?

19. What is a heel and how can a heel affect loaded product?

20. Why are pre-transfer and post-transfer inspections made on cargo vehicles?

21. Explain why it is important to wait 15 minutes before lowering a sampling device into an acid railcar?

CHAPTER 14

Material Handling II: Bulk Solids

Learning Objectives

After completing this chapter, you should be able to

- *List four categories of bulk solids.*

- *List four characteristics of bulk solids.*

- *Explain how particle size affects flow through a conveyor.*

- *Explain how flowability affects material handling.*

- *Describe two common material flow problems.*

- *Explain why dust is a hazard.*

- *List four methods of mechanical conveying.*

- *Explain the principle difference between dilute phase and dense phase material transport.*

- *Explain the difference between volumetric and gravimetric feeding.*

- *Describe a generic bagging operation.*

INTRODUCTION

Many processing plants produce a product that is a solid, such as a powder, pellet, or granule. Their product may have to be dried, screened, polished, conveyed to storage, loaded into bags at the bagging house, or loaded directly into railcars and barges. A processing site may receive bulk solids as reactants in their process. In this case, solids must be unloaded and conveyed either to storage or directly to the processing unit. Materials can be transported by a conveyor belt or bucket conveyor, or powdered and granular solid materials can be conveyed through pipe or ducts with compressed air using a technique called pneumatic conveying. These are versatile techniques for material handling, which are efficient, environmentally acceptable, and cost effective. They are used to move material into storage, between process operations and storage facilities, within a storage area, and into railcars and trucks for shipment.

Safety is a high concern during the processes just described. Dust concentrations in the air can be high enough to fuel a dust explosion. Inhaled particulates may lead to health problems, plus the particulates may damage rotating equipment by contaminating oil and grease reservoirs or clogging filters.

PORTABLE CONTAINERS

Before going directly into the subject of handling of bulk solids, this section will discuss the characteristics of portable containers used for any state of chemical material. Small quantities of materials are transported or stored in portable containers, such as tote bins for solids and liquids, super sacks for solids, portable tanks for liquids, and high-pressure cylinders for gases. These portable containers are discussed in the following paragraphs.

Tote Bins

Tote bins are used for both solids and liquids. Most have a loading valve, a removable lid on the top, and an unloading valve on the bottom. Construction can be of steel, galvanized steel, stainless steel, or aluminum. Gravity is the preferred method for unloading, and a conical bottom improves material flow for easier discharge of solids. A vent valve vents the bin as it is loaded or unloaded, and a relief valve or rupture disc prevents overpressure. Tote bins range in volume from 200 or more gallons and hold about 1 ton of material with an approximate density of water. They have channels in the base and lifting lugs attached to the top so that they can be moved by fork truck or crane. Low-pressure tote bins are rated for 15 to 20 pounds per square inch gauge (psig); high-pressure bins have a working pressure of 150 psig. See Figure 14-1 for an example of a tote bin.

Sack Containers

Sack containers, such as *SuperSacks*™, are containers made of heavy, woven cloth fabric, usually Nylon or polypropylene, that range in size from a few cubic feet up to 100 or more cubic feet in volume. Bulk bags typically hold between 500 and 2,000 pounds of material, and the large woven ones are referred to as flexible intermediate bulk storage (FIBC) bags. The basic model is a four-sided sack with lifting loops, a filling inlet, discharge outlet, and usually a liner. Lifting loops attached to the top allow the bags to be hung in loaders or unloaders and handled by a fork truck or crane. Material does not bridge in sacks and they are easily unloaded by gravity. Filled sacks form easy-to-stack cubes for storage. When empty, they can be folded for storage. Sacks are made of non-conductive materials and can develop static electrical charges if material is rapidly dumped from them. This has been a cause for dust explosions. Some sacks have metallic wire woven into the fabric to provide a ground wire that can carry away the static charge. See Figure 14-2 for an example of an FIBC sack.

Figure 14-1   Tote Bin

Figure 14-2   FIBC Sacks

Portable Liquid Tanks

Portable liquid tanks that meet Department of Transportation (DOT) specification are frequently used to store and transport smaller quantities of hazardous liquids, such as diesel fuel and gasoline. These are steel cylinders and tanks with capacities from a few gallons to over 1,500 gallons and working pressures up to about 200 psig. They usually have a relief

Figure 14-3   Portable Liquid Tank

valve and/or fusible plug overpressure protection and a protective cap or a recessed well to protect the unloading valves. The larger tanks are equipped with forklift truck pockets and crane lifting lugs for handling. They have a vent valve and an unloading valve located on the top of the tank. The unloading valve is attached to a dip tube that extends to the bottom of the container. The tanks can be unloaded using nitrogen pressure to push the liquid up the dip tube and out the unloading valve. The procedure for doing this is very much like the procedure used for unloading railcars and tank cars. See Figure 14-3 for an example of a portable liquid tank.

High-Pressure Gas Cylinders

High-pressure gas cylinders are used to store and transport a wide variety of gases. They are normally made from carbon steel, unless the gases are corrosive or reactive. They range in size from about 3 to 150 gallons. The largest can hold 1,000 pounds of a gas like chlorine. The normal operating pressure of a high-pressure gas cylinder is about 2,000 psig when full, but some cylinders are designed to operate at pressures up to 4,000 psig or greater. As a safety measure, the Compressed Gas Association (CGA) has persuaded the various gas manufacturing industries to adopt a system of different threaded fittings for the more common gases. The different thread configurations prevent the opportunity for serious accidents, such as a nitrogen cylinder being connected to a breathing air system.

Technicians should keep in mind the amount of energy stored in high-pressure bottles. Cylinder valves are usually constructed of brass, a soft metal that fractures easily. It is important

Figure 14-4   High-Pressure Gas Cylinders

to protect the brass valves from harsh impact. If a valve breaks off or if the cylinder wall is punctured during handling, this stored energy is released and the cylinder can take off like a rocket and smash through bricks walls and buildings. A ruptured, full cylinder is a severe safety hazard. When cylinders are mishandled they may explode, release their contents, or become dangerous projectiles. Move large cylinders with a cylinder cart and secure them with a chain. To protect the valve threads and the valve, make sure a cylinder cap is in place when the cylinder is not in use. If a cylinder is dropped, knocked over, involved in a fire, or potentially damaged, take it out of service and send it to the manufacturer for inspection.

- Never roll a cylinder to move it.
- Never carry a cylinder by the valve.
- Never leave a cylinder unsecured.
- Never force the improper attachments on to the wrong cylinder.

Store cylinders in areas designated and marked only for cylinders. Cylinders should be stored in compatible groups. When storing cylinders, separate

- Flammables from oxidizers
- Corrosives from flammables
- Full cylinders from empties
- All cylinders from corrosive vapors

See figure 14-4 for an example of a high-pressure gas cylinder.

Boxes, Bags, and Drums
Paper, cardboard, plastic, and metal boxes; fiberboard drums; and paper, plastic, and cloth bags are common containers for most of the solid materials that are used in smaller quantities

by the chemical industries. Likewise, metal, plastic, and lined fiberboard drums are used extensively for storing and transporting liquid materials. Handling materials in small containers requires manual lifting.

MATERIAL HANDLING OF BULK SOLIDS

Now, we will discuss in detail the material handling of bulk solids. Bulk solids can exist as pellets, powders, flakes, granules, and slurries. They are moved by (1) gravity flow, (2) mechanical conveying, (3) hydraulic conveying, and (4) pneumatic conveying. In *gravity flow*, the bulk solid moves vertically (falls) due to the pull of gravity. This type of conveyance can be seen in silos and hoppers which are shaped to utilize gravity flow. This is such a simple concept I don't think more than this is needed. In *mechanical conveying*, equipment such as conveyor belts and bucket belts carry the solid from one point to another. *Hydraulic conveying* mixes a solid with a liquid to form a slurry that is pumped to the desired location. In *pneumatic conveying*, solid particles are pushed through a pipe by air or inert gas flow in much the same way that the wind pushes dust.

Characteristics of Bulk Solids

Bulk solids have certain physical characteristics, such as particle size, density, flowability, ability to be fluidized, and moisture content. These characteristics affect how the solids are handled.

Particle Size and Shape. *Particle size and shape* determine the type of conveying systems used. Solid particles can be relatively large, rock-sized objects, grains, pellets (polypropylene) or fine powders (aluminum chloride). The size of the particle basically determines how the material should be handled. For example, materials consisting of particles that are too large or heavy to be carried pneumatically (by air pressure) may be carried on unenclosed conveyor belts or bucket belts. However, materials consisting of small, light particles may have to be moved in enclosed conveying systems like pipes, not just because they are light enough to be conveyed by air pressure, but as a safety factor. The smaller or finer the particles, the easier it is to generate dust even at slow conveying speeds. Dust can be a serious safety (explosion) and health hazard (lung diseases). Occasionally, we hear on the news that a grain elevator has exploded. The grain didn't blow up, rather the grain dust did. The dust became a fuel that formed a combustible mixture with air. Safety and health hazards can be avoided by using enclosed conveying equipment to prevent or minimize dust generation.

Particle size and shape are factors in determining whether a particle flows through or gets caught in a filter, plus they are factors in determining whether a particle moves up and out of a cyclone separator or falls down and through the separator. *Particle size distribution* affects how a material must be handled. A material may be predominately very small particles but have a few large particles. Air can be used to convey this material. As long as there is enough air velocity to keep the large particles moving along with the small ones, there is no problem. However, if the air velocity is too low, the large particles will not move properly and will form clumps that act as dams and clog the system or restrict material flow. This will result in maintenance problems and reduced efficiency.

Density. *Density* refers to the amount of space occupied by a particle of a certain weight. The smaller the amount of space occupied for a given weight, the greater the density of the particle. Density is a factor in determining how a solid material is handled and determining how much air velocity is needed to move a material through a pneumatic conveying system. Density, along with particle size, also influences the size and type of mechanical conveyors used to move a material.

Bulk density refers to the weight of a quantity of solid material per unit of volume, such as pounds of plastic pellets per cubic foot in a container. Knowing a material's bulk density helps a technician determine if there is enough space in a storage vessel to hold all of the material to be unloaded.

Flowability. *Flowability* is a measure of a material's ability to flow steadily and consistently as separate particles. Flowability is important for handling material in storage vessels. Materials that will not flow on their own from storage vessels may require vibrators or feeders to help move them. **Feeders** are mechanical devices that stimulate movement of solid particles and meter bulk material out of storage. Feeders keep materials flowing evenly and discharge them in controlled amounts. Flowability can also be impacted by the particles of some materials that tend to lump together. In this case, the use of a mechanical device called a **lump breaker** will aid the flow of the material. Rotating cutter teeth inside the lump breaker break up the lumps and enable the material to be moved without clogging up the conveying systems.

Some very fine solids can be **fluidized** by mixing them with a gas, such as air or nitrogen. The fine solid (dust-like solid) will flow just like a gas or liquid. A good example of this is the catalyst movement in a fluidized cat cracker. A fluidized solid is easy to convey pneumatically but may have to be fed to the pneumatic system by a feeder because conveying it mechanically might create a dust hazard.

Other Characteristics. Other characteristics affecting solid material handling are *abrasion*, *friability*, and *hazardous nature*. An abrasive material acts like sandpaper and the equipment and systems carrying that material must be designed to withstand the abrasion. Friable materials break easily and must be handled by equipment designed to minimize damage to the fragile particles. Some bulk solids are hazardous materials and require special handling. A hazardous material may be corrosive, toxic, or explosive. Very fine particles similar to dust can be carried deep into the lungs and create respiratory problems or diseases in later life. Technicians must be familiar with all of the safety precautions and procedures associated with handling hazardous materials.

MATERIAL HANDLING EQUIPMENT

Bulk solids handling systems typically have some pieces of equipment in common, such as blowers, pumps, compressors, and hoses. Conveying hoses are rated to match specific materials and operating conditions. It is important to use the correct hose for each bulk solids transfer. Friction, abrasion, and chemical reactions can damage or deteriorate the hoses. Material flow is controlled or directed by valves throughout material handling systems. Slide gate valves are commonly used to control flow from storage vessels and processing equipment. Some valves, such as diverter valves, are designed to redirect the flow of material within a system. **Diverters** are used to divert the flow of material away from one process unit into a line leading to another process unit.

Raw materials and finished products require storage vessels. These vessels are typically either rectangular bins or cylindrical silos. A bin or silo usually has a hopper section above the vessel's outlet and the hopper will have at least one sloping side. **Hoppers** channel the flow from the silo to the discharge valve. The storage vessel usually has a high-level alarm system to warn technicians that the vessel is nearly full. See Figure 14-5 for an example of a silo with a hopper.

Material flow problems can occur in storage vessels, conveying systems, and transfer points. Two common flow problems are (1) solids clinging to the interior of a vessel and (2) solids forming blocks or restrictions at hopper outlets. Feeders are used to assist and control material flow. The outlet for a storage vessel normally connects directly to a feeder or some other flow-assisting or metering device. The feeder controls the discharge of material at the desired rate for a process. Flow-assisting devices also help give uniform flow from funnel-flow storage vessels and can be used to restart blocked hoppers.

Material Flow

One of the most common types of feeding devices is called a **rotary valve** (see Figure 14-6). A rotary valve is basically a rotor with vanes that turns inside a housing. It is like a paddle wheel and each paddle (vane) gathers a certain quantity of material and feeds it into the system. Driven by a slow-speed gearmotor through a roller chain drive, the volume of material delivered to the air stream in the pipeline can be closely controlled. An inlet chute leads to the rotary valve which is mounted below the hopper. The rotary valve feeds the material to a discharge chute and a discharge line carries the material to the next stage of processing. A rotary valve is typically driven by an electric motor and is connected to the motor by a drive chain. The shaft that connects the motor to the drive chain generally has a shear pin as a protective device. If material forms a blockage in the rotary valve, the pin is sheared off and the feeder stops operating.

Figure 14-5 Silo with a Hopper

Figure 14-6 Rotary Valve

General Safety Concerns

Safety is a primary consideration of a technician handling bulk solids. Accidents and injuries are avoided by planning and following proper procedures. Before beginning a material transfer involving bulk solids, a technician should consider the

- Characteristics of the material
- Starting location and destination for the transfer
- Method of transfer and the equipment required
- Period of time required

Every material has its own physical characteristics that affect how it behaves under different conditions, which in turn dictates how it should be handled. If a material is hazardous, approved safety procedures for handling it must be followed and the appropriate personal protective equipment (PPE) worn. Technicians should know where the nearest fire fighting and first aid equipment is located and how to use them.

Safely and correctly transferring a material requires a technician to know where the material is located and where it should be moved. Sending the wrong material to a process unit could result in unsafe conditions and an interruption in production. The technician also must know what transfer method to use—hydraulic, mechanical, pneumatic—and what equipment is required for making that type of transfer. Auxiliary equipment and safety gear such as pumps, blowers, dust collection equipment, and fire fighting gear must be available and in good working order. Finally, the technician must know the period of time in which the transfer must be completed in order to meet process demands and product delivery schedules.

Specific Safety Issues

The four most common and most serious hazards associated with handling bulk solids are *dust, dust explosions, moving equipment*, and *high pressures* or *vacuums*. Moving equipment with its attendant guards and emergency shutdown devices will not be discussed, nor high pressures and vacuums, because these subjects have been discussed in earlier chapters and should be familiar to students of this higher level course.

Dust. Dust is a hazard that endangers the health of personnel, damages the environment, and initiates explosions and fires. As we mentioned earlier, dust is a very fine particulate matter and can be carried deep into the lungs where it is difficult to remove. Breathing in dust can cause emphysema and, depending on the dust material, cancer. Skin contact with dust from certain materials can be harmful to workers. Also, dust from product transfers and processing can add to air pollution.

Bulk solids should be handled in equipment that minimizes the generation of dust. Dust-prone solids should be carried in enclosed systems, such as pneumatic conveyors or screw conveyors. Provisions must also be made for dust collection and removal. This is done with dust collectors which function to remove solids from the air. Cyclones are used to collect large solid particles while smaller particles can be separated from air in spray and mist scrubbers, filters, or bag houses. Although dust collectors reduce the hazard of breathing dust, they do not prevent dust from accumulating. Dust collectors slow down the rate of accumulation. To keep dust levels low, flat surfaces should be vacuum-cleaned regularly to remove accumulated dust.

Dust Explosions. Dust can be as explosive as gunpowder. Dust consists of tiny particles of material that expose a large surface area to atmospheric oxygen. If the dust particles are combustible, there is the potential for an explosion. If a dust explosion occurs, the explosion may create a cloud of dust and this dust cloud may also explode.

Dust explosions are a hazard with many bulk solids. On average, the gross loss for a dust explosion in a U.S. processing plant is about $400,000, according to statistics published in *Data Sheet 7-76* by FM Global, a major industrial property insurer. Fire and explosion due to dust are potentially deadly risks in any process operation that employs powders or other bulk solids. Such operations are widespread throughout the chemical process industries, not only in plants that produce bulk solids, but also in units that handle raw materials, intermediates, byproducts, catalysts, or additives in powder form. Death, injury, equipment damage, productivity losses, and sharply increased insurance premiums are only some of the consequences of unsafe bulk solid plant operation or design. See Table 14-1 for the explosibility levels of some types of dust.

Engineers working with bulk solids and powders must anticipate and prevent dust fires and explosions. First, they must make the proper choices during process design and plant layout, selecting and specifying the right equipment. For instance, processing systems may be equipped with explosion suppression systems and explosion vents. In addition, engineers must be fully aware of all possible ignition sources, and some of these sources aren't too obvious. Even a material that is normally non-explosive can explode if it is mixed with another material that is explosive. The chance of an explosion also increases if excess solvent is carried over with dried solids from a slurry process.

Eliminating ignition sources helps prevent dust explosions. In dust-prone work areas, technicians may be required to use non-sparking tools and explosion-proof electrical devices and equipment. Equipment used in dusty areas should be kept well lubricated to help keep

Table 14-1 Explosibility of Dust

Material	Ignition Temp. of Dust Cloud, °C	Explosive Concentration (oz/ft^3)	Relative Explosion Hazard
Aluminum	650	0.045	Severe
Coal	610	0.055	Strong
Copper	900	——	Fire
Epoxy Resin	530	0.020	Severe
Iron	420	0.100	Strong
Magnesium	520	0.020	Severe
Silicon	——	0.110	Strong
Tin	630	0.190	Moderate
Zinc	600	0.480	Moderate

Source: Based on laboratory test results by the U.S. Bureau of Mines (USBM) on dried samples of fine dust (passing 200 mesh sieve). See USBM Investigations No. 5624, 5753, 5971, and 6516.

dust from getting into bearings. Remember, dust (particulate matter) is the number one cause of rotating equipment failure because it causes bearings to fail. Also, failing bearings may overheat and become a source of ignition.

Static electricity is always a potential safety hazard because it is source of ignition. The conveying gas and the material being conveyed can build up static electrical charges while in motion and release the charge as a spark. To control static electricity, pneumatic conveying systems have grounding devices.

Overpressure of a conveying system creates a hazardous condition by causing equipment or piping to rupture, thus damaging equipment and possibly harming workers. Conveying systems typically include a variety of safety features, such as high-pressure alarms and high-pressure shutdown switches. Pressure relief devices may also be included as safety backups in case the high-pressure shutdown fails.

Environmental Considerations

Air pollution caused by dust exhaust in many areas is not tolerated by the Environmental Protection Agency (EPA). Where this condition cannot be tolerated and complete separation—100 percent visible retention of dust and material from the exhaust—is required to meet EPA requirements, a combination filter-receiver is used. The **filter-receiver** will have a cyclone bottom section and an upper section equipped with a multiplicity of cloth or synthetic filter bags to retain fine dust particles (similar to a vacuum cleaner bag).

MECHANICAL CONVEYING

Mechanical conveying can move solids horizontally, vertically, or at an angle. In addition to transporting bulk solids from one location to another, mechanical conveyors are used for feeding, discharging, metering, and proportioning materials. Most mechanical conveyors have drive motors and a speed reducer that converts the high speed of the drive motor to a lower speed for efficient conveyor operation. Four of the most common types of mechanical conveyors are

- Belt conveyors
- Vibrating conveyors
- Screw conveyors
- Bucket elevators

The different types of conveyors meet a variety of operating needs. For example, belt and screw conveyors are typically used to transport material along paths that are not more than about 30 degrees from horizontal. Bucket elevators are commonly used if the material flow path requires a vertical lift.

Belt Conveyors

The components of a **belt conveyor** usually include a drive mechanism, a belt, belt idlers, head and tail pulleys, loading chute, a guard, and a belt cleaner. The belt on a belt conveyor may be flat or troughed, depending on the type of material and the amount of material being conveyed. A troughed belt can carry more material than a flat belt. See Figure 14-7 for an example of a conveyor belt.

Several kinds of idlers are used on a belt conveyor to support and guide the belt. **Idlers** guide, support, or carry the belt. Carrying idlers support the belt in the section of the conveyor that carries the material. Impact idlers prevent damage to the belt at loading points. The grooved molded rolls or springs of these idlers bear the impact of the material being loaded and protect the belt. Training idlers help keep the belt trained (tracked) on the conveyor, causing the belt to remain centered along the length of the conveying path. Belt cleaners remove sticky or built-up material from the belt and help keep the conveyor operating smoothly. When a belt conveyor's drive motor is started, the belt's pulleys begin

Figure 14-7   Conveyor Belt

turning. The belt travels from the tail pulley to the head pulley. Material falls from the loading chute onto the belt and is carried towards the head pulley. As the belt travels across the head pulley, the material is discharged. During the return run to the tail pulley, the belt is cleaned by the belt cleaner.

Vibrating Conveyors

A *vibrating*, or *oscillating*, *conveyor* can handle a wide range of materials from granular, free-flowing materials as well as hot, abrasive, and lumpy materials. A typical vibrating conveyor has a drive motor and a metal trough. The motor supplies the vibrating action needed to move material through the trough in a uniform, continuous flow.

Screw Conveyors

Screw conveyors are often used when material must be transferred at very low speeds. They are also used for injecting additives, extracting samples, and as mixers and stirrers to blend dry or fluid ingredients. During the operation of a screw conveyor, a drive motor turns the screw mounted inside an enclosed trough. The material is conveyed through the trough by the rotating action of the screw. A loading chute directs the material into the trough. At the other end of the trough, a discharge chute routes the material to the next stage of processing. Adjusting the speed at which the screw rotates varies the rate of material flow.

Bucket Elevators

Bucket elevators are used for transporting material vertically or along an inclined path. A bucket elevator is a conveyor consisting of a series of buckets mounted on a chain or belt inside a casing. A drive motor attached to a head wheel drives the bucket elevator. A belt runs under a foot wheel at the base of the bucket elevator. During operation, material enters through the inlet chute at the bottom of the elevator and the buckets scoop up material as they travel around the foot wheel and carry the material to the top of the elevator. The buckets turn upside down as they pass over the head wheel and dump the material into the discharge chute. See Figure 14-8 for an example of a bucket elevator.

Figure 14-8 Bucket Elevator

HYDRAULIC CONVEYING

In addition to gravity flow and mechanical conveying, other methods of moving bulk solids include hydraulic conveying and pneumatic conveying. **Hydraulic conveying** suspends solid particles in a liquid to form a slurry. A slurry is a liquid containing solids or insoluble material. The solid particles may be fairly large and heavy or they may be relatively fine and light. In either case, a pump is used to move a slurry from one location to another.

PNEUMATIC CONVEYING

Pneumatic conveying uses a gas to transport bulk solids. During pneumatic conveying, solid particles are moved through a hose or a pipe by the flow (pressure) of a gas such as air or nitrogen. Nitrogen or another inert gas is used in a pneumatic conveying system when the product is adversely affected by humidity or by contact with air. Pneumatic conveying can be used for feeding raw materials to processes or loading and unloading products. To perform these different tasks, pneumatic conveying systems are designed in a variety of ways. A pneumatic conveying system may use positive pressure, negative pressure (vacuum), or a combination of the two.

In a typical positive pressure system, a blower located at the feed end of the system introduces gas that moves the solid particles through the system. In a negative pressure system, the blower is located at the discharge end and suction created by the blower draws the solid particles through the system. These vacuum systems allow multiple product inlets through the use of simple diverter valves. However, it becomes costly to have multiple destinations because each must have its own filter receiver with partial vacuum capability. Vacuum systems are also more "distance sensitive" than pressure systems due to the maximum pressure differential of 5.5 to 6.0 psig. Pressure-vacuum operation (utilizing both methods) is sometimes ideal for a given conveying setup, such as the unloading of a standard railcar. Since the cars cannot be pressurized, air is pulled from the outside, through the car (carrying solids with it), and to a filter. Then after passing through the filter, a blower can be used to forward the solids to the final receiver.

Most pneumatic conveying systems are open systems. In an open system, the conveying gas follows a once-through path in and out of the equipment and is released to the environment. Closed systems capture the conveying gas and recirculate it.

Powdered and granular solid materials can be conveyed through pipe or ducts with compressed air. This is a versatile technique for material handling, which is efficient, environmentally acceptable, and cost effective. It is used to move material into storage, between process operations and storage facilities, within the storage area, and into railcars and trucks for shipment. Two different techniques are used; (1) a dilute phase transport and (2) a dense phase transport. The choice between dilute and dense phase operation is typically dependent on the solid's properties. For example, the lower velocity bulk phase operation is popular for transporting highly abrasive products or for those that degrade easily, such as kaolin clay.

Dilute Phase Transport

Dilute phase transport uses low-pressure, high velocity conveying air to completely suspend the solid particles in the conveying pipe. These systems may be either positive pressure and push the material or negative pressure and pull the material through the pipe. The conveying line pressure ranges from a negative pressure to about 15 psig, while the conveying velocity ranges from 2,000 to over 5,000 feet per minute. Dilute phase systems have lower initial investment but higher operating costs than the dense phase transport system. In addition, because of the high conveying flow rate, they are not as well suited for conveying abrasive or fragile materials. At these high velocities, the abrasive materials erode the conveying pipe, plus the fragile material may be damaged in the conveying process.

In a dilute phase system solid particles are continually fed through a feeder into the conveying line. The particles mix with gas from a blower and are carried to the next stage of processing. The ratio of particles to gas is lower in a dilute phase system than in a dense phase system. The widely dispersed (diluted) particles are carried along in the gas stream at a high velocity. However, they are carried at a lower pressure than in a dense phase transport system. See Figure 14-9 for a schematic of this process.

Dilute phase pneumatic conveying is extremely cost effective and is well suited for handling non-abrasive, non-fragile, or light materials. The key benefits of the dilute phase material handling system are

- Smaller capital investment
- Lower installation costs
- Minimized dusting
- Flexibility in routing

Dense Phase Transport

Dense phase transport uses high-pressure, low velocity air to push slugs of solid material through the conveying pipe. Conveying line pressure is typically 45 to *50* psig while the conveying velocity will range from *50* to *500* feet per minute. Initial investment of a dense phase system is higher. These systems are built to withstand higher pressure and use

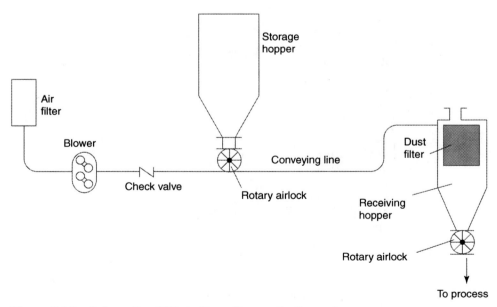

Figure 14-9 Schematic of Dilute Phase Pneumatic Conveying

positive displacement compressors rather than a low-pressure blower or fan to generate the conveying air. They require a separate transporter vessel and, in some cases, booster air injection nozzles along the length of the conveying pipe. However once these systems are installed, they provide lower operating costs than dilute phase transport systems. In addition, they experience much less pipe erosion and material degradation. Due to the low conveying velocity (50 to 500 feet per minute) produced by a dense phase system, materials are gently pushed through the conveying line in controlled slugs. As a result, dense phase is preferred for moving mixed batches or materials that may be abrasive, fragile, heavy, or hygroscopic in nature.

A typical dense phase system operates on a fill and discharge cycle. In this type of system, solid particles are fed into a transmitter vessel, the vessel's discharge valve is opened, and the material is discharged into the conveying line. Compressed gas pushes the material through the conveying line. This system moves a fairly large mass of material at a low velocity but under a high pressure. See Figure 14-10 for a schematic of dense phase transport. The key benefits of dense phase material handling are

- Gentle handling of materials
- Reduced product degradation and segregation
- Reduced component/system wear
- Minimized dusting
- Flexibility in routing

Overview of a Dilute Phase Pneumatic Conveying System

Dilute phase pneumatic conveying systems are widely used in process plants. The system described here operates under positive pressure. The blower is the heart of this pneumatic

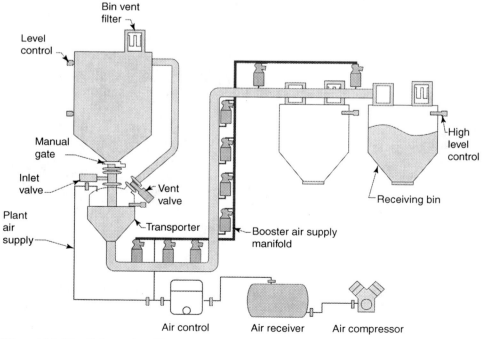

Figure 14-10   Schematic of Dense Phase Pneumatic Conveying

conveying system and is located at the feed end of the system. It supplies the gas that keeps the material circulating through the piping. The feed is contained in a silo. When a slide gate on the silo's hopper is opened, feed enters a rotary valve. The amount of material fed into the conveying system is determined by the speed of the rotary valve. The faster the rotary valve turns, the more material is fed into the system. The material is conveyed by the gas introduced by the blower and flows through piping to a process. Farther downstream, diverter valves direct the flow of material to specific parts of the system. Dust collection equipment separates the feed material from the conveying gas before the gas is released or returned to the system. The dust collection equipment minimizes the escape of feed material into the atmosphere where it could harm the environment and be a health and safety hazard.

TECHNICIAN CHECKS FOR CONVEYING SYSTEMS

Routine walk-by inspections of pneumatic conveying systems are part of the normal rounds of a technician. Inspections should include checks for

- Unusual sounds, vibration, odors, oil spills, or process leaks
- Excessive heat, moisture, vapors, and dust generation
- Abnormal pressure gauge readings
- Indications that the blower filter needs cleaning or replacement
- Hot bearings that can ignite a dust explosion

Some checks are general to all conveying systems while others depend on the type of conveyor.

WEIGHING AND METERING BULK SOLIDS

Most plants use a computerized control system to monitor the operation of metering and feeding systems and to enter or change operating data. Bulk solids are usually weighed and metered (fed at a certain rate) so that material is supplied to the processing unit at the proper rate. The two basic methods of weighing and metering bulk solids are *volumetric feeding* and *gravimetric feeding*. Several technician tasks are associated with volumetric or gravimetric feeding. Technicians should know how to change the feed rates for the feeders. In particular, they should know how to enter the necessary data into the control system and verify the accuracy of the scales.

Volumetric Feeding

Volumetric feeding introduces material into a process at a desired rate. This type of feeding delivers a set volume of material per unit of time, such as cubic feet per hour. The rotary valve is an example of a volumetric feeder. To change the output of a volumetric feeder and achieve a different feed rate, the technician has to reset the feeder manually (or from the control board) or adjust the speed of the feeder's variable-speed driver. This type of feeder is most accurate when a material's bulk density and flow characteristics are fairly constant.

Gravimetric Feeding

A **gravimetric feeder** maintains a set feed rate by weight per unit of time, such as pounds per hour. Two types of gravimetric feeders commonly used are weigh-belt feeders and loss-in-weight feeders.

Weigh-Belt Feeder. A **weigh-belt feeder** weighs solids on a moving belt. Material is withdrawn from a hopper, transported past a shear gate onto the belt, weighed, and then discharged to the process unit. During the operation of a weigh-belt feeder, a controller continually receives data that compares the current feed rate to the desired feed rate. The belt speed is automatically adjusted to compensate for any difference and maintain the desired feed rate at all times.

Loss-in-Weight Feeder. A *loss-in-weight feeder* works by keeping track of and controlling the amount of weight lost from the feeder container. A typical loss-in-weight feeder includes a scale and a controller. Typically, the feeder is a hopper with paddles on two of its sides and a rotating screw at the outlet. The paddles agitate the hopper to help create a uniform bulk density in the material and aid the flow of the solids into the screw. The rotating screw then discharges the desired amount of material to the process unit. During operation, the scale weighs the hopper and the material in it at all times. The controller receives signals from the scale that reflect the current weight loss and compares this weight loss with the desired weight loss. The screw speed is then automatically adjusted to control the rate of weight loss and to compensate for any difference between the current and the desired feed rates. Loss-in-weight feeders must be recharged periodically. On some feeders, an automatic refill system is activated when the feeder senses that its supply of material is about to be depleted.

BAGGING OPERATIONS

Some operations, such as polypropylene production, produce huge quantities (many tons per hour) of bulk solids. These solids, small pellets of plastic, are pneumatically or mechanically conveyed to huge silos adjacent to the plant railroad tracks. The pellets are funneled from the silos into hopper cars. Since polypropylene is a basic commodity used extensively in manufacturing industries, it is not uncommon for a string of railcars to leave the plant each day. However, some process industries may ship their product in smaller containers, such as 50-pound bags or in FIBC sacks. This will require technicians to work bagging operations.

Bagging operations is similar to drumming operations (previously discussed in Chapter 13) with the exception that the technician is loading 50-pound or heavier sacks or bags of solids, such as nitrate-based fertilizers, seeds, feed, powdered chemicals, etc. Let us look at a generic bagging operation. Assume a technician has an order for thirty 500-pound FIBC bags of a pelletized product. The technician will operate a filling station (see Figure 14-11) that consists of a metal rectangular frame about 7 feet tall containing the combination hydraulic lift/weighing table. Running through the station is a slated conveyor belt, and to the side of the station is the operating control panel. The station is located under a product silo. The technician lowers the lift table, attaches an empty bag's lifting loops to hooks on the loading frame, and then tares the bag. Next, the technician verifies the bag's discharge outlet is closed and attaches the bag's grounding strap to ground. The bag's filling inlet is then attached to the feeder's filling spout, which consists of two concentric tubes, one for product flow into the bag and the other for air and dust return to the top of the hopper or a dust collector. When the correct weight of material is in the bag, the feed valve is stopped, the bag's filling inlet tied off, and the bag lowered to the slated conveyor. The bag is carried to the end of the conveyor where a lift truck lifts the bag and carries it to storage. The technician then prepares to load a second bag.

Figure 14-11 Filter Receiver with Bulk Bag Filler. Photo courtesy of Flexican Corporation

In many processes bulk bags are unloaded and fed by either pneumatic conveying or screw conveyors directly into the process or into silos that feed into the process. An example of a bag unloader is shown in Figure 14-12.

SUMMARY

Many processing plants produce a product that is a solid, such as a powder, pellet, or granule. Or, their process may require a solid as a reactant. In either case, solids must be handled, conveyed, packaged, and stored. They are moved by (1) gravity flow, (2) mechanical conveying, (3) hydraulic conveying, and (4) pneumatic conveying. In gravity flow, the bulk solid moves vertically (falls) due to the pull of gravity. In mechanical conveying, materials can be transported with a conveyor belt or bucket conveyor, or powdered and granular solid materials can be conveyed through pipe or ducts with compressed air using a technique called pneumatic conveying. Hydraulic conveying uses a liquid to carry solids from one point to another.

Bulk solids have certain physical characteristics, such as particle size, density, flowability, ability to be fluidized, and moisture content, which affect how the solids are handled. Bulk solids handling systems typically have some pieces of equipment in common, such as blowers, pumps, compressors, and hoses. Conveying hoses are rated to match specific materials and operating conditions. It is important to use the correct hose for each bulk solids transfer. Friction, abrasion, and chemical reactions can damage or deteriorate the hoses. Material flow is controlled or directed by valves throughout the material handling system. Slide

Figure 14-12   Bag Filter Feeder with Pneumatic Lines. Photo
courtesy of Flexicon Corporation

gate valves are commonly used to control flow from storage vessels and processing equipment. Diverter valves are designed to redirect the flow of material within a system.

The four most common and most serious hazards associated with handling bulk solids are *dust*, *dust explosions*, *moving equipment*, and *high pressures* or *vacuums*.

REVIEW QUESTIONS

1. List four portable containers for transporting process material.

2. The _____ is the weakest part of a compressed gas cylinder.

3. Explain how tote sacks are designed to prevent accidents caused by static electricity.

4. List four categories of bulk solids.

5. List four characteristics of bulk solids.

6. Explain how particle size affects flow through a conveyor.

7. Explain how flowability affects material handling.

8. Describe two common material flow problems.

9. Explain the function of a diverter valve.

10. Explain why dust is a hazard.

11. List four methods of mechanical conveying.

12. Screw and belt conveyors are used as long as the transport path is not greater than _____ above horizontal.

13. Explain the tasks screw conveyors are commonly used for.

14. What is a *slurry*?

15. List the two pneumatic systems.

16. List five checks technicians make on conveying systems.

17. Explain the principle difference between dilute phase and dense phase material transport.

18. Explain the difference between volumetric and gravimetric feeding.

CHAPTER 15

Material Handling III: Oil Movement and Storage

Learning Objectives

After completing this chapter, you should be able to

- *Describe the various types of storage tanks.*

- *List the four conditions that affect the evaporation of storage tank contents.*

- *Give three reasons for tank gauging.*

- *Describe how tanks are automatically gauged.*

- *Describe how tanks are manually gauged.*

- *List four major responsibilities of a tank farm technician.*

- *Discuss several technician safety concerns about tank farms.*

- *Describe the routine inspections and duties of tank farm technicians.*

- *Discuss the precautions a technician should take before sampling a tank.*

- *Define the terms* innage *and* outage *as they relate to storage tank liquid level.*

- *List the four major steps of a material transfer.*

INTRODUCTION

Process facilities use aboveground storage tanks to meet a variety of operating needs. A large refinery or petrochemical plant may have over 100 storage tanks of various sizes and limitations. Technicians who work with these tanks need to know what their responsibilities are and how to carry them out safely. This chapter discusses the types of aboveground storage tanks and tank farm technician responsibilities for routine inspections, material transfers, sampling, and gauging. Though a tank farm is not as complex as a large processing unit, it can seriously affect safety, health, the environment, quality, and profitability if wrongly operated. A technician's action in the tank farm can affect processes and activities in other parts of a plant and the environment in the plant and outside the plant. To ensure that work is performed safely and that tank farms and equipment operate properly, technicians must follow their plant's procedures and observe all appropriate safety precautions. Poor judgment and improper actions at tank farms can cause polluted rivers and large-scale evacuations of local communities.

TANKAGE

Tanks of various types are used for the storage of raw materials and finished products. Since the stored material represents a large concentration of value, the protection and correct operation of material transfer into and out of storage tanks is important. Technicians responsible for storage tanks and material transfer must be very familiar with the various tanks and related equipment. Improper operation can cause serious accidents resulting in injury to personnel, damage to the environment, loss of revenue, and costly damage to equipment.

There are various types of tanks, each with their own advantages and disadvantages. The type of tank to be used is generally determined by the material to be stored. As a rule, tanks can be divided into two general divisions:

      1. Atmospheric tanks (cone roof, floating roof, and umbrella roof tanks)
      2. Pressure storage tanks (drum, sphere, and spheroid tanks)

Atmospheric Tanks

Atmospheric tanks operate very close to atmospheric pressure. Pressures that exceed 0.5 pounds per square inch above or below atmospheric pressure can rupture atmospheric tanks. Atmospheric tanks are good for storing most non-volatile materials and some hydrocarbons.

A **cone roof tank**, a type of atmospheric storage tank, has a fixed cone-shaped roof with one or more internal support columns (see Figure 15-1). It has a flat bottom, cylindrical wall, and is used for the storage of low vapor pressure liquids. This type of tank is designed to operate within a range of about 1.5 inches of water pressure to 1.5 inches of water vacuum. The welded joint where the roof joins the shell is purposely made weaker than other joints so that if pressure causes a cone roof tank to rupture, it will rupture there to relieve pressure. The tank's contents will still be contained because the walls will not have ruptured. See Figure 15-2 for a detailed schematic of a cone roof tank.

Figure 15-1   Cone Roof Tank

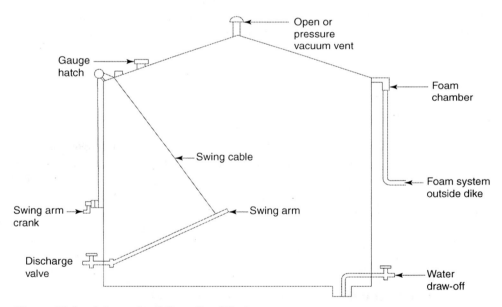

Figure 15-2   Schematic of Cone Roof Tank

A *floating roof tank (FRT)* may have an open top or be covered, and a pan-like structure that floats on the top of the oil and moves up and down inside the tank as the liquid level changes (see Figure 15-3). A close clearance is maintained between the roof and the shell of the tank with the opening sealed by means of a flexible curtain-like fabric attached to the roof and to steel-bearing surfaces called shoes. Reducing product evaporation and improving safety and overall operating economy remain the primary reasons FRTs are accepted as

Figure 15-3   Floating Roof Tank

a worldwide standard for volatile product storage. Product evaporation is not only a significant environmental and economic concern but is also a major consideration when improving safety. Four conditions that affect the evaporation rate of storage tank contents are

1. Liquid temperature
2. Vapor space above liquid
3. Vapor space ventilation
4. Available liquid surface area

For any liquid stored in a container with an open vapor space, a portion of that liquid will evaporate until equilibrium is achieved. Once equilibrium is established no further evaporation will occur unless conditions are altered. Many refined petroleum products are naturally volatile and will readily evaporate under normal operating conditions and produce combustible vapors with air. Floating roof tanks can reduce product evaporative emissions by more than 98 percent and greatly improve operating safety. See Figure 15-4 for a detailed schematic of a floating roof tank.

Pressure Storage Tanks

A **pressure storage tank** is used to store volatile liquids that have a Reid vapor pressure greater than 18 pounds per square inch gauge (psig). The three types of pressure storage vessels are drums, spheres, and spheroids. They are mainly used for the storage of liquefied petroleum gases (LPG), such as butane and propane. *Drums*, also called *bullets*, are cylindrical vessels with hemispherical ends built to withstand a given internal pressure (see Figure 15-5). Usually, a drum is supported in the vertical position on a concrete pad or in the horizontal position on two or more concrete piers. *Spheres* are pressure vessels shaped like a ball and supported above grade on tubular columns (see Figure 15-6). A sphere 65 feet in diameter will have a volume of 25,000 barrels. Spheres normally have deluge systems that spray the surface of the sphere with water to cool off the shell of the sphere if it becomes too warm. A warming or hot shell may cause the liquefied butane or propane to

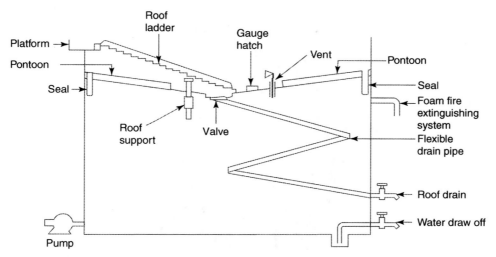

Figure 15-4 Schematic of Floating Roof Tank

Figure 15-5 Drum Storage Tank

begin to vaporize and increase pressure within the sphere to hazardous levels. See Figure 15-7 for a detailed schematic of a sphere storage tank.

TANK GAUGING

Tank gauging is the generic name for the static quantity assessment of liquid products in bulk storage tanks. Two methods of assessment are recognized:

1. A volume-based tank gauging system that determines quantity based on level and temperature measurement.
2. A mass-based tank gauging system that determines quantity based on hydrostatic pressure of the liquid column measurement.

Figure 15-6   Sphere Storage Tank

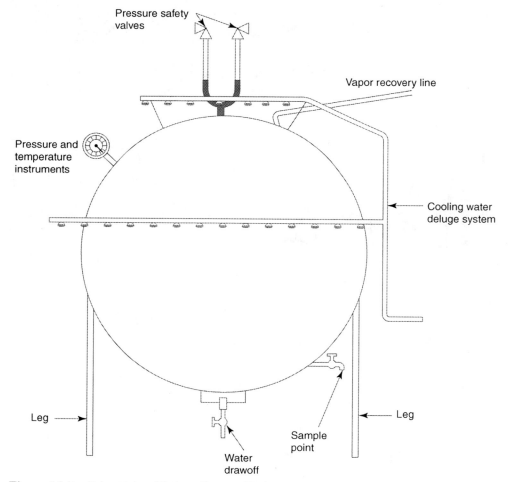

Figure 15-7   Schematic of Sphere Storage Tank

Whatever method is used, a high degree of reliability and accuracy is of great importance when the data is used for inventory control or custody transfer purposes. Tank gauging is essential to determine the inventory of liquid bulk storage tanks. Typical capacities of bulk storage tanks range from 6,300 barrels (bbl) to more than 755,000 bbl. A level uncertainty of only 1 millimeter (0.04 inch) or 0.01 percent in a 10-meter (33 feet) tall tank (315,000 bbl) equals 31 bbl. This means we are uncertain about 1,302 gallons of product. Hence, accuracy is very important for good inventory management. However, accuracy is only one aspect involved in tank gauging. Reliability of the measurement system to prevent product spills and safety of the environment and personnel is equally important. Some general requirements for a tank gauging system are safety, accuracy and repeatability, reliability, operator friendliness, and low maintenance.

The Need for Tank Gauging

Tank gauging is required for the assessment of tank contents, tank inventory control, and tank farm management. Tank gauging system requirements depend on the operation. The types of operation can be categorized as

- Inventory control
- Custody transfer
- Oil movement and operations

Inventory Control. Inventory control is an important management tool for any processing plant, terminal, or storage company. Inventory represents a large amount of company assets. Tank inventory control is either based on volume or mass. In-plant accuracy requirements for inventory control are often non-critical and measurement uncertainties do not result in direct financial losses. Reliability and repeatability are much more important to in-plant material transfer. Independent storage companies and terminals that strictly store and distribute products owned by their customers cannot operate without an accurate inventory control system. To do so could invite legal repercussions.

Custody Transfer. Many plants use their tank gauging system for the measurements of product transfers (*custody transfer*) between their production site and the external customer. Where custody transfer or assessment of taxes, duties, or royalties is involved, the gauging instruments and inventory control system are required to be officially approved and certified for this purpose. Inventory management that requires information related to taxes, loss prevention, and custody transfer requires great precision. It requires accuracy to better than 1/8 inch. (± 3 millimeters) as well as precise corrections for possible volume inaccuracies caused by high or low temperature, high humidity, thermal shock, etc.

Oil Movement and Operations. Generally, tank content measurements for day-to-day operational use (oil movement and operations), for scheduling purposes, and for blending programs do not require the same accuracy as custody transfer operations. However, measurement reliability and repeatability are still important. Accurate level alarms are necessary to operate safely. When level measurement supports a process and has no other purpose, a certain amount of imprecision is allowable and expected. Thus, the level measurements suitable for processes require the precision necessary to start or stop a pump, trigger a warning or an alarm, etc. The calculated volume of the contents of a vessel may be off a little, but the tank is not going to overflow or become empty and starve the process.

Manual Versus Automated Tank Gauging

Traditional methods for gauging and oil-water interface detection such as dip tape and water-finding paste are labor-intensive and difficult to use accurately. There are automatic tank gauging systems designed for atmospheric tanks and low-pressure tanks that continuously measure level in bulk storage tanks. They are referred to as *float and tape tank gauges* due to the manner in which the product level is measured (or gauged). A large float follows the level of the product as it moves up and down in the tank. The float is connected to a tape or wire that in turn is connected to an indicator on the outside of the tank (see Figure 15-8). This simple design has changed very little over the past 50 years due to its reliability and wide industry acceptance. It allows the gauge to perform with minimal maintenance throughout its working life. Automatic tank gauges like the float and tape do not require power or communications wiring to provide local display of the product level on the side of the tank or the tank roof.

Manual tank gauging is often the only method accepted for custody transfer of bulk liquids and is also used for the calibration of automatic gauges. In some cases, a gauging paste or a gauging chalk must be applied to a gauge tape before the tape is used to manually gauge a tank. When the material in the tank (either water or product) contacts the gauging paste or chalk, the paste or chalk either changes color or dissolves. This action produces a mark on the gauge tape. When the tape is withdrawn, technicians can read the mark on the tape's measuring scale to determine tank level. Whether a chalk or a specific type of paste is used depends on the material that is being gauged. Gauging chalks are generally not suitable for gauging certain oil products because the products tend to *creep* on a chalked tape and do not leave a clear mark. Certain gauging pastes are used to gauge water levels, other pastes for oil levels.

The importance of accurate tank gauging and temperature measurement within the petroleum industry cannot be overemphasized. As crude oil and petroleum products

Figure 15-8   Automatic Tank Gauge

have relatively high thermal expansion coefficients, any error in temperature measurement will result in a significant error in the standard volume measurement of stock or transfer quantities. In general terms, a 1°C error in the temperature of most petroleum products will equate to an error of approximately 0.1 percent in the standard volume calculation. The error becomes more significant as the hydrocarbon density decreases, so a 1°C error can equate to a volumetric error of approximately 0.3 percent for liquefied petroleum gas. Hence, the calibration of temperature measuring equipment is critical.

In recent years, there has been growing concern over the release of volatile organic compound (VOC) vapors with regard to both air pollution and occupational exposure during manual gauging of a tank. When a technician lifts the gauge hatch to gauge a tank, they might expose themselves to escaping vapors from the tank. New vapor control systems that require closed operation to avoid VOC emissions have been developed, and gauge manufacturers have responded by introducing gas-tight gauges that allow gauging to be carried out without releasing any vapors.

One of the newer technologies used for process level measurement and inventory tank gauging (ITG) is non-contact instrumentation based on microwave and radar. Non-contact instrumentation is less susceptible to corrosion, fouling, and mechanical problems. ITG is the use of level measuring for inventory storage and custody transfer applications, rather than in process control. Independent storage companies and sites with large numbers of storage tanks use instrumentation and distributed control systems (DCS) to monitor the level, temperature, and pressure of each tank. Otherwise, the task would require a large force of technicians.

TANK FARM TECHNICIAN RESPONSIBILITIES

The major responsibilities of a tank farm technician include

- Safety concerns
- Performing routine inspections of aboveground storage tanks
- Performing routine inspections of auxiliary equipment and systems
- Sampling
- Gauging duties
- Material transfers

With each responsibility there are certain basic tasks that a technician must perform without deviation according to the facility's standard operating procedures (SOPs).

Safety Concerns

Safety concerns in a tank farm are related to

- Tank and stored material compatibility
- Head pressure
- Fire
- Hazardous characteristics of stored materials
- Weather conditions
- Leaks

Tank and Stored Material Compatibility. Technicians must know whether the tank is an atmospheric tank or a pressure tank, what functions the tank is designed to serve, and what types of materials the tank can safely handle. They must also know whether the tank is equipped with any special features, such as a liner, a blanketing system, or a floating roof. Before filling a tank, technicians must ensure that the material and the tank are compatible. Storing material in a tank that is not designed to handle that material can cause serious problems. An incompatible material could damage the tank, degrade product quality, cause a process unit shutdown, or pollute the environment. An example would be storing styrene monomer in an uninsulated tank. If the tank heats up in hot weather, the styrene monomer may polymerize into a solid or semi-solid or degrade so badly it cannot be sold. If the styrene monomer polymerizes into a semi-solid, it will not be able to be pumped with the tank's normal pump. And pumping is impossible if the monomer turns into a solid.

Technicians must be familiar with all of the *auxiliary equipment* and systems associated with the tanks in their facility and may be responsible for minor maintenance on some equipment. Examples of this equipment include motor control centers (switch racks), pumps, and other equipment at pump slabs or pump pits; vapor recovery units; flare systems; and fire safety equipment.

Head Pressure. Pressure exerted by the column of liquid in an aboveground tank, often referred to as **head pressure**, is a concern for tank farm technicians. The pressure of this column of liquid creates positive suction (head pressure) for the tank's centrifugal pump. Head pressure ensures there is adequate net positive suction head to the pump and prevents cavitation. However, that same head pressure can promote an uncontrolled flow of product from a tank if valves are accidentally left open or a piping section ruptures. Also, negative pressure created by partial vacuums can cause tank roofs or the entire tank to buckle and fold inward. If a technician is on the roof at that time, they are in serious trouble. It is the same situation as the example of negative pressure collapsing a railcar: too many square inches of tank collecting one or two pounds of pressure and the total pressure becoming too great for the tank to withstand. As an example, assume the vacuum breaker on a tank has become defective. The tank has a diameter of 100 feet. As the tank is pumped out, a negative pressure of 2.7 psig develops in the tank. The top of the tank has a surface area of 1,130,976 square inches and when multiplied by 2.7 psig equals more than 3 million pounds of pressure exerted on the tank's roof!

Fire. Fire is another safety concern. Many tank farms are huge storage depots of flammable and combustible materials. Overheated auxiliary equipment or maintenance work on auxiliary equipment or tanks may initiate fires. To eliminate the possibility of electrostatic sparking, tanks and their associated equipment are bonded and grounded.

Today, fires in storage tanks are rare events. During the development of the oil industry, however, tank fires were common occurrences. These experiences led to improvements in better codes and guidelines for handling storage tank fires. Floating roof tanks were adopted for petroleum products and proved to be a very reliable method of protection against losses due to fires. The less volatile products with higher flash points continue to be stored in welded, steel-fixed cone roof tanks. As the frequency of storage tank fires decreased, the size of a fire, when it happened, became larger due to the increase in tank size and capacity. It is quite common to see tanks with diameters in excess of 328 feet (100 meters) capable of storing a million barrels of product.

One practical method to protect flammable liquid storage tanks from fire is with a fixed or a semi-fixed foam fire protection system. When engineered, installed, and maintained correctly, these systems give years of reliable service. The foam system can be used for fire prevention, control, or direct extinguishment of any flammable or combustible liquid fire within a tank. A *fixed system* is a complete installation piped from a central foam station, discharging through fixed discharge devices into the tanks. Most tanks with flammable materials have fixed systems for fire fighting and fire suppression.

Hazardous Characteristics of Stored Materials. Stored materials contained in aboveground storage tanks generate many safety concerns. Every material has hazardous characteristics—flammable, toxic, corrosive—that dictate how it must be handled. Flammable or combustible vapors inside a tank or in the area around a tank are serious fire and explosion hazards. Some materials have toxic vapors (hydrogen cyanide, hydrogen sulfide). To avoid the danger of hazardous materials, safe work practices must be followed at all times. For example, technicians should never enter the diked area around a tank or descend onto the roof of a floating roof tank until they have gas tested the area and determined it does not contain hazardous levels of gases and vapors. These areas can act like confined spaces and collect heavy vapors. Before opening a gauge hatch on a tank, technicians should position themselves so that they are upwind of the hatch to avoid the vapors that exit the hatch when it is opened. Hazards may be indicated by National Fire Protection Association (NFPA) material identification stencils on tanks and by warning signs posted in the tank farm.

Weather Conditions. Weather conditions create other safety concerns. Windy, wet, or icy conditions increase the hazards of slips, trips, and falls on access ladders, stairways, and walkways. Electrical storms and the threat of lightning make it unsafe to perform certain tasks, such as working on top of a large metal tank. Technicians should monitor the weather in their area and be aware of expected sudden weather changes, such as thunderstorms.

Leaks. Lastly, technicians should be aware that leaks are a constant hazard in a tank farm. They will detect leaks or incipient leaks when they make their rounds. High head pressure in a leaking large storage tank can eject hundreds or thousands of barrels of product into the environment, plus create very hazardous situations. Several times a year we hear of a river or estuary being covered with an oil slick from hundreds or thousands of barrels of a hydrocarbon that leaked out of a shore tank.

Performing Routine Inspections of Aboveground Storage Tanks

Routine inspections of aboveground storage tanks and their related auxiliary equipment and systems are a major responsibility of a tank farm technician. Technicians are generally required to perform a walk-by inspection as part of their daily rounds. They may also be required to make detailed, documented inspections of in-service tanks at regularly scheduled intervals. As part of a routine inspection, technicians should look for obvious signs of damage, vandalism, or abnormal conditions. The tank shell should be checked for corrosion, especially at the bottom of the tank and along seams or welds.

Tanks with cathodic protection systems should be checked to ensure it is functioning. Cathodic protection systems protect the tank's metal from galvanic corrosion. Galvanic corrosion refers to corrosion damage induced when two dissimilar materials are coupled in a corrosive electrolyte. The dissimilar metals can be where valves, piping, or instruments join the metal of the tank. It occurs when two (or more) dissimilar metals are brought into electrical contact under water or in fluids contaminated with water. When a galvanic couple forms, one of the metals in the couple (the corrosion coupon) becomes the anode and corrodes faster than it would all by itself, while the other (the tank shell) becomes the cathode and corrodes slower than it would alone.

Technicians should check for leaks of vapors or liquids, particularly at manways, valves, gaskets, and flanges. Gas detectors are used for leak detection of vapors. Abnormal pressure readings may be caused by leaks in a tank or tank piping system. Monitoring and familiarity with pressure readings for individual tanks helps to detect leaks.

Technicians may be required to inspect the flame arrestor on a tank. A **flame arrestor** is an in-line device installed under equipment vents to prevent flame propagation through the line. The small passages in a flame arrestor element may plug from corrosion or foreign objects (dust, wasp nests, etc.).

Gauge hatches are provided in the roofs of atmospheric tanks to enable its contents to be measured, but an important secondary function of a gauge hatch is to provide some emergency pressure relief if excess pressure builds up inside the tank. The tank literally *burps* through the gauge hatch. Gauge hatches should be kept closed to prevent the loss of vapors, fire hazards, and entry of rain. But they should not be weighted or otherwise restricted from opening as this eliminates their function as a pressure relief device.

Blanketing systems protect the tank's contents from air-borne contaminants (moisture, oxygen) and, in some instances, prevent the occurrence of a flammable mixture within the tank. The vapor space in tanks storing materials of a low vapor pressure at ambient storage temperature is usually too lean to explode. Tanks storing very volatile materials are usually too rich to explode. In some tanks, however, the vapor space would nearly always be in the explosive range if air were allowed to enter. In general, gas blanketed tanks are similar to other fixed roof tanks except that they are equipped with a supply line for adding a gas, usually nitrogen, sometimes natural gas (see Figure 15-9). A pressure regulator limits the gas supply to a pressure slightly lower than the exhaust set point of the conservation vent. This causes the tank to be under a slight positive pressure at all times and prevent air from entering. Blanketing systems should be checked to ensure they are working properly. Technicians should check pressures on both the supply side and the backpressure side of a blanketing gas regulator.

Tank temperature is checked because temperature has a direct effect on pressure. Temperature is checked by way of local mounted temperature gauges or console screens. Checking tank levels is an important part of preventing overflows caused by high liquid levels.

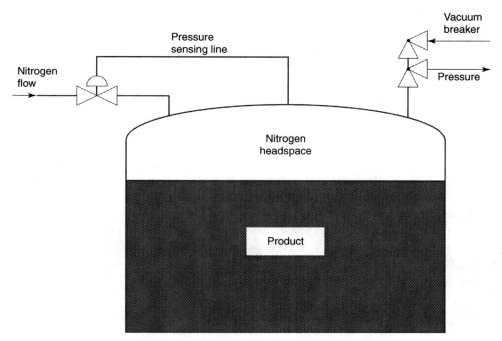

Figure 15-9 Tank with Nitrogen Blanketing System

Level checks are also another way to detect leaks or theft. Assuming that manual gauging is done correctly, automatic or electronic gauge levels that do not match the manual gauged levels may indicate leaks or theft. Level readings that do not match may also indicate that electronic gauges are malfunctioning. Unusual changes in levels are often signs of abnormal conditions and operating problems. Level checks also help protect transfer pumps from being damaged as a result of insufficient head pressure. For instance, on many tanks it may be SOP not to lower tank level below a certain height because below that level insufficient net positive suction head will damage centrifugal pumps. Level gauges and alarm systems associated with tanks should also be checked for proper operation.

If conditions permit, a good practice is to manually operate devices such as vent, fill and suction, bleeder, motor-operated, and drain valves and automatic gauge tapes. Regular manual operation helps keep the devices from seizing or sticking. The normal and emergency vents on a tank should be checked periodically. The vents prevent a vacuum from being pulled when the tank is pumped out and prevent excessive pressure from building up when the tank is filled.

Policies concerning the operation of roof drain valves for open-top or external floating roof tanks vary. In some facilities, roof drains are left open to handle water runoff as it occurs. In other plants, the roof drain valves are normally kept closed, primarily to prevent leaks from the drainage system from contaminating material inside the tank. The drain valves are then opened periodically to remove water from the roof. Roofs should be drained after heavy rains to ensure that they do not sink under the weight of accumulated water. Storm water runoff that collects in the diked area around the tank is usually removed through a dike drainage system that directs the water into a facility's industrial sewer or is drained directly into the environment if analytical tests indicate the water is within compliance for release. Normally, the discharge valve is kept closed. Technicians manually open a dike discharge drain valve to remove storm water runoff.

The specific routine tasks for tank farms vary, depending on the features of the individual tanks. Tanks with external floating roofs have several features that require regular inspection to ensure that the roof is safe and in good working order. Technicians may also have to check instrumentation and steam traps associated with tanks requiring internal heating by steam coils and mechanical mixing devices (stirrers, stirring motors, jets). They will also open water draw valves provided at the lowest point on the tank bottom to drain to process sewer water that has settled to the bottom of the tank.

Performing Routine Inspections of Auxiliary Equipment and Systems

Routine tank farm inspections include the examination of auxiliary equipment and systems. In general, auxiliary equipment commonly includes the piping system, motor control centers (MCCs), pumps and other equipment located on pump slabs or pump pits, and fire safety equipment.

Motor control centers (MCC) route electricity that drives the auxiliary equipment located on pump slabs and in pits (Figure 15-10). All of the equipment served by a given MCC is labeled on the rack. MCCs for the auxiliary equipment should be checked to ensure that there are no electrical problems.

Fire safety equipment is another function of routine inspections. Depending on the tanks and the materials involved, fire safety equipment may be a foam injection system, fire monitors, or a water deluge system. To protect against fires at pump slabs, manually activated fire monitor stations are often used. Technicians should verify the fire safety equipment is in good working order and that they know how to use it when it is needed.

Figure 15-10   Motor Control Center for Tank Pumps

Sampling

Another routine duty of tank farm technicians is collecting tank samples for laboratory analysis. Samples are analyzed to determine whether the tank's material meets specifications as feed for process units or product for sale. Verifying on-specification tank product is a preliminary step before material is transferred from a tank to a process unit or to a product distribution site (via pipeline or loading rack). In chapter 8 we stated that sample analytical results are no better than the sample submitted. Restated, bad samples yield bad results. Technicians must (1) provide samples that are representative of the entire contents of the tank and (2) avoid contaminating samples during sample collection.

Before sampling a tank, technicians must determine if the tank has been static for a prolonged length of time. A **static tank** is neither receiving product nor releasing product. An appreciable inflation of the tank's content can occur due to air inclusion resulting from the circulating or receiving action. A settling time of 2.5 hours for all products is considered normal for some sites before gauging tanks. Depending on the product, static tanks may stratify into layers of different densities. If the product has a tendency to stratify the tank will have to be *rolled*, an operations term that means mixed, usually by bubbling nitrogen through the tank or circulating its contents with a pump. If the tank has recently received product, it is important to allow the contents of the tank adequate settling time to discharge any electrical charge generated by the flowing liquid through a grounding conductor. Before collecting samples, the technician should review the material safety data sheet (MSDS) about the material in the tank, then collect the necessary sampling equipment and the required personal protective equipment (PPE).

Before climbing to the top of the tank to collect a sample or to gauge it, the technician should make a visual inspection of the area around the tank for signs of abnormal conditions. After climbing to the top of the tank, to protect against static electricity hazards, the technician should ground his body by touching his bare hand to the steel stair rail before opening the gauge hatch. Next, the technician should check a windsock to verify they are positioned upwind of the gauge hatch before opening the hatch and allowing the initial release of vapors from the tank to clear the area. During the sampling procedure, the technician keeps the chain of the **sample thief** (sample collection device) in contact with the metal hatch to prevent any discharge of static electricity from the tank's liquid through the sample thief. When the sample thief reaches the bottom of the tank, the technician pulls the chain on the sample thief to open the bottle. As the technician pulls the bottle and sampling device upward through the column of liquid, the open bottle fills with sample. The filled sample bottle is withdrawn from the tank, sealed, labeled, and put in a carrier basket.

Gauging Duties

Material inventory and available storage space in storage tanks is determined by gauging tank levels. Gauging storage tanks also provides a means for

- Detecting material losses resulting from leaks or theft
- Preventing overfills and spills
- Checking for damage to transfer pumps due to loss of suction head

Tank farm technicians are responsible for taking level readings as part of routine inspections of aboveground tanks and equipment. Besides gauging, level readings can be determined from local level indicators and remote indicators on control boards.

Various devices can be used to gauge aboveground storage tanks. The specific device that is used in a particular situation depends primarily on the type of tank and the material it contains. There are two basic methods used to gauge storage tank liquid levels. One is **ullage** or **outage**, the amount by which a vessel falls short of being full. The other is **innage**, the amount by which a vessel is full. With either method, it is critical that accurate, reliable measurements are obtained because what would seem to be a very small change in liquid level, for example one half inch, can significantly affect material inventory calculations. For example, in a vertical cylindrical tank that is 100 feet in diameter, a half-inch change in liquid level equates to nearly 2,500 gallons of material. That much material selling at 90 cents a gallon is a misrepresentation to somebody (buyer or seller) of $2,250.

Tanks may be gauged using long reels of tape similar to measuring tapes used by carpenters. At the top of the tank, the weighed tape is attached to a stainless steel cable. The tape is reeled in and out of the storage tank by springs and counterweights in response to changes in liquid level. The reel mechanism is housed in a weatherproof housing at the base of the tank, where the technician can conveniently read the measurement, typically given in feet. The technician may later refer to a strapping table for the tank to convert the measurement of feet to its equivalent in terms of gallons or barrels.

Technicians may also gauge a tank with a *tape and bob*, a hand-held gauging tape that has an attached weight. Several feet of the tape may be coated with a water indicating paste. Then the tape is lowered into the tank through the gauge hatch. The level of both liquids in the tank, water level at the bottom and oil level above the water layer, is determined.

Generally, inventory data is recorded daily and reconciled monthly. To keep daily inventories, technicians are required to obtain and record level measurements from automatic, pneumatic, and electronic gauges at regular intervals. Inventory figures that do not reconcile at the end of the month indicate a problem in the operation of the tank farm. Discrepancies will result in an investigation to determine if the discrepancy is caused by an undetected leak, theft or malfunctioning level gauges.

Gauging Safety. Several safety concerns are associated with gauging external floating roof tanks. Before descending down onto the roof of a floating roof tank the technician should be aware of factors that affect vapor loss from the tank and the safety precautions that should be taken.

In general, the higher the temperature, flow rate, and volatility of the material entering the tank, the more vapors are created. This creates the potential for more vapors to escape from the tank. More vapors are also created when the contents of a floating roof tank are mixed or agitated, and more vapors are likely to escape when the liquid level is not high enough to float the roof. These conditions make it unsafe to go onto the roof to sample and manually gauge the tank unless technicians use a respirator or other PPE.

In some situations, descending onto a floating roof is considered the same as entering a permit-required confined space and is subject to all of the applicable safety precautions. Technicians should keep in mind that the atmosphere above a floating roof could contain hazardous levels of toxic, flammable, or combustible gases and vapors. Specific company SOPs should be strictly followed. For instance, some facilities require the use of a standby, or buddy, system. Using this system, one technician remains at the tank's sampling platform and stays in communication with another technician who goes down onto the floating roof. Before anyone goes onto the roof of a floating roof tank, the area should be tested for hazardous levels of gases and vapors. No one should go onto the roof unless the test indicates that the area is safe to enter.

After the area has been gas tested, the floating roof and its auxiliary equipment should be inspected for signs of corrosion and for obvious damage and wear. Roofs that have been seriously weakened by corrosion may not be strong enough to support personnel. In some cases, a technician may have to wear a safety belt and harness or other fall protection equipment during the inspection of a floating roof. Next, the roof seals should be verified to be in good condition to keep vapor emissions from the tank at or below allowable levels. After the inspection is complete, the tank can be gauged. Technicians can perform an innage gauge through the gauge hatch at the tank's sampling platform and need not go out onto the roof. Before opening the gauge hatch, technicians should touch the metal cover of the hatch to ground their body, and then position themselves upwind of the hatch before opening it. The gauge tape is lowered until it reaches the bottom of the tank, then it is withdrawn. The technician then reads the mark left by the tank's material on the tape's measuring scale. The reading indicates the level of product in the tank in feet. For accuracy's sake, the technician makes a second level measurement and then reports the reading to the control room.

Gauge measurements reported to the control room can be converted to gallons or barrels using a computerized strapping table. The manual gauging results can then be compared to the recorded results for the tank's automatic gauge. If the figures match, this indicates that the inventory is accurate and that the automatic gauge is functioning properly. Gauging tank levels is a basic procedure for keeping track of material inventory. It is also an integral part of another major technician responsibility, which is making material transfers. Generally, before material is transferred into or out of a tank, a measurement known as an *opening gauge* must be made. An opening gauge determines the liquid level or volume in a tank before transferring more fluid into it. Then, after the material is transferred, technicians must make a *closing gauge* measurement. A closing gauge is a determination of the liquid level or volume of a tank after pumping into it has ceased.

Material Transfers

Material transfers are basically a matter of moving material from one location to another. When a technician transfers material from a process unit to a storage tank he/she must coordinate with the control board technician and/or with personnel at the delivery end (where material is coming from) or the receiving end (where material is going) of the transfer. In general, a material transfer from storage tanks can be broken into four steps: (1) making preparations, (2) lining up valves and initiating the transfer, (3) monitoring the transfer, and (4) completing the transfer.

Step 1: Making Preparations. To prepare for the material transfer, the technician reviews relevant safety information and operating procedures. Next, the tank's opening gauge shown by the tank farm's computerized level monitoring system is recorded. Then the technician checks the available space in the receiving tank and verifies that the tank can hold the amount of material that is to be transferred. The technician will estimate the amount of time needed for the transfer. Careful planning before the transfer helps to ensure that overfills and spills do not occur.

Step 2: Lining Up the Valves. Next, the technician must do a valve lineup, which is the task of opening certain valves and closing others in a specific sequence so that material will move along the desired flow path. A general rule of thumb for opening valves in a lineup is to begin with the valves on the receiving (low-pressure) end of the transfer and work toward the delivery (high-pressure) end. Correct lineups are important because:

- An incorrect lineup can be costly in terms of both productivity and workplace safety. For example, as little as one barrel of the wrong material entering a tank with a 1000-barrel capacity can contaminate the contents of the tank and result in an off-spec product tank.
- Sending material to the wrong tank, such as a tank without enough available space, can cause a spill or overfill, with all of the associated safety and environmental problems.
- The wrong material entering a process unit may cause a unit upset. In some cases, this material may react violently with normal unit materials, resulting in a fire or explosion.

To prevent the preceding situations from occurring, valve lineups should be double-checked before a material transfer is initiated. After double-checking the lineup, the technician starts the transfer pump, or if the transfer pump is board controlled, they radio the board person to begin the transfer. The technician also alerts process technicians receiving the material that the transfer is about to begin. The initial flow rate should be kept relatively low, since a high initial flow rate and splashing during a fill could generate static electricity.

Step 3: Monitoring the Transfer. Now, the tank farm technician begins the third step, which is monitoring the transfer. The transfer must be monitored to prevent leaks, spills, or overfills, and to detect abnormal conditions. Temperature and level must be checked at regular intervals. Good communication among the outside technician (field technician), process unit technicians, and inside technician (board man) is important. All are capable of affecting the material transfer process. The process unit technicians at the delivery end of the transfer depend on the tank farm technician at the receiving end to notify them when flow must be diverted or stopped.

Step 4: Completing the Transfer. When product delivery is stopped, the tank farm technician rearranges the valve lineup as necessary to block in the tank. After allowing time for the tank contents to settle, the technician takes a closing gauge reading and collects samples of the product for the quality control laboratory. A final inspection of the tank and its auxiliary equipment is made and then the details of the transfer in the tank farm's logbook are recorded.

Some storage tanks are *dedicated tanks*, that is, the tank is used to store only one type of material or product. This is normally true of process units that have their own tanks. Using dedicated tanks avoids problems of cross-contamination of materials. Because the tank is a dedicated tank, a technician may not need to have samples from the tank analyzed and approved by the quality control laboratory before more of the same material can be transferred into the tank from the process unit.

Making a material transfer should be an orderly, step-by-step process. However, a material transfer can be quite complex and involve multiple tanks, intricate piping arrangements, and numerous valves. For instance, the transfer of material from a process unit to multiple storage tanks can take place over several hours and involve workers on different shifts. For jobs of this type, maintaining good communication, paying close attention to details, and double-checking work before proceeding are key factors for a safe, successful transfer.

SUMMARY

A large refinery or petrochemical plant may have over 100 storage tanks of various sizes and limitations. Technicians who work with these tanks have the responsibilities of inspections, material transfers, sampling, and gauging. Though a tank farm is not as complex as a large processing unit, it can seriously affect safety, health, the environment, quality, and profitability if wrongly operated. Tanks of various types are used for the storage of raw materials and finished products. Since the stored material represents a large concentration of value, the protection and correct operation of material transfer into and out of storage tanks is important. Improper operation can cause serious accidents resulting in hazards to personnel, damage to the environment, loss of revenue, and costly damage to equipment.

Tank gauging is required for the assessment of tank contents, tank inventory control, and tank farm management. Tank gauging system requirements depend on the operation. The types of operation can be categorized as inventory control, custody transfer, and oil movement and operations. Tank gauging is performed by automated gauging using the float and tape gauge, and by manual gauging, using a tape and bob.

The major responsibilities of a tank farm technician include safety concerns, performing routine inspections of aboveground storage tanks and their auxiliary equipment and systems, sampling, gauging tank levels and taking material inventory, and material transfers. With each responsibility are certain basic tasks that a technician must perform without deviation according to the facility's standard operating procedures (SOPs).

REVIEW QUESTIONS

1. Name two types of atmospheric tanks.

2. List the four conditions that affect the evaporation of storage tank contents.

3. Give three reasons for tank gauging.

4. Describe how tanks are automatically gauged.

5. Describe how tanks are manually gauged.

6. What is meant by the term *custody transfer*?

7. In general terms, a 1°C error in the temperature of most petroleum products will equate to an error of approximately _____ in the standard volume calculation.

8. Describe the potential harm that can result from transferring a product into the wrong storage tank.

9. List the five major responsibilities of a tank farm technician.

10. Explain why pressure is a safety concern on tank farms.

11. How does a gauge hatch act as a pressure relief device?

12. Give two reasons why some tanks have nitrogen blankets.

13. Explain why weather is a safety concern on tank farms.

14. Describe the routine inspections and duties of a tank farm technician.

15. List three reasons why level checks are important on storage tanks.

16. Discuss the precautions a technician should take before sampling a tank.

17. List the four major steps of a material transfer.

18. How is the water level in the bottom of oil-bearing tanks determined?

19. What is the purpose of a water draw valve?

20. What does an innage gauge measure?

21. What does an outage gauge measure?

22. Give three reasons why correct valve lineup is important.

CHAPTER 16

Process Unit Shutdown

Learning Objectives

After completing this chapter, you should be able to

- *List five reasons that may require a unit shutdown.*

- *List five expenses of a unit shutdown.*

- *State when the greatest number of injuries and accidents occur on a process unit.*

- *List five hazards associated with a unit shutdown.*

- *Explain why training is important to a successful unit shutdown.*

- *Discuss the planning and preparation required for a unit shutdown.*

- *List three methods of purging equipment.*

- *Explain the purpose of blinding.*

- *Explain the purpose of gas testing a purged vessel.*

INTRODUCTION

Many of the chemical plants and refineries in operation today run continuously, 24 hours a day, seven days a week. Shutdowns and startups of processing units can be costly and time consuming. As an example, a styrene unit running wide open may gross $200,000 a day. Think about how much money is lost when the unit comes down unexpectedly for a week! A process unit makes money only when it is running. When it is down, workers are still being paid wages and benefits but no profit is being generated. Thus, unit shutdowns are very infrequent and are scheduled as planned shutdowns once every 18 to 36 months or longer, depending on the type of unit. If a unit comes down before a planned turnaround, it is normally because of an emergency situation, such as an explosion, fire, or a hurricane nearing landfall. Unit shutdowns are usually very costly due to the reasons revealed in Table 16-1. Additional reasons for shutdowns are

- *Scheduled turnaround:* A processing unit is subject to entropy, a way of saying it begins to wear out and become more subject to process disruptions. It requires periodic maintenance and repairs due to corrosion, erosion, wear, and equipment fatigue. Depending on the type of unit, there is a schedule of planned shutdowns for maintenance and repairs. A planned shutdown of this type is called a **turnaround (TAR)**. A turnaround may last for 15 to 30 days or longer, depending on the extent of repairs needed.
- *Unit inspections:* Unit inspections that determine vessel integrity and corrosion rates may reveal severe enough piping and vessel weakness to mandate a shutdown for emergency repairs. Such inspections are usually done under the umbrella of a unit turnaround, however it is not unusual that corroded or stressed piping and equipment have necessitated a partial unit shutdown if not the complete unit.
- *Major equipment problems:* Most major equipment (i.e., pumps) is spared but not all. The spares are either on standby or stored in maintenance. Compressors are very expensive and it is uncommon to have compressors spared. If key critical equipment fails between scheduled turnarounds, it will be necessary to shut down the unit or parts of the unit while a hot-shot order is placed for spare parts. It might be possible to keep parts of the unit up and in circulation. If the failure is sudden and unexpected, quick repair becomes an urgent situation.
- *Emergency situations:* Emergencies such as a major fire, explosion, or a complete electrical power failure may require the shutdown of a unit. Once the

Table 16-1 Monetary Costs of Process Unit Shutdown

Shutdown Action	Reason for Expense
Flaring of unit materials	Waste of unit streams
Off-specification products produced	Waste of manpower, utilities, energy
Additional manpower	Increased payroll
Loss of production	No unit gross income during shutdown
Equipment not properly cleaned, purged, or isolated	Damaged equipment, accidents, injuries, accident investigations
Additional environmental compliance	More leaks, spills, slop material
Increased accidents and injuries	Contract labor, pressure to meet deadlines

unit is down and the emergency has been resolved, bringing the unit back up becomes top priority. If major damage occurred during the emergency situation then there will be an emergency budget meeting and contractor notification.

- *Business downturns:* Sometimes production exceeds product sales because of business downturns (recessions). In this situation, process units reduce production rates or, if necessary, circulate for a period of time. In the worse case scenario, the unit is shut down and its operators are sent to other units or the maintenance division until the business climate becomes more favorable.

HAZARDS OF UNIT SHUTDOWN

In general, the highest number of accidents and injuries occur during a unit shutdown or startup. This is because during these situations there is a tremendous amount of pressure to get repairs done quickly and the unit back up and making money. Everyone is in a hurry. Opening lines and equipment increases the risk of chemical exposure, plus, there are a large number of contractors on the unit.

Experience is a slow and painful teacher, sometimes a deadly teacher. It isn't a good idea to learn by experience when you work around flammable and toxic chemicals, of which some are under high pressure or at high temperatures. Preparation and training for shutdowns and startups is critical to avoid accidents and incidents. When hazards cannot be eliminated they (1) must be recognized and all personnel must be trained on hazard avoidance and prevention, and (2) procedures must be created to include hazard avoidance or hazard avoidance must be included in existing procedures. The hazards encountered most frequently in shutdowns are detailed in Table 16-2.

Table 16-2 Hazards of Unit Shutdown

1. Mixing of air and hydrocarbons to form explosive mixtures
2. Water contact with hot oil
3. Freezing of residual water in pipes
4. Contact with corrosive or toxic liquids and gases
5. High pressures
6. Vacuums
7. High temperatures
8. Poorly trained contract workers
9. Anxiety to meet deadlines
10. Poor communication
11. Pyrophoric materials
12. Uncontrolled sources of energy

NORMAL SHUTDOWN PROCESS PLANNING

It is hard to completely separate shutdown planning from turnaround planning. Often you are doing both at the same time because there is a gray area where they overlap. It is not uncommon to bring part of a unit down and begin turnaround work on it while the rest of the unit is still coming down.

A normal shutdown should be well planned. Manpower, material, and resources must be estimated and made available when needed so that schedules are adhered to. Critical paths in the planned work will have to be determined. A *critical path* is a sequence in the planned work events, which if not completed, may delay other work. A critical path is like a highway leading up to a river. There better be a bridge across the river or there will be a long wait to get one built. Critical paths are predetermined and closely monitored so that one job does not delay work on several other jobs.

Turnaround Work Scope Planning

An operations mechanical work scope is developed by the operations department. It includes any known mechanical repair needed to enable the process to safely function at the desired ability for the length of time the following run is planned, say for three years. This work scope should include only items that are required to meet these criteria. Technicians learn through their normal workday what pieces of equipment have begun to need repair or replacement and include this information in logbooks and repair lists. Also, additional equipment problems will be discovered during the shutdown and clearing process and must be immediately and clearly reported to ensure these late discoveries are put on the repair list. Technicians must realize there are limitations (time and money) on the work to be done. Not everything suggested or listed will be corrected. Management makes the final decisions on repair work and upgrades during the turnaround based on safety, economics, operability, and legal issues.

Since scheduled shutdowns are not frequent events, technicians hired as long as two or three years ago may never have experienced a shutdown. More experienced technicians may not remember all of the steps in safely shutting a unit down since it is not a common occurrence. Proper planning and preparation for a shutdown will need to include technician training. Someone will have to verify all shutdown procedures are up-to-date. What things have changed? Was a procedure modified but not updated in the procedure manual and only a few technicians know about the modification? Technicians will either review hard copies or review the electronic copies on the unit computer one to two weeks prior to the unit shutdown and attend unit shutdown planning meetings.

Operations/Product Group Turnaround Planning

Changes in the unit's consumption of raw materials and utilities will have to be communicated to those suppliers. Steam for process heating, natural gas for fired heaters, cooling water, and electricity consumption will be greatly reduced. Nitrogen consumption will increase because it is used to purge inert process equipment. Suppliers of raw materials and process utilities will need to be notified of these changes several weeks in advance so that they can adjust their operations to the reduced or increased demand. Plans for raw materials and unit feed for startup will have to be made.

Since the unit will no longer be making product, either enough product must be in inventory for customers or plans prepared for purchasing product on the market. Most companies want to maintain or increase their market share, so careful thought and planning is given to building an inventory of product(s) to provide for customer needs or contracting with another site to supply customers. It is not uncommon for two competing companies to assist each other when one has a unit on turnaround. For example, Plant A's styrene unit shuts down for a planned turnaround but it still must supply rail cars, tank trucks, and barges of styrene daily for shipment to customers. Plant B, a competitor of Plant A and located several miles down the road, has a styrene unit. Plant A contracts with Plant B to buy their styrene for its customers. Tank trucks and rail cars are routed to Plant B, and barges to Plant B's docks. Plant A keeps its customers by selling them Plant B's styrene. Plant A will respond in kind when Plant B shuts down its styrene unit for a turnaround. Both plants may be competitors but they are not enemies. They cooperate for mutual benefit.

Operations/SHE Departments Turnaround Planning

As a rule, the highest number of accidents and injuries occur during a unit shutdown or startup. I quote again: *Experience is a slow and painful teacher, sometimes a deadly teacher.* Preparation and training for shutdowns and startups are critical to avoid accidents and incidents.

Environmental issues are a concern during a shutdown because there is a greater potential for chemical spills as process piping, equipment, and vessels are drained and opened. The potential for unwanted environmental releases increases because changing pressures and temperatures can cause joints to leak. Also, as systems are opened, materials that were not completely drained and purged from the system may contaminate the environment. The first line of defense against emissions is to prevent spills and/or recover oil and other materials at the process unit before they have a chance to become a larger problem. An emergency action plan for environmental releases should be in place, and an adequate supply of absorbents and materials for diking or otherwise confining spills should be readily available. Some waste materials generated during process equipment cleaning will be classified as hazardous wastes and must be disposed of according to environmental regulations. Plans for storage of these materials and their eventual disposal must be developed prior to the shutdown.

When a unit is shutting down, hazards not normally present during routine operation will exist. Equipment temperature may change by hundreds of degrees, especially when fired furnaces are involved. Rapid temperature changes may result in equipment damage or leaks due to thermal stress. Shutdown procedures involving temperature control and cooling rates should be followed closely to prevent such problems. As systems are cooled rapidly and vessels are emptied, a vacuum can be created if changes proceed too quickly. The vacuum can damage equipment not designed to withstand a vacuum or pull air into an unpurged system that is not vacuum tight, creating a hazardous situation. Changes must be made slowly to avoid creating vacuums that damage equipment or create hazardous conditions, and system pressure should be monitored closely. When steam is used to purge a system during clear up, a vacuum can easily be created as the system cools and the steam condenses. Vents should be opened to admit air (or some kind of gas) at a rate sufficient to displace steam as it condenses or water is drained from a vessel.

Other hazards may be created by pressure surges resulting from water added to a system with a temperature greater than the boiling point of water. If there is enough water, the vaporization of water to steam can generate enough pressure to rupture equipment and harm personnel. There is also the danger of air entering through an open vent or leaking flange and mixing with flammable materials. An ignition source, such as static electricity, could result in a fire or explosion. Procedures and training during shutdowns can prevent such situations.

GENERAL OUTLINE FOR UNIT SHUTDOWN

A general outline for a unit shutdown would be similar to the following:

Phase 1: Preparation

1. Management determines what repair work and upgrades are to be done during the turnaround.
2. Engineering planning, diagrams, and so on are completed on repair and project work for the turnaround. Plant departments and maintenance are notified of the shutdown and planned work.
3. Logistics of materials, supplies, and tools needed for shutdown are determined.
4. A list is prepared of equipment to be inspected.
5. The shutdown schedule is continually reviewed to determine ways to reduce turnaround time.
6. Shutdown procedures, blind lists (discussed in detail later in the chapter), and check-off lists are verified as up-to-date.
7. Additional operations and maintenance personnel are scheduled for the shutdown.

Phase 2: Unit Shutdown

1. Plans, including training, are prepared to minimize the flaring of materials, equipment damage, and accidents.
2. Unit feed and temperatures are slowly decreased to prevent damage from thermal shock.
3. Unit vessels and piping are deinventoried as much as possible to recover all unit hydrocarbons.
4. After all hydrocarbons have been pumped out or drained from piping and vessels, purge with an inert gas.
5. Insert blinds, aerate, and use gas detectors to check the lower explosive limits (LEL) of each vessel after equipment is hydrocarbon free.
6. Check vessels daily for LEL before allowing entry and check hot work jobs while maintenance and contractors are doing turnaround work.
7. Engineering and operations personnel do a final inspection before equipment is closed up.

Generally, process units are shut down in an orderly way. First, heated unit charge (feed) is gradually reduced in volume and temperature to avoid thermal stress. Then process material is salvaged and stored in designated tanks. Waste tanks, also called slop tanks, are available to accept unit material designated as slop. **Slop** is material that is not an intermediate stream or final product. If the unit cannot rework the slop when it comes back up, it will send the slop to a unit that can utilize it to make product. Once all material has been pumped out, it may be necessary to purge to flare or atmosphere to eliminate any remaining hydrocarbons or other process materials in piping and vessels. All process materials are purged out until every piece of equipment has been thoroughly cleared. Bypass lines, level chambers, dead end lines, and other parts of the system that may be difficult to purge should be given special attention. Hard to purge piping and equipment not adequately purged are often the cause of hazardous situations once the opening of piping and equipment begins.

Technicians will be busy preparing equipment for maintenance personnel once the unit is down. It is very discouraging for maintenance crews to show up ready to begin turnaround work when only half of the equipment is ready for them. This wastes time and money and disrupts the tightly planned turnaround schedule.

DE-INVENTORY AND EQUIPMENT SHUTDOWN

After the process materials have cooled to acceptable temperatures, they can be transferred to storage tanks if they are recoverable. If they are not worth the time and effort to recover, the materials are flared or sent to slop tanks. If centrifugal pumps are used for de-inventory, they must be watched to prevent pump damage from loss of suction and running dry. Reciprocating pumps or steam turbine-driven pumps give the operator better control of the pump-out rate and are not as easily damaged.

Depending on the nature of the process materials within the unit, they will be sent to various storage or slop tanks. The unit may have one or two slop tanks, one used for off-specification product and the other for slop that cannot be reworked by the unit. Nothing will be wasted. All hydrocarbons, even slop, can be rerun in some unit to make a salable product. If nothing else, slop can be blended with other hydrocarbons to make a furnace fuel. Gases will normally be depressured and purged to the plant fuel gas system or flare system for destruction. When all of the process materials have been pumped out, the pumps are shut down.

Next, unnecessary auxiliary systems and equipment will be shut down or blocked in. This may include cooling water systems, compressed air systems, steam systems, lube oil systems, pumps and compressors, air cooled fans, etc. Water-cooled condensers and coolers should have the waterside blocked in and drained with the vents and drains left open. Leaving vents and drains open prevents damage caused by residual water flashing to steam if these units are purged with steam. Electric equipment requiring maintenance must be locked out and tagged out.

UNIT CLEAR UP

Once the unit has been safely shut down, the major part of an operating team's shutdown work is just beginning. Securing of equipment begins as furnace fires are cut out, feed is discontinued, and product and reflux pumps are shut down. Some areas that may need special attention during purging and clearing up are summarized in the following sections.

Pumps. Pump seals may be damaged during the product pump out if the pumps are allowed to run dry. It is important to shut down the mechanical seal pumps immediately before liquid levels are depleted. The pumps may have to be restarted after liquid on the tower trays drain down in order to pump out the majority of the hydrocarbons from the tower draft trays and tower bottoms. After all available liquid has been pumped out via a product pump, battery limit charge and product rundown block valves must be closed to prevent the backing of hydrocarbons into the unit.

Furnaces and Boilers. Furnace and boiler fuel lines must be blinded after fuel is shut off for protection against a possible firebox explosion from inadvertently introducing fuel into still hot fireboxes. Isolating valves must be closed between systems of greatly varying pressure to prevent over pressuring lower pressure systems and popping relief valves within the system. When shutting down furnace fires, the master fuel valve must be closed and one of two burners must remain lighted in order to burn the fuel remaining in the furnace fuel line as the fuel system is depressured. All remaining burner block valves and pilots are then closed. This prevents leakage of fuel trapped in the burner headers into the firebox.

Steam Purging Precautions. Water-cooled exchangers left full of water can be pressure damaged during steam purging since the water can flash to steam when it is heated. All water should be drained from such heat exchangers and the vents should be opened before purging. Some equipment may be damaged during steam purging because of the high temperature and pressure involved. Instrumentation, turbine meters, and pump seals are examples where steam should not be used for clear-up purging. Other areas to watch include salt beds, catalyst beds, and molecular sieves. These areas may be damaged by steam or steam condensate.

Process Chemical Injection Points. One area often overlooked by operating personnel is process chemical injection points. Continued injection of process treatment chemicals or additives into equipment that has been shut down not only wastes valuable chemicals but could be hazardous during draining procedures when a technician encounters a line completely full of an unexpected, hazardous, and sometimes flammable, material.
Be thorough in preparations and inspections before turning equipment over to repair or inspections groups. Do not place them in harm's way through your inattention to detail.

CLEARING PROCESS EQUIPMENT BY PURGING
To *clear* equipment means to remove traces of flammable, toxic, or corrosive materials from the equipment. Though the equipment is drained, it may still contain residues of hazardous materials. Some process vapors and residual liquids or solids may be trapped in hard to reach locations. These materials must be removed before piping and vessels are opened up for maintenance or inspection. They are removed by purging with steam, nitrogen, or by flooding the system with water.

A *purge* is a procedure in which an inert gas, such as nitrogen, is used to remove undesired gases or vapors from a vessel so as to reduce the concentration of the undesired gas to a safe and acceptable level. Purging can be one of the most hazardous operations in a plant as the possibility of a combustible mixture of air and flammable gases or vapors is always present. Often it may be necessary to purge several vessels connected to one another. It may be possible to introduce gas into the first vessel and vent out the last one but this method is not recommended. It lengthens the time that a combustible mixture can be present in the system. Each vessel should be purged separately. Also, welding slag and other metal debris should be removed from vessels before closing them up. Purge gas introduced too fast might hurl the debris against the metal vessel and cause a spark. Some purge gases introduced too rapidly may cause a build up of static electricity that may spark.

Purging with Steam

Steam can be used for most purging applications. It is ideally suited for purging waxes, sticky materials, highly viscous materials, heavy crude oils, and polymers from process systems. When steam is used to clear up a system, purging actually does not begin until the steam heats the metal piping and equipment of the system hot enough to keep the steam from condensing on it. This is because hot steam contacting ambient temperature (cool) metal condenses into water, thus there is no gas (steam) to purge hydrocarbon vapors or heat hydrocarbon residues into vapors that can be purged. Steam must be introduced at a rate that allows the metal to slowly expand without thermally stressing the equipment. It should be introduced slowly at the bottom of the vessels or piping and the flow gradually increased until a good, steady flow enters the system. Steam flow rate should be dictated by procedure. If the procedure does not specify a flow rate then a good flow can be determined by the sound of the steam going through the valves and the increase in heat of the piping.

Steam purging accomplishes the following:

- Removal of air
- Evaporation of water trapped in equipment
- Removal of hydrocarbon liquids and gases
- Revealing plugged drains and vents

Since steam is a gas, when it is injected under pressure into a vessel it will force air out. The steam used for purging will be much hotter than the boiling point of water, thus it will evaporate any pockets of water trapped in the system. The hot steam will also vaporize hydrocarbon liquids and drive these gases out of the system. Finally, opened drains and vents that do not have steam plumes exiting them reveal plugged drains and vents.

Equipment should be steam purged according to unit procedure. If no time is designated in the procedure, equipment should be steamed until the blowdown line is hot. Technicians should periodically drain steam condensate from low points in the system to prevent liquid buildup. Once the system is thoroughly heated and steamed per procedure, the system must be tested to ensure hydrocarbon gases or liquids have been reduced to the required minimum levels before opening the vessels to atmosphere. When opening to atmosphere, open the vents at the highest point of the vessels or piping. Check the equipment with a pressure gauge when opening the vents to ensure the vessel pressure is greater than air pressure, thus preventing air from entering the equipment and creating an explosive mixture. If steam does not immediately flow out of the vents, close them off until the system is hot enough to establish a good visible flow from all vents.

If steam does not flow from vents and drains they may have accumulated rust or other debris that prevents their proper operation. This may require rodding, inserting a slim metal rod (similar to a welding rod) into the drain or vent to break up the material blocking the drain. Technicians should dress with appropriate personal protective equipment (PPE) for this task. Trapped fluids, such as hydrocarbons and water, will be very hot because of the steaming and may suddenly erupt from the drain, creating a hazardous situation.

Instruments and pump seals can be damaged by the high temperature and pressure associated with steaming. They should be valved out of the system before starting steam flow and purged with an inert gas such as nitrogen.

Once purging is completed, the system should be blinded, opened, and gas tested to verify that purging has removed all traces of process materials. As a general rule for systems that contained hydrocarbons, the system is considered to be safe to open if gas testing indicates that hydrocarbons are less than 1 percent of the vented purge stream. If the gas test indicates that process material concentration is still greater than 1 percent, additional purging will be required. Process vessels that have been purged with steam should not be allowed to cool before they are opened because of the hazard of creating a vacuum when the steam cools and condenses.

Purging with Inert Gases

Steam cannot be used to purge certain areas of a process unit because it may heat the equipment or instruments above design temperatures or because steam condensate may react with water-sensitive process materials such as catalysts. When equipment can't be purged with steam, inert gases such as nitrogen or carbon dioxide may be used. Inert gas purges work well to displace process vapors and free flowing liquids from the system but do not work well with highly viscous liquids or sticky materials. An inert gas is any gas that is non-flammable, chemically inactive, non-contaminating for the use intended, and oxygen deficient to the extent required.

Inert gases should be introduced from the top of a vessel and allowed to flow down to the lowest drain valve and be safely vented away from personnel. The use of nitrogen as an inert atmosphere is used by a wide number of industries. Nitrogen gas is colorless, odorless, tasteless, non-toxic, and a poor conductor of heat and electricity. It comprises about 78 percent of normal air and is slightly lighter than air at standard atmospheric pressure and temperature. However, cold nitrogen gas is considerably heavier than air and will accumulate in low areas. Commercial nitrogen gas is extremely dry, with a dew point of less than 800°F.

Purging with Nitrogen

The nitrogen purging method used is dependent upon several factors such as

- The time allowed
- The desired oxygen concentration
- The type of vessel to be purged (i.e., atmospheric or pressure vessel)

Though there are a variety of different methods and techniques used for purging of vessels and equipment, three fundamental methods are commonly used:

1. Pressure purging
2. Sweep purging
3. Vacuum purging

When equipment is placed in service, the air in the equipment is displacement with nitrogen. The nitrogen is then displaced with the subsequent introduction of a combustible gas into the vessel without the formation of an explosive mixture. Similarly, when equipment is taken out of service for inspection or repairs, the displacement of the combustible gas vapor with a suitable quantity of nitrogen allows for the safe introduction of air after the purging procedure.

Pressure Purging. *Pressure purging* is a technique used primarily on pressure vessels. As an example, tank trailers carrying liquefied petroleum gases (LPG) are normally transported under pressures of about 100–150 pounds per square inch gauge (psig). Once the gas is off-loaded into storage, a residual gas remains in the cargo tank. The purging procedure in this case is simple. The vessel is pressurized with nitrogen allowing for a thorough mixing of the nitrogen and residual LPG. Once sufficient time for mixing has elapsed, the mixed gases are bled to a flare line and the procedure is repeated until the vented gases test at 1 percent or less hydrocarbon. At this point, the remaining gas in the vessel should be below its LEL.

The most efficient method of pressure purging is pressurizing to two atmospheres (approximately 29 psig) per cycle and repeating the cycles until an LEL of 1 percent or less is obtained. The least number of volume changes and lowest nitrogen requirement will be achieved at this rate. However, this technique is very time-consuming. Most pressure purging applications are performed at higher pressures determined by the maximum vessel operating pressure. As an example, consider an LPG tank car carrying butane and pressurized to 150 psig. Hot work needs to be performed, and the vehicle must be taken out of service. The tank car is off loaded to storage but a butane gas residual remains in the tank car at one atmosphere. Pressurization of the tank car with nitrogen to 2 atmospheres will dilute the residual of the original atmosphere to 50 percent of its original value. After venting, the procedure is repeated. This time the residual is reduced to one half of 50 percent, or 25 percent of the original concentration. When repeated four more times, the original concentration is finally reduced to a level below the LEL of butane, which is 1.9 percent. By pressurizing to 2 atmospheres, six vessel volume changes are needed in six pressuring cycles. The same resulting concentration could have been achieved by pressurizing to 10 atmospheres in the first cycle. The original concentration would be reduced to below the LEL in just two cycles.

Sweep Purging. *Sweep purging*, also known as *dilution purging*, is the most common purging technique. In this process, the gas or vapor to be purged is diluted or displaced with nitrogen, causing the gas to be vented from the vessel. Nitrogen is introduced into the vessel through the vessel vapor line, forcing the vapor residual out of the vessel through the liquid withdrawal line and into a flare line. The nitrogen purge is maintained at a constant rate until the vented gas is below a predetermined LEL. The volume of nitrogen required must be determined by a test based on the time required and the flow rate used to reduce the residual gas content to the required percentage. The slower the purge (with a maximum amount of mixing), the more efficient the use of nitrogen.

As a rule, purging an air-filled vessel a minimum of four volume changes with nitrogen will reduce the oxygen content to approximately 3 percent or less; five volume changes are required to purge a vessel to less than 1 percent. Sweep purging is less effective than pressure purging because sweeping nitrogen through a vessel will miss hard to get to pockets of chemicals. Sweeping is best used for long runs of piping and tubing. Figure 16-1 illustrates the necessary volume changes for desired oxygen concentration.

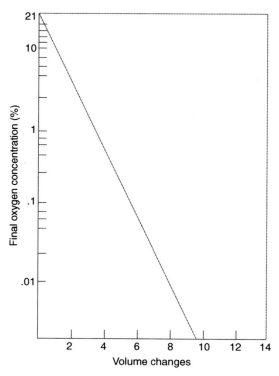

Figure 16-1   Sweep Purging Chart

Vacuum Purging. *Vacuum purging* is similar to pressure purging except that it is used on vessels that can withstand a vacuum. Once the vessel to be purged has been pumped out, a vacuum pump is used to evacuate the vessel. Injecting nitrogen then breaks the vacuum, time is allowed for mixing, and the vessel is vented. An explosion meter or oxygen analyzer is used to determine the hydrocarbon concentration. The cycle is repeated until the desired concentration is reached. Vacuum purging is generally more efficient than pressure or sweep purging in terms of volumes of nitrogen required; however, it is a more difficult procedure, and it takes longer to evacuate a vessel than to pressurize it.

Purging by Water Flooding

Purging by *water flooding* works well if the vessels and lines contained water-soluble materials. Examples of water-soluble materials are ammonia, alcohol, and acids and bases. Using this purge method, water is introduced from the lowest level of the system or equipment and allowed to flow up and out the highest part of the system or equipment. Water flooding can also be used to remove some hydrocarbons and other materials that do not react with water, but water flooding is not the best purge method for systems that contained hydrocarbons. Water may not be sufficient to gas free the system if there are scale or sludge deposits. Hydrocarbons may remain under the scale and sludge and degas after the system is drained. Steam purging may be needed to gas free the system. Usually, water purging is performed with water heated to a temperature of 100°F because warm water does a better job of removing hydrocarbons than cool water. Water flooding should only be used if the equipment and foundations of the system to be purged are designed for the stresses that the added weight of the water will impose. Water is on average about 25 percent heavier than most hydrocarbons.

Figure 16-2 Frac Trailer Tank

Before water flooding is started all rundown, feed, and gas lines should be blinded at the battery limits to prevent water from entering them. Then water should be added to the bottoms of towers, vessels, and other equipment until they overflow. Continue overflowing for approximately two hours or until the water at the drains is free of process materials. When inspection shows that the system is free of process materials, drain all water from the system. Draining water from a vessel can pull a vacuum on the vessel and may damage it, so air or nitrogen should be injected into the vessel to prevent this from occurring. The water drained from the system during water flooding may be hazardous and should not be dumped to grade or the unit storm water system. Instead it should be routed to the process sewer or, if conditions warrant, to Frac tanks for temporary storage before being hauled off to a waste disposal site. **Frac tanks** are wheeled trailers that serve as temporary storage tanks that can provide up to a 20,000-gallon capacity storage per tank (see Figure 16-2). Many FRAC trailer tanks are a rectangular straight-wall design that allows adjacent tanks to be positioned only inches apart. Many are insulated and contain valves, piping, and internal electric heating.

Water that has been analyzed for total organic carbon, pH, and oil and grease and meets environmental requirements may be dumped to the unit storm water system.

Purging with Natural Gas or Fuel Gas

Sometimes equipment is purged with natural gas or propane because it contains a catalyst or some material that would be harmed by nitrogen or steam. This is not as hazardous as it seems because purging with either will create an environment too rich to ignite. Since natural gas is lighter than air, it will rise above air. Thus, when purging with natural gas, the gas must enter from the top of the equipment and the air withdrawn from the bottom. Propane, being heavier than air, should be introduced at the bottom of the equipment and the air withdrawn at the top.

If a vessel is open to the atmosphere and full of air, it contains 14.7 pounds per square inch (psi) of air and the pressure gauge (psig) reads zero. If that same vessel is closed to the atmosphere and gas is introduced until the pressure gauge reads 14.7 psig, the vessel will contain a mixture of 50 percent air and 50 percent gas, a mixture too rich to burn. Thus, when purging with gas it is often best to pressure a vessel to about 14.7 psig for safety purposes.

BLINDING, OPENING AND GAS TESTING FOR ENTRY

Vessels and equipment must be blinded, opened, and inspected before work can be begun or workers allowed inside vessels. *Workers doing the gas testing of vessels should wear a breathing apparatus prior to opening any vessel that may contain hydrocarbons or hydrogen sulfide gas.*

Blinding

Blinds must be installed to keep unwanted materials out of the recently purged systems. **Blinds (blanks)** are metal plates that completely block flow in piping and vessels. All blinds are not created equal. Certain lines or vessels may require blinds of a different thickness or made of special alloys or they may fail. Blinds prevent process materials from inadvertently being released into a vessel where maintenance or contract workers may be working and causing injury or death. If the entire unit has been shut down, blinds should be installed at the battery limits in the main process and utility lines (those not necessary for the coming turnaround) supplying the unit. In addition, individual pieces of equipment subject to human entry should be blinded and all energy sources should be locked and tagged out before permitting entry.

Blinds may be kept stored in unit buildings or the warehouse on blind boards (sheets of plywood with a hook and number for each blind). If figure-eight blinds are used, they are not stored but are located on the equipment to be blinded. Figure-eight blinds are two blinds in one (see Figure 16-3). One of the circles comprising the figure eight is solid metal and is the blind that stops flow. It is slid into place and bolted in when piping or vessels are to be blinded. When flow is to be returned to the piping or vessel, the bolts are loosened, and the other circle of the figure eight, which has the center cut out to allow flow, is slid onto the equipment and bolted in place. Now flow can resume (see Figure 16-4). Some sites attach bright Day-Glo plastic chains to their figure-eight blind handles to make it easy to spot the blind location.

Figure 16-3   Figure Eight and Paddle Blinds

264

Figure 16-4   Figure Eight Blinds on a Contactor

Units have blind lists to keep a record of installed blinds. Later, this same blind list will be used to verify all blinds have been removed. A **blind list** is just what its name implies, a list of all the blinds used in an area of the processing unit. A processing unit may have one or more blind lists. The blind list lists all the blinds, the size of each, and assigns them a number. The blind list is important to ensure that

- All required blinds have been installed before anyone enters equipment
- All blinds are removed before unit startup

Keeping track of blinds is important both for safety purposes and for economic purposes. As a safety factor, blinds prevent the accidental release of material into a vessel occupied by workers. Valves won't work. Most valves leak to some degree or another. As an economic factor, blinds can prevent a unit from successfully coming up because a blind was overlooked and left in line, preventing fluid entry into a vessel. Removing the blind will allow air back into the system, thus maintenance workers or contractors will redo work that had already been paid for. They will seal, purge, and gas test again. Depending on the size of the vessel or system to be purged again, the cost could be in the thousands or tens of thousands of dollars. A blind list is a document of critical importance. An example of a blind list to be used for blinding a distillation tower for repair work, such as tray replacement, is shown in Table 16-3.

A special blind to be aware of is a **running blind**. This is a blind with the function of blocking certain lines, vents, or drains at all times when the unit is running. They are called running blinds because they blind equipment while the unit is running. They are removed during shutdowns and turnarounds for purging and draining purposes.

Table 16-3 Distillation Tower Blind List

DEBUTANIZER TOWER (T-203) BLIND LIST					
No.	Service & Location	Date in	Initial	Date out	Initial
1	Tower bottom to tankage				
2	Tower bottom to reboiler				
3	Reboiler to tower				
4	Overhead to reflux drum				
5	Reflux to tower				
6	Liquid feed to tower				
7	Level gauge connection, LG-201 (top)				
8	Level gauge connection, LG-201 (bottom)				
9	DP instrument connection, DPI-201 (top)				

Note: The above example is a partial blind list.

Opening Equipment and Testing for Entry

After equipment and vessels are blinded, they must be vented and fully opened and then tested to ensure they are free of hazardous materials. Testing requires that vessels be tested for explosive mixtures, toxic gases, and oxygen levels. Blowers and/or air movers should be setup on equipment to be entered. Blowers are huge fans that force air into a vessel—such as a reactor or wash drum—to ventilate the vessel and replace the inert purge gas with air. Air movers are pieces of equipment that operate on a venturi design and create drafts in vessels. A distillation tower may have several blowers at varying heights on the tower to ensure a constant flow of air moves through the tower. Another approach to insuring a safe atmosphere is to use the push-pull effect created by using both ventilation and exhaust fans. Nomograph charts based on the volume of the space and the capacity of the blower can readily give the minimum purge time. It is important to assure that the source ventilation air is safe to use. In one situation, the air on the suction side of the blower became contaminated with the exhaust of the gasoline motor powering the blower. Fortunately, by properly using both atmospheric testing and forced air ventilation, the problem was discovered before anyone was exposed to high levels of carbon monoxide.

Although nitrogen is a very common purge gas because it is non-toxic, it does cause frequent fatalities. It is a very dangerous gas because it is odorless and colorless and widely used in plants. If a vessel is purged with nitrogen and not well ventilated, workers who enter it will die from a lack of oxygen. Their senses cannot warn them of the high volume of nitrogen that has reduced the normal level of oxygen in a vessel. They will become drowsy or tired and sit down to rest for a moment, and then will die of **hypoxia**, a lack of oxygen.

Turnaround work can be started once all systems have been properly blinded, ventilated. and gas tested. Constant monitoring of all confined spaces permitted for entry must be maintained at all times.

SUMMARY

A process unit makes money only when it is running. When it is down, workers are still being paid wages and benefits but no profit is being generated. Thus, unit shutdowns are very infrequent and are scheduled as planned shutdowns. If a unit comes down before a planned turnaround, it is normally because of an emergency situation, such as an explosion, fire, or a hurricane is nearing landfall.

In general, the highest number of accidents and injuries occur during a unit shutdown or startup. This is because of the (1) time pressure created by these situations, (2) the large amount of work being done, (3) the increased potential for chemical exposure and accidents, and (4) the large number of contractors on the unit.

Manpower, material, and resources must be estimated and made available when needed so that schedules are adhered to. *Critical paths* in the planned work will have to be determined. Critical paths are pre-determined and closely monitored so that one job does not delay work on several other jobs.

To *clear up* equipment means to remove traces of flammable, toxic, or corrosive materials from the equipment. Though the equipment is drained, it may still contain residues of hazardous materials. Some process vapors and residual liquids or solids may be trapped in hard to reach locations. These materials must be removed before piping and vessels are opened up for maintenance or inspection.

Vessels and equipment must be blinded, opened, and inspected before work can be begun or workers allowed inside vessels. Workers doing the gas testing of vessels should wear a breathing apparatus prior to opening any vessel that may contain hydrocarbons or hydrogen sulfide gas.

REVIEW QUESTIONS

1. List five reasons for a unit shutdown.

2. Explain why flaring is a shutdown expense.

3. List five expenses of a unit shutdown.

4. State when the greatest number of injuries and accidents occur on a process unit.

5. List five hazards associated with a unit shutdown.

6. Explain why training is important to a successful unit shutdown.

7. Discuss the planning and preparation required for a unit shutdown.

8. Explain why there is a greater potential for environmental releases during a shutdown.

9. Discuss how a process unit ensures its customers continue to receive product while the unit is shut down.

10. Define *purge* and *clear up*.

11. List three methods of purging equipment.

12. Steam purging accomplishes what four objectives?

13. The high temperatures and pressures associated with steam can damage _____ and _____.

14. Explain the purpose of blinding.

15. Explain the function of each end of a figure eight blind.

16. What is the purpose of a blind list?

17. Explain the purpose of gas testing a purged vessel.

18. A system is considered to be safe for opening when gas testing reveals that hydrocarbons are less than _____ of the vented purge stream.

19. Why is nitrogen considered a killer gas?

CHAPTER 17

Process Unit Turnaround

Learning Objectives

After completing this chapter, you should be able to

- *Discuss the purpose of unit turnarounds.*

- *List four factors that affect the time between unit turnarounds.*

- *List the four common goals of a turnaround.*

- *Explain why each of the turnaround goals is important.*

- *Discuss several factors that would justify a turnaround.*

- *Explain the importance of a turnaround work list.*

- *Explain the importance of creating milestones.*

- *List five turnaround planning activities.*

- *Describe the job of the inspection group.*

- *Explain five factors that contribute to accidents during a turnaround.*

- *Discuss why the quality of repair is important to a turnaround.*

- *List several pre-startup activities.*

- *Describe how a turnaround is evaluated.*

INTRODUCTION

Unit shutdown, turnaround, and startup are not distinct events, meaning one does not start until another ends. Part of a unit may still be up while turnaround work has begun on a section of the unit that has been brought down. Midway through the turnaround the logistics of coming up is being planned and prepared. Sections of a unit that have completed turnaround work may start up and re-circulate until the rest of the unit is ready to come up.

Process unit turnarounds are major undertakings that have a significant impact on a plant's annual maintenance budget and future operating and maintenance performance metrics. A turnaround is the scheduled shutdown of a processing unit for major repairs and/or upgrades. The name comes from the concept of coming down, making fast and furious repairs, then turning right around and coming back up. A successful turnaround is one where safety, environmental compliance, cost, and duration are within expectations and benchmark performance is achieved. To be successful, turnarounds require careful planning and scheduling. Skilled and experienced personnel utilizing proven practices typically perform these duties. Since the interval between turnarounds is usually long, lessons learned from past turnarounds are sometimes lost because of personnel transfers and poor documentation.

The cost of a turnaround is the single biggest maintenance expense a process plant can expect to encounter. The cost of a single turnaround can vary from less than one million dollars to tens of millions, depending on the work scope. On an annual basis, a turnaround can take up more than half of the maintenance budget for a site. Turnarounds carry serious safety concerns because of the large number of contractors working in the plant, many of whom may be minimally trained in safety procedures.

PROCESS UNIT TURNAROUNDS

The modern chemical complex cannot run indefinitely without taking some time out to make repairs and improvements. Piping and vessels erode and corrode. Rotating equipment wears out. Technology changes. The purpose of a turnaround (TAR) is to open, inspect, and repair all equipment in a process unit that cannot be worked on while the unit is running. You can normally spot a process unit undergoing a TAR by the large number of cranes in the area (see Figure 17-1). Typically, major unit TARs occur in cyclical patterns, taking place approximately every 3 to 6 years. The time between TARs is determined by

- Equipment reliability
- Mechanical integrity
- Economics
- Special projects/technology upgrades

Process units use TARs for (1) improving or retrofitting their equipment to achieve optimal performance, (2) inspecting vessels and piping for erosion or corrosion that could lead to failure, and (3) checking the operation of safety equipment such as pressure relief devices and safety interlocks. Think of all the equipment—piping, valves, heat exchangers, pumps, drums, distillation towers, tanks, compressors, turbines, etc.—that a large unit may have that have been in service for three or more years. A TAR is the opportunity to fix a lot of things right away. TARs also provide the opportunity to install and implement the latest

Figure 17-1 Cranes Signify Turnaround Activity

technologies for reaching or exceeding government mandates (i.e., achieving emission targets).

Fall and spring are the traditional TAR seasons because this avoids shutting down production during the height of the summer gasoline season. TARs are expensive since no product is made while the unit is down, and the work being done during this time requires large investments in manpower, materials, and equipment. Conflicting forces lead to tension and frustration. The sales, marketing, and operation groups are interested in getting the unit started up as soon as possible so they can make money. The maintenance group is interested in keeping the unit shut down long enough to do the planned work properly. These two opposing interests require that TARs be carefully planned to get the work done at minimum down time. A well-planned TAR may take 20 to 30 days to complete. During that time hundreds (perhaps thousands) of planned jobs will be completed using a workforce made up of the operations, maintenance, and technical groups, plus a large number of contract craft groups (welders, pipe fitters, electricians, insulators, etc.).

The complexity and expense of a TAR can be illustrated by the facts and figures of a debottlenecking TAR of a 2.1-billion-pounds-per-year ethylene plant. The capital work included compressor overhauls, extensive piping modifications, several complete tower retrays, installation of numerous pieces of equipment, electrical work, and instrumentation work. Plus, a significant amount of accumulated maintenance work had to be completed during the downtime. Over 1,200 work items were included in the TAR. A five-week schedule was prepared to accomplish the work. The entire scope required 1.3 million TAR work hours to be executed in that five-week window by 3,200 craftsmen. Imagine the capital expense for new equipment and piping, and contractual expense for several thousand people for 1.3 million work hours (many on overtime), all the while the unit is not making one penny!

Also, there is a tremendous safety initiative involved with 3,200 craftsmen working on a process unit. Imagine the safety concerns of management and safety personnel! It is a nightmare scenario, 3,200 people on one unit working feverishly to accomplish 1.3 million man-hours of work in five weeks! The unit process technicians will have to keep a careful eye on the craftsmen as they perform their jobs because part of the final "grade" assigned to a completed TAR comes from safety. Too many accidents and injured people can fail a TAR even though it was finished on time and within budget.

Process technicians will spend most of their time in the field during a TAR doing the checkout (punch list-audit) of the repair and maintenance jobs, overseeing contractors, and standing fire watch. This is a critical responsibility because once their unit is up and running they would like their normal job in their area of responsibility to be as hassle-free as possible. For that to happen, they have a personal interest in the thoroughness and quality of the work done during the TAR. Failure to be diligent with their punch list may mean they will experience some frustrating and miserable days after the unit is up and running, solving problems that resulted from poor maintenance work.

GOALS AND OBJECTIVES

Every TAR may have special goals and objectives, which may differ from time to time, depending on a complex set of circumstances. Goals and objectives are usually determined by the best compromise of time, cost, quality, and safety. The most critical objectives are to

- Complete the TAR in the shortest possible time.
- Complete the TAR at the lowest possible cost.
- Achieve the best quality of workmanship.
- Execute the TAR with the highest possible safety record.

It is impossible to achieve all four of these objectives concurrently because they are conflicting (self-excluding). To complete the TAR in the shortest time, overtime costs may soar and quality and safety may be impacted. To complete the TAR at the lowest cost, the duration and quality may be affected. To achieve the best quality, duration and cost may be increased. To achieve the highest safety, cost may be impacted. Time and cost are always critical and require special attention. These two important objectives can be controlled through detailed planning and scheduling.

TURNAROUND JUSTIFICATION

Since TARs can cost millions of dollars, the decision to shut a large unit down for a planned TAR is always weighed carefully. The various reasons for scheduling a TAR are

- Old or worn out equipment is causing excessive downtime for unscheduled repairs.
- Old or worn out equipment is causing serious problems with maintaining product quality.
- New equipment designed to improve operations or replace obsolete equipment is scheduled for installation.

If any of the three preceding reasons, or any combination of them, is reason for a TAR, during the TAR management will also schedule (1) piping and equipment inspections needed to comply with mechanical integrity programs and (2) the inspection of safety interlocks and pressure relief devices according to the Occupational Safety and Health Administration's (OSHA's) mandated safety programs.

Many other items needed to maintain the profitable operation of the unit could be added to the list. The unit manager, with help from the unit's operating team, must weigh the value of these benefits against the cost of the TAR and decide when to schedule the TAR. The manager must also establish goals to be accomplished during the TAR. These goals should list the specific improvements that are expected as a result of the TAR work. They might include such things as

- Increasing production capacity
- Improving product quality
- Creating greater energy efficiency
- Reducing unit downtime due to equipment problems
- Reducing operating or utility costs

Once justification and goals for the TAR have been established, a comprehensive TAR plan is formulated.

Turnaround Planning Process

A lot of man-hours and personnel are involved in TAR planning and several documents may be created. Some of these documents are the milestone plan, TAR worklist, and activities timeline.

The milestone plan is a time chart that identifies all key TAR planning activities needed prior to execution. It is a high level timeline and is generally called "planning the plan." The milestone plan is the first document prepared since it establishes the timeline for strategic TAR activities and their interrelationship. Many jobs can be started at the same time, but some have to wait until a particular job is completed first. As an interrelationship example, if job A doesn't get done first, it delays the start of jobs B and C. The milestone plan can take several forms but some of the most common are tables and Gantt charts. The key is that the plan outlines the major tasks involved in planning against a timeline. The milestone plan provides an overview of strategic activities in the TAR and communicates the planning progress. Typically, 14 different areas are represented on a milestone plan. Some of the areas are work scope, contracting, inspection, cost control, scheduling, health/safety/environment, etc. Table 17-1 is not a milestone plan, but is very similar.

The time range used for the milestone plan and starting time generally depend on TAR man-hours (size of the TAR). Table 17-2 gives some reasonable time ranges in months based on estimated man-hours.

Turnaround Work List

The TAR work list is the key to a successful TAR since it is the basic foundation upon which budgeting, operations, and maintenance plans must be built. Items for the TAR list come from many sources. The work that did not get completed during the last TAR provides a starting work list for the next TAR. Also after the last TAR during unit startup, additional items that need correction or repair will be discovered and added to the list. As time goes by, all groups continue to add to the list. The operations department will always have a list of work that needs to be accomplished to maintain efficient operations. The processing unit's operations and maintenance engineers are aware of equipment and process conditions that will need attention during the TAR or modifications to meet new safety or environmental

Table 17-1   Key Turnaround/Project Activities Timeline

Pre-Turnaround Estimated Timeframe				Post T/A
>20 months	>20–14 months	>14–8 months	>8–5 months	+1–3 months
Turnaround Activities				
	• Appoint T/A manager • T/A objectives • Milestone plans • Contracting strategy	• Determine work scope • Initial cost estimate • Determine critical paths • Milestone plans • Contractor plans	• Freeze work scope • Integrated schedule • Critical path plans • Subcritical path plans • Mobilization plans	• Critique • Update files • Update P&IDs • Update procedures
Project Activities				
• Appoint team • Objectives • Estimate budget • Contracting strategy	• Design basis • Cost estimate • P&IDs • Procedures	• Appropriation • Engineering details • Procurement • Construction	• S/D plans • S/U plans • Turnover plans • Testing • Commissioning	• Commission • Evaluate

Table 17-2   Turnaround Time Ranges

Turnaround Man-Hours	Milestone Planning Time (months)
>300,000	−24 to −18
100,000 to 300,000	−18 to −14
>100,000	−14 to −8

Source: R. J. Motylenski, "Proven Turnaround Practices," *Hydrocarbon Processing* 82 (April 2003): 37–42, Table 1.

regulations. Engineering and technical groups will be working on projects authorized to be installed during the next TAR. The inspection groups will be concerned with inspections required for continued safe operations (safety valves, interlock systems, etc.). Some of these may be mandated by company safety policies or federal and state regulations. All of these items will be added to the list. Once a complete list of suggested TAR items has been compiled, it is critically reviewed to eliminate items that should not be on the TAR list. Such items expand the TAR work list to such an extent that it interferes with performing necessary TAR work while adding unnecessary time to the length of the TAR. Such items are not ignored but are scheduled as regular maintenance items.

The preliminary TAR work list is arranged in a logical manner for ease of planning. For example, similar repair items may be grouped by equipment (heat exchangers, pumps) or by type of craft (pipefitting, insulation). An addendum to the normal preliminary work list may be used to separate items requiring special attention, such as special projects for the engineering and/or maintenance divisions. The preliminary TAR work list is typically generated as a joint effort between the unit manager and the entire operating team.

Once prepared and issued, the work list is used as a guide to expand the TAR planning. This includes the following four steps:

1. Preliminary TAR costs will be estimated using the work list and past TAR costs as a guide.
2. The TAR budget will be established using these estimates.
3. A TAR schedule will be created and will include an estimate of the number of days that the unit will be shut down and establish the dates for the TAR.
4. Materials that require long delivery time will be ordered early to ensure that they are available when the TAR begins.

A good preliminary TAR work list should be completed at least a year before the shutdown is planned. This work list will be put into final form at least three to six months before the work starts. At this time, the preliminary TAR work list is reviewed, updated, and issued as the final list. Changes after the final update will require special approval from the unit manager.

TURNAROUND PLANNING

A unit TAR, regardless of scope, does not just happen; it must be carefully planned. A successful TAR must meet the following requirements:

- Safety
- Schedule

- Budget
- Completed work must accomplish the expected goals

Planning to achieve these requirements begins as soon as the last TAR has been completed. The work that did not get completed during the TAR provides a starting work list for the next TAR. This work list becomes the key to planning a successful TAR. All work that needs to be done while the unit is shut down should be on the list. Likewise, any work that can be done while the unit is running should not be on the list.

Planning continues, starting out low-key, but increasing in intensity as the next TAR date approaches. Final planning should be completed two to three months before unit shutdown. Schedules for all the planned work should be complete. These schedules should have the work organized by crafts and work areas. They should include information on manpower requirements and the optimum sequence of work activities. The schedules provide an estimated time required to do the work and give realistic starting and finishing dates for each job. The goal of the schedulers is to get all the work completed in the allotted time while keeping costs within budget. Detailed engineering design for any special projects should be completed during these last few months before the actual shutdown. Prefabricated piping work should be underway and, as the date for the TAR approaches, tool trailers for the various crafts should be stocked and ready to move into place.

Operations personnel are responsible for developing

- Detailed work lists
- Work order initiation
- Checklists for lockout/tagout
- Blind lists
- A utilities plan to coordinate outages, such as nitrogen, plant air, etc.
- Lockout pre-planning (which valves, switchgear, blinds, etc.)

Final Details

Activities during the final three months before TAR are numerous and sometimes chaotic. Final planning details are the responsibility of the operations, maintenance, and mechanical inspection group and they will be holding meetings to ensure that everything is ready. The TAR work list should be in final form by this time. All design work should be complete and work orders for these activities should be issued as soon as the unit is down and cleared for work. Table 17-3 provides a partial list of some of the activities that need to be completed during this time frame. A detailed list might contain as many as 50 or more items.

During the last few weeks before shutdown, much of the emphasis shifts from planning to the actual TAR work. This involves coordinated work between operation and maintenance groups. Where permissible, scaffolding needs to be erected. Blinds and gaskets need to be spotted in the field at those locations where they will be used. Pipe that will be removed during the TAR needs to be marked in some manner for easy identification. A list should be made that identifies the vessels or equipment that must have insulation removed before work on them can begin. Defective or inadequate insulation should also be identified and

Table 17-3 Final Turnaround Planning Activities

Finalize TAR work list.
Review critical path schedules.
Ensure short delivery materials are available and in adequate quantity.
Verify long delivery items have been ordered.
Review environmental compliance procedures.
Set TAR housekeeping policies.
Schedule and conduct pre-TAR meetings.
Review milestone plan.
Review policy for working with and overseeing contractors.

listed. De-insulation work should begin as soon as possible. Equipment that can be shut down early should be shut down, cleared, and locked out.

Staging is an important logistical function. **Staging** is the critical placement of temporary buildings, power, air, tool cribs, field parts and materials warehouses, and runoff containment barriers. How and where these items will be located will have an impact upon the efficiency of the TAR execution. Maintenance materials and equipment should be accounted for in tool trailers (cribs) at appropriate places throughout the area to minimize worker travel distance (and time wasted). Leased dumpsters to hold all the trash and debris should be spotted at various sites around the unit to facilitate housekeeping.

ACTIVITIES DURING TURNAROUND

Assume that all the various process areas of the unit are shut down, purged, isolated with blinds, and locked or tagged. As each area of the unit is ready, it is turned over to the maintenance group and the TAR work begins. The planned work can be divided into several categories:

- Testing and inspection
- Preventative equipment maintenance
- Modifying existing equipment
- Installing new equipment
- Completing major project work

Testing and Inspection

A TAR provides a good opportunity to function check safety and process interlocks to verify that they work properly. Function checks can be actual tests if the variable can be run to interlock conditions during shutdown activities. If not, then a simulated load can be applied to the interlock system while the unit is down and the response to that load can be checked.

This is also the time for testing and inspecting pressure relief and safety valves by contract companies that specialize in these tasks. The actual pressures where relief occurs will be determined and compared with design set points. The valves will also be inspected for corrosion or erosion that may prevent proper operation and for foreign materials such as rust or caked process materials that could plug inlet or discharge ports, preventing proper operation. When all of this is completed, the valve will be rebuilt and placed back in service on the unit.

The plant mechanical inspection group is responsible for ascertaining the integrity of the unit's piping and equipment. They will inspect tanks, piping, and other equipment for evidence of corrosion, erosion, mechanical fatigue, or other signs of wear which could lead to system failure. The electrical group will be looking for evidence of insulation breakdown in large power cables and inspecting electric motors for signs of needed repair.

Preventative Equipment Maintenance

TARs are also the time to do those preventative maintenance jobs that could not be done while the unit was running. The commutator brushes in large electric motors need to be changed out. Bearings, packing, rings, and other parts that wear with time need to be inspected and replaced on rotating and reciprocating equipment. Unit storage tanks, after being completely emptied and cleared, will be inspected. Tank insulation, lining, and floating roof seals may be replaced. Equipment to be repaired or cleaned are

- Heat exchangers that are fouled or require retubing
- Cooling towers with fouling problems
- Coked furnace tubes
- Polymer separators coated with gel
- Sumps filled with residue
- Corroded trays in distillation towers
- Centrifugal pumps with worn impellers

Contractors may be brought in for the majority of the cleaning and repair work because many plants have cut back on their maintenance personnel and rely more and more on contractors. Plant maintenance personnel will be primarily engaged in items of critical concern or assisting in overseeing contractors.

Modification of Existing Equipment

Sometimes installed process equipment doesn't work as envisioned in the original design. There are also times when original equipment is no longer adequate for the job. In either case a major modification of equipment may be required to fix the problem. Examples include rotating equipment with undersized motors and distillation columns with nozzles too small for the required flow. Whatever the problem, the unit must be shut down to make the modification, and modifications are usually made during the scheduled TAR.

Installing New Equipment

Frequently new equipment is waiting on site for a unit TAR so it can be installed. Sometimes, much of the preliminary work for the equipment can be done while the unit is still running so all that remains for TAR work is the final tie-ins of the equipment to the unit. Other times, the work that needs to be done during the TAR may be extensive. Special planning may be required if a large process vessel needs to be brought in over the highway or additional handling equipment may be needed for a heavy equipment lift.

SAFETY CONCERNS DURING THE TURNAROUND

More accidents occur during a TAR than at any other time. Factors that contribute to this rash of accidents are

- Piping and equipment are opened, providing opportunities for flammable mixtures to form or for toxic materials to leak out.
- A large number of contractors are involved in the TAR work. Many have a minimum knowledge about the unit, its processes, or its chemicals.
- Maintaining the TAR schedule creates a *hurry up and get it done* work attitude that leads to accidents.
- Working conditions are crowded and confusing as the different crafts work in the same area at the same time.
- Overtime is frequently needed to remain on schedule, resulting in a higher level of fatigue among all workers.

Safety must never be compromised during a TAR. It must be a prime consideration as TAR jobs are planned and executed. Safety guidelines play an important role in the planning and scheduling of a TAR. The two basic safety concerns that must be addressed are

1. Safety of all personnel
2. Safety of plant equipment

Technicians will spend the majority of their time involved in permitting (confined space, hot work, lockout/tagout) and directing and monitoring contract personnel and in-house maintenance personnel on their unit.

Ensuring a safe TAR will require additional manpower. Technicians from other units still running may assist in some TAR activities, such as fire watch and monitoring contractors. Technicians on the unit undergoing the TAR will be working seven days a week, twelve hours a day (7/12s) either with no day off until the TAR is completed and the unit back up and running, or one day off every two weeks. A lot of people will be paid overtime and sixth and seventh day pay (special pay for working a long string of days with no time off), though not all companies have this pay arrangement. Despite the fact that the majority of unit technicians are at first excited about all the overtime pay, after the first few weeks they are often just tired and want the TAR to end. Definitely, they do not want it to extend beyond the planned time limit. At this time they have realized that money isn't everything. Some safety items that require the expenditure of additional manpower are listed in Table 17-4.

Vessel Entry Procedures (Confined Space)

Confined spaces such as towers, columns, tanks, drums, and sumps must be made safe for inspection and repair personnel to enter. The vessel must be purged and then isolated by installing blinds at all vessel connections. Blinds prevent the accidental introduction of hazardous materials. Operations must ensure that all blinds are installed before a vessel is opened. Once opened, air movers must be installed to ventilate the vessel, then technicians must gas test the vessel interior for a breathable (oxygen) atmosphere free of flammable or toxic material. Next, unit personnel will issue and post a valid entry permit before entrance of the vessel is allowed. On very large vessels, the technician must wear a self-contained

Table 17-4   Safety Tasks Requiring Manpower

Scaffolding
Obtaining permits
Tagging equipment for entry to work
Hot work in process areas
Blinding
Fire watch/hole watch
Lockout/tagout
Heavy equipment lifts
Housekeeping
Testing (hydrostatic, X-ray, etc.)
Installing air movers
Monitoring contract workers
General cleanup

breathing apparatus or be supplied an air respirator before entering to test all locations (top, middle, bottom, all corners, etc.) in the vessel.

Blinding Procedures

Shutdown blinds must be installed after unit shutdown in all process lines entering or leaving the unit at the battery limits (except for utility lines such as steam, water, air, etc.) as soon as possible. They prevent the backing of hazardous materials into the unit during the TAR. All blinds must be tagged out. Blind handles can be painted a bright color to permit easier location. This will also help ensure that all blinds are recorded on the blind list. All blinds installed or removed must be logged in the unit blind list. This includes shutdown blinds and all equipment blinds installed during the TAR. Usually, an individual technician on each shift is assigned the responsibility of keeping this list up-to-date. During unit shutdown for a TAR, running blinds are also removed. Running blinds are in place while the unit is running and their function is to block certain lines, vents, or drains at all times the unit is running. They are reinstalled prior to startup.

Hot Work Procedures

A lot of hot work is involved in most TAR jobs. Anything that can generate sufficient heat for combustion must be considered hot work. This includes mobile equipment, chipping, sanding, electric tools, and welding or open flames. Precautions must be taken to prevent ignition of flammable materials by hot work. It is always assumed that equipment will still contain some process materials even though the unit was thoroughly purged during shutdown. Inevitably, pockets of such materials are encountered throughout the TAR, sometimes in the most unexpected places. Sewers and underground piping, for example, are not process vessels but can be a source of process materials that can ignite and cause an accident.

Authorized hot work areas, as well as the equipment to be worked upon, must be gas tested and current hot work permits issued before hot work is allowed. This is the responsibility of process technicians. The TAR maintenance group, including contractors, is responsible for notifying operations in advance of when and where hot work permits will be required. Both groups must police hot work areas for conditions that may require hot work be stopped.

Lockout Procedures

The supervisor or lead technician is responsible for proper lockout and cannot delegate this duty to a subordinate. Craftsman working on locked out equipment must also properly document and sign all tags according to the established site lockout procedures.

Heavy Equipment Lifts

Heavy equipment lifts that involve massive loads require prior review and written approval from the appropriate plant authorities. The unit supervisor must be informed when a heavy equipment lift is scheduled on the unit. He must investigate for conditions which may interfere with the lift procedure or which may cause it to become unsafe (underground voids, high-tension electrical conduits, etc.).

Housekeeping

Good **housekeeping** (i.e., keeping a work area and equipment in a safe, clean, and usable condition) is critical for accident prevention during TARs (see Figure 17-2). During a TAR, good housekeeping is difficult to maintain because contractors do not have the training and mindset of process technicians. They will leave tools, hoses and equipment strewn everywhere and these become trip hazards. The process technician will have to police contractors carefully. Some things to closely monitor are

- Fire lanes should be kept open at all times.
- All refuse is to immediately be placed in waste containers.
- All spills, no matter how small, must be cleaned up immediately.
- Material should not be allowed to accumulate and interfere with unit traffic.
- Sewer drains clogged or choked increase the danger of fires and should be cleared.
- Safety equipment must be kept unobstructed.
- Equipment and material necessary for the TAR work should be maintained in an organized, orderly manner.

Figure 17-2   Housekeeping Sign on Process Unit

Scrap material such as pipe, insulation, wiring, and metals should be promptly removed from the unit and disposed of in the leased dumpsters. Excessive clutter is not only unsafe but also slows down work and results in lost time and increased cost.

PIPING AND EQUIPMENT INSPECTION

During the TAR, the inspection teams will be busy checking the many pieces of piping and equipment that could not be inspected while the unit was running. They will also inspect newly fabricated piping and newly installed equipment. Their work will include

- Inspection of process piping and equipment for signs of corrosion, erosion, or chemical attack
- Inspection of tanks for damaged linings or corroded areas
- Inspection of pressure vessels for corrosion that could affect pressure rating
- Hydrostatic testing of piping, equipment, and completed systems
- Inspection of completed jobs to ensure that everything has been assembled properly

Inspectors may be plant personnel or contractors. Records will be made of X-ray tests, corrosion allowances, metallurgy tests, dye tests, etc. Inspectors are also responsible for hydrostatic inspection of all new welds.

The process technician does visual inspections to ensure equipment cleanliness and job completion. Because all technicians involved in testing, inspection, and approval of maintenance work must operate the process equipment after the TAR, they should be very conscientious about ensuring that all jobs are completed satisfactorily. Every job should be closely inspected for loose flanges, proper gaskets, bolting, blinds, and unplugged vents and drains.

As inspection proceeds, the team will generate a ***punch list*** (work list) of incomplete maintenance items and forward it to the TAR superintendent, allowing sufficient lead-time for completion of these items during the TAR. Punch list items are rechecked before startup to verify they were completed. Operations personnel should also witness any pressure test required before the job is completed. A close visual inspection by operations personnel at this point can prevent many headaches during startup.

QUALITY OF REPAIR WORK

The quality of the repair work has an impact on time and cost. Quality guidelines should be prepared and issued to establish the minimum acceptable level desired. Operating personnel should have access to these minimum levels. Poor quality can result in accidents, rework, and equipment failure. Quality can be specified in the TAR planning by the amount of testing required. Some tests required might be X-rays, pressure tests, and ultrasonic/infrared inspection. Four basic concerns governing the acceptable level of quality are

1. Safety (preventing equipment failure)
2. Production (desired productive run lifespan for the unit)
3. Cost (lowest cost for the useful lifespan of repairs)
4. Schedule (extra time needed for welding and inspection)

PRE-STARTUP ACTIVITIES

As the TAR work nears completion, steps must be taken to ensure that all scheduled work has been completed satisfactorily and that the unit will be ready for startup. Consideration must also be given to preparing the technicians for the startup of the unit and for the operation of newly installed equipment. Startup meetings and training are usually held about once a week prior to the scheduled startup time to familiarize technicians with the unit startup procedures and to implement any technician training necessary for running new or modified equipment installed during the TAR. Post-TAR unit performance expectations are also discussed to make technicians aware of what operating improvements are expected.

Unit process equipment must be given a final inspection for cleanliness and work completion. This is usually a joint effort of the inspection team and the unit process technicians. The team will generate a punch list of incomplete maintenance items and forward it to the TAR superintendent. Punch list items are rechecked before startup to verify that they were completed.

On most TARs, work on individual unit systems that make up the processing unit is rarely all completed at the same time. As system equipment becomes available, the equipment is purged, pressure-tested, and then maintained at purge system pressure until needed for the unit startup. In some cases individual sections of the unit may be started up before mechanical work is completed on other sections. This may be necessary, for instance, if TAR work on one section extends far beyond the time required to complete repairs to the major portion of the unit.

EVALUATION OF THE TURNAROUND

Once the TAR has been completed, a post-TAR work list should be created showing needed follow-up items (rental equipment returned, items returned to warehouse, etc.). An evaluation should be made to determine if the TAR successfully reached expected goals. This evaluation should consider the goals of *costs*, *timing*, *safety*, and *unit performance* after the TAR.

First, was the TAR cost on budget or below budget? If the answer is yes, congratulations are in order for implementing effective TAR cost controls. If not, the TAR budgeting and control plan should be thoroughly reviewed to improve future TAR cost planning. Was the budget missed due to inaccuracies in the original cost estimate, poor cost control during the TAR, or unexpected and unbudgeted major expense repairs encountered during TAR? Answers to these kinds of questions will help budget for the next TAR.

Second, did the actual TAR length meet or exceed the scheduled time objective? Did work items have to be deleted to meet the TAR deadline? Was the schedule missed because of poor time estimates, poor manpower management, or excessive unplanned work? Again, the contributing cause must be determined for consideration in future TAR planning.

Third, no TAR can be considered a complete success if people are injured during the TAR. Even though unit performance, budget, and timing goals were met, accidents and injuries will reduce the TAR success score. TAR safety records and accidents or near miss situations that occurred during TAR should be reviewed for causes and future prevention. Were they a result of not following procedures, unsafe working conditions, or an unsafe act? This information should be published widely throughout the company so other units can avoid similar situations.

Fourth, unit performance must be measured to determine if the improvements that were planned as part of shutdown justification were actually achieved. This will involve collecting actual operating data, and comparing these data with expected performance based on projected TAR improvements. Comparisons will be made based on production rates, product quality, operating costs, etc. If the TAR was successful, the projected improvements should be seen in these parameters. If they were not achieved, an in-depth study of the reasons for the shortcomings should be initiated.

Finally, critiques should be held for the various aspects of the TAR. Critiques may be in the following categories:

- Schedule—a review of the events that impacted the schedule and ways to improve
- Major machinery—maintenance engineers and supervisors to review machinery overhauls
- Materials—a review of the availability of parts and supplies and problems with availability and delivery
- Process unit—a meeting of maintenance and operations personnel meet to discuss common problems encountered during the TAR
- Safety—a review of any incidents and significant safety problems
- Contractor—a review of contractor performance

SUMMARY

Plant and process unit turnarounds are major undertakings that have significant impact on the plant's annual maintenance budget and future operating and maintenance performance metrics. A *turnaround (TAR)* is the scheduled shutdown of a processing unit for major repairs and/or upgrades. A successful turnaround is one where safety, environmental compliance, cost, and duration are within expectations and benchmark performance is achieved. To be successful, turnarounds require careful planning and scheduling.

Every turnaround may have special goals and objectives that are usually determined by the best compromise of time, cost, quality, and safety. The most critical objectives are to complete the TAR in the shortest possible time, complete the TAR at the lowest possible cost, achieve the best quality of workmanship, and execute the TAR with the highest possible safety record.

A unit TAR, regardless of scope, does not just happen; it must be carefully planned. A successful TAR must meet the requirements of safety, schedule, and budget, and the completed work must accomplish the expected goals. Once the unit is down, several critical activities begin such as testing and inspection, clearing equipment, repairing equipment and piping, modifying existing equipment, and the installing new equipment.

Safety must never be compromised during a TAR. It must be a prime consideration as TAR jobs are planned and executed. Safety guidelines play an important role in the planning and scheduling of a TAR. The two basic safety concerns that must be addressed are the safety of all personnel and the safety of plant equipment.

Once the TAR has been completed, a post-TAR work list should be created showing needed follow-up items (rental equipment returned, items returned to warehouse, etc.). An evaluation should be made to determine if the TAR successfully reached expected goals. This evaluation should consider the goals of *costs*, *timing*, *safety*, and *unit performance* after the TAR.

REVIEW QUESTIONS

1. Discuss the purpose of unit TARs.

2. Unit TARs occur about every _____ to _____ years.

3. Why are fall and spring chosen as TAR seasons?

4. List four factors that affect the time between unit TARs.

5. Discuss the job of a process technician while the unit is undergoing a TAR.

6. List the four common goals of a TAR.

7. Explain why each of the TAR goals is important.

8. Discuss several factors that would justify a TAR.

9. Explain the purpose of a milestone plan.

10. Explain the importance of a TAR work list.

11. List six TAR planning activities.

12. Explain the importance of *staging*.

13. Describe the job of the inspection group.

14. Describe some of the repairs that will be made on process equipment.

15. Two basic safety concerns that must be addressed during a TAR are the safety of _____ and _____.

16. Describe some activities that would be considered hot work.

17. Explain five factors that contribute to accidents during a TAR.

18. Describe what a process technician will be looking for when they are ensuring housekeeping is not neglected.

19. Discuss why the quality of repair is important to a TAR.

20. Discuss how a TAR is evaluated.

CHAPTER 18

Process Unit Startup

Learning Objectives

After completing this chapter, you should be able to

- *List five activities of unit personnel when preparing the unit for startup.*

- *Describe two activities of unit personnel when actually starting the unit up.*

- *Describe several duties technicians must perform on their equipment before unit startup.*

- *List the plant auxiliary systems.*

- *List the plant utility systems.*

- *Explain why it is important that the flare system be operational before process materials are brought into the unit.*

- *List the various water systems on a process unit.*

- *Explain the three problems steam presents after purging equipment.*

- *Describe the inspection duties of technicians regarding piping and equipment support.*

- *Explain the advantages of* running in *equipment.*

- *Explain the importance of a post-startup review.*

INTRODUCTION

Startups must be as carefully planned as shutdowns. Like shutdowns, startups are not events that occur frequently so preparation for a startup is similar to that for a shutdown. Planning and preparation for a startup should include technician refresher training on startup procedures and following the startup check sheet for each system. The re-introduction of unit feed, raw materials, and utilities will require that suppliers of these things be notified in advance when to expect startup activity. Startups present as much opportunity as shutdowns for chemical spills and accidents. Also, when feed is introduced to the unit there will be a lot of off-specification material before the unit lines out. This material will go into the unit slop tank for reprocessing later. Material that is badly contaminated, for whatever reason, and cannot be reworked will be routed to a plant slop tank where hydrocarbons are skimmed off for some use, either as feed to another unit or as fuel. Nothing will be wasted.

Too many fires, equipment damage, and injuries occur during startup, principally because unit procedures were not up-to-date, were not followed, or verification of turnaround repairs were not thorough. Emergency shutdown procedures should exist and be up-to-date for the failure of any utility and major equipment failure. In Chapter 4 we discussed the importance of management of change. It is during startups and shutdowns that the management of change of procedures is so critical to preventing accidents or costly unnecessary expenses. Assume a unit is delayed by two hours from coming up because of a procedure that had been modified but not updated in the procedure manual. Remember Murphy's Law: *If there's any way it can go wrong, it will go wrong.* Supposedly everyone was told about the change and knew about it—except the technician working the board. He was on vacation at the time of the change. So, what was the cost of the two-hour delay? $30,000? $50,000? How will this affect the startup review?

Additional operating and supervisory personnel are available during startup because of the potential for unexpected events. Extra manpower, to the extent of double coverage on all shifts, is not uncommon during the first few days of startup. Extra pairs of eyes and hands are always needed. Because the unit is coming up, most of the instruments will be in manual mode. (Note: There are software programs available that can literally bring a unit up without human help.) This means that many tasks usually completed by automated equipment must be done manually until the unit is lined out. Data readings may have to be taken more frequently, pumps started and stopped more often, technicians sent out to troubleshoot an anomaly, and instruments may have to be operated manually. Maintenance personnel will be scheduled on standby in case they are needed for the urgent repairs of failed equipment and leaks. Everyone will be under a lot of tension and pressure because they know it is important to get the unit back up and making money and the psychological pressure may lead to hasty actions that result in accidents. All the extra personnel, many on overtime, add to the high cost of a startup and the need to get the unit up and making money. See Table 18-1 for a list of advantages and disadvantages of coming up too slow or too fast.

Since a process unit may only be started up once every few years, the steps in a startup must be carefully planned so that nothing is overlooked. This is a requirement of OSHA's Process Safety Management regulation. Operators must familiarize themselves with the startup procedures to be used and follow the startup check sheet for each system. These may be reviewed on computers or printed out as hard copies. They should verify the procedures are accurate and up to date. That is their responsibility, not the unit engineer's or unit supervisor's. Startup meetings will be held to ensure everyone has the same information.

PRE-STARTUP EVALUATION OF NEWLY CONSTRUCTED OR MODIFIED PROCESS UNITS

After the construction phase of a newly constructed or a modified unit is complete, two questions must be answered when the unit is turned over to the operations personnel:

1. Is construction truly complete?
2. Can the unit be operated as designed and operated safely?

These questions need to be answered as completely as possible during the preparation and planning stages of the unit startup and then verified after the unit is lined out.

The design team and the construction groups of the modified or newly constructed process unit will have conducted operation and safety studies to ensure that nothing was overlooked in the unit design. The construction groups will use similar checks to ensure that everything was installed as designed. A pre-startup evaluation should be conducted for all process or equipment modifications and a process hazard analysis conducted for all major changes and all new construction.

A team of local experts, such as the unit operators and engineers, should do the pre-startup evaluation. The Process Safety Management (PSM) standard mandates pre-startup checks and that operators be included in this activity. The evaluation team should do a thorough inspection of the newly installed or constructed equipment. A checklist should be used as a guide for this inspection. The local experts should start at one end of the modified area and work their way to the other end, inspecting everything as they go. Two lists should be developed during the inspection. One list should include all problems identified during the inspection that need to be completed before the unit can safely start up. The other list should include those items that need to be fixed, but will not prevent the safe start up of the unit. Items on the first list need to be worked on and completed as part of the startup preparations. Items on the second list need to be scheduled for completion once the unit lines out.

Before the startup phase of new equipment can be considered complete, the unit will have to be at design specification. The engineering firm that installed new equipment or made major modifications to a system must perform certain tests while the equipment or system is operated under its supervision. The engineering firm will conduct a performance test that verifies the new equipment or system meets agreed upon specifications. The specified performance conditions are usually kept simple—capacity, yield, and critical utility consumption. The length of the performance run is pre-set, usually not less than two days.

PROCESS UNIT STARTUP

The actual steps involved in starting up a processing unit are similar for many processing units. A generic outline for a unit startup would be similar to the one that follows.

Phase I: Preparation
1. Inspect equipment for closure.
2. Verify good unit housekeeping.

Table 18-1 Startup—Advantages and Disadvantages

Coming Up Too Fast	
Advantages	**Disadvantages**
Save time	Rapid temperature and pressure changes might cause leaks
Save money	Overloading of equipment
	Difficulty of coordinating work
	Promotes operational errors
	Increases probability of accidents
Coming Up Too Slow	
Advantages	**Disadvantages**
More time to detect unsafe conditions	Extended off-stream time
Better coordination, less operational errors	Excess slop production
Less stress	Difficulty in bringing the entire unit up at the same time
	Unstable operation of the unit over an extended period of time

3. Verify all shutdown blinds are pulled and running blinds in place.
4. Verify flanges are tightly bolted.
5. Commission utilities (electricity, air, water, fuel gas, steam, and nitrogen).
6. Verify auxiliary systems available (cooling water, flare, heat transfer fluids, refrigeration).
7. Ensure systems have been inerted with nitrogen or steam for oxygen removal.
8. Ensure water has been eliminated from all hydrocarbon systems.
9. Verify all systems are air-free and pressure tested.

Phase II: Start Up the Unit

1. Introduce process materials into the unit.
2. Eliminate purge gas from vessels and piping by backfilling.
3. Bring process conditions to normal operating ranges.
4. Verify unit product is on specification.

A lot of critical communication outside of the process unit must take place before startup. Utilities and raw material suppliers should be notified to prepare for an increased demand of their products or services. Product storage tanks should be ready to receive product. The shipping department must be notified to arrange for pipelines, rail tank cars, barges, and tank trucks for shipping unit product. Engineers responsible for newly installed and designed equipment must be on hand to ensure their equipment is operated correctly and meets design specifications.

TURNOVER AND INSPECTION OF EQUIPMENT

Operations must communicate with maintenance to determine when the last few pieces of equipment to be repaired or refurbished will be ready for turnover to operations. Large electric motors must be *megged* before they are started up. Electric motors that have remained idle for the weeks of the turnaround can collect condensate from humid air. If they are operated at high voltage the condensate may act as an electrical conductor and short out the motor when it is started, causing added expense and repairs. Measuring the electrical resistance between the armature and field coils before attempting to start the motor can prevent this damage. This resistance is measured in Megohms, hence the term **megging.** If the resistance is less than a critical value then hot air or nitrogen should be circulated through the windings to dry out the motor. All large electric motors should be megged at least 24 hours prior to the unit startup. This allows time to dry out motors with a low meg reading.

Before technicians start up a unit that has been shut down for turnaround activities, they should verify that maintenance personnel have completed the work on the turnaround list. Technicians should inspect all repaired equipment and repair work performed to determine if the required work has been completed and completed correctly. If you are a technician, you should not take the contractor's or maintenance person's word that everything is good to go. It is your unit and your area of responsibility; *you inspect and you verify*. You will be the one signing off on the equipment. Inspection should verify that all flanged joints have been properly reassembled, all pressure relief devices and control valves removed for inspection or repair are back in place and installed correctly, and all instrument control loops are reassembled and working properly. This is an important responsibility and should not be taken lightly.

Towers, vessels, and other equipment that were opened and worked on should not be closed up before personal inspection by the technician assigned to that area. The technician should check the vessels and equipment for foreign materials left inside by maintenance and contract crews. It is not uncommon for hard hats, leather gloves, ropes, and other equipment used by contractors and maintenance personnel to be left inside of vessels. In one case a startup failed due to a length of rope left behind that plugged a tower bottom discharge drain. The tower had to be blinded, drained, purged, opened up, and the problem located and removed. Then the tower had to be closed, purged, and blinds removed. How many hours did that take? How expensive was this failure to properly inspect equipment? $50,000? $100,000?

During a turnaround it is common for oyster shell, dirt, rust, and metal fragments to get into piping and vessels that have been opened. Piping is often laid on the ground and stuff gets kicked, blown, or knocked into them. This debris will have to be flushed out or it will cause problems. The proper operation of valves, steam traps, heat exchangers, and instruments can be affected. When this check is completed, the equipment should be closed and bolted to keep the system clean. Newly installed facilities should be checked to determine that the work has been installed as designed. New work should also be reviewed for safety and operating concerns.

Equipment used by maintenance or construction groups during the turnaround should be removed before startup activity begins. Operators should have uncluttered access to all towers, pumps, and other process equipment as they prepare for unit startup. Maintenance or contract personnel should remove scaffolding, boards, tools, buckets, ropes, and other maintenance equipment remaining after the turnaround. Such equipment cluttering the unit area is a safety hazard and violates good housekeeping policy. It is not uncommon when maintenance or contract personnel fail to respond in a prompt manner to remove their materials that the equipment is thrown into a dumpster.

COMMISSIONING AUXILIARY AND UTILITY EQUIPMENT AND SERVICES

Auxiliary equipment and services should be checked out or started-up before they are needed. Never assume that the auxiliary systems will start up trouble free. Auxiliary systems typically consist of the flare, cooling water, lubrication, heat transfer fluid, and refrigeration systems. Utility systems typically consist of water, steam, air, nitrogen, gas, and electrical systems (see Figure 18-1 and Table 18-2). The steps for commissioning each utility and auxiliary system should be systematically set out because it is important for setting the tone of the startup. A systematic set of steps tests the startup team's discipline in following routine procedures. (A systematic approach is following a procedure step-by-step

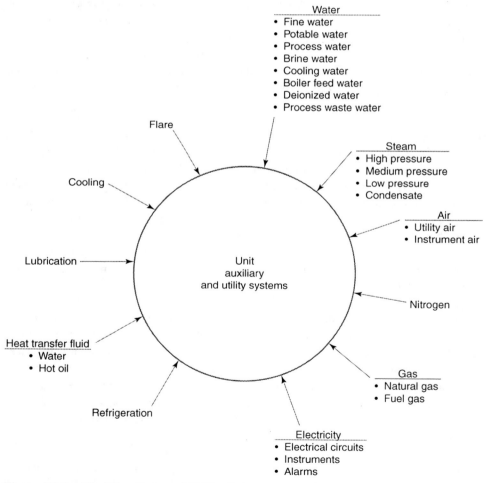

Figure 18-1 Plant Auxiliary and Utility Systems

Table 18-2 List of Process Auxiliary and Utility Systems

Utility Systems	
Plant air	125 psig
Instrument air	105 psig
Nitrogen	400 psig (or greater)
Natural gas	250 psig
Steam	—
Condensate	—
Auxiliary Systems	
Cooling water	75 psig @ 90°F
Flare	—
Lubrication	—
Hot oil	—
Refrigeration	—

and line-by-line and checking off each step as it is accomplished.) The following broad guidelines apply to the commissioning of all utilities:

- Check the supply pressure of all auxiliary and utility services up to the block valves at the entrances to the unit headers.
- Open drains and vent valves at the most distant points and purge until all fluids come out clean and debris free.
- Check that the instrument air is clean and dry and at the right pressure.
- Circulate water to the sewers until the water lines are clear.
- Blowdown the steam distribution system until it is clean.
- Check the operation of steam traps and drain condensate to the sewer until it is clean.

Auxiliary Systems

Flare System. The unit flare system, a critical safety system, must be in operation before any process materials are brought into the unit. This is mandated by law. This requires that the flare header and connecting piping systems all be completely made up and checked for leaks. The system is then purged with an inert gas, such as nitrogen, to remove any air. A sweep gas, such as methane (natural gas), is then started at rates specified in the operating procedures and the flare pilots are lighted. The flare system must be in operation whenever process materials are present in the unit since its primary function is as a safety device. Consider the alternative if the flare is not available and your unit has a reactor overpressuring with several thousand pounds of highly flammable and volatile material.

Heat Transfer Fluid Systems. Heat transfer fluid systems vary but often only steam or hot water is used as heat transfer fluids. However when operating temperatures in excess of about 600°F are required, other heat transfer fluids must be used. These oils are manufactured from a variety of chemicals to replace water that is limited by its freezing and boiling point, and air that is a poor heat transfer medium. Heat transfer oils have many advantages over the steam that was formally used in heat transfer applications:

- The product does not flash.
- No boiler blowdown.
- No high pressure. This means it is not only safer but also tends to leak less.
- No licensed boiler operator needed.
- These oils are excellent in systems that are water/steam sensitive.
- There is less corrosion in the system.

Among the popular heat transfer oils are Dowtherm from Dow chemical, Therminol from Monsanto, and Syntrel from Exxon.

Utility Systems

Water Systems. Most units will have several different sources of water for varying functions. These may include cooling water, process water, brine water, potable water, firewater, and specially treated waters, such as boiler make-up water. The various water systems must be verified as ready for use. Cooling water is water from a cooling tower used in heat exchangers. Process water is river water that has suspended solids removed and is typically used for washing down pads. It is not fit to drink. Firewater may be similar to process water or it may be municipal water, depending on the plant's location. As a rule, the unit's firewater system, even during shutdowns and turnarounds, is always available. Brine water is special water that may contain salt or another chemical to lower its freezing point so the water can be used in chilling operations.

Each unit will have procedures for putting these systems in service. It might be nothing more than opening the block valves at the battery limits, then starting up pumps, venting air from the system, and checking to see that the various water supply headers reach normal operating pressures. In some cases it may involve starting up cooling towers or cooling pond operations and sampling to determine if corrosion inhibitors or other additives are at the proper concentrations. In the case of chilled water or brine water systems, it may involve starting up a refrigeration or chiller unit.

Steam Systems. If steam is supplied by plant utilities, they should be notified in advance of the unit's plans for startup. Steam should be brought into the unit steam distribution system slowly to gradually warm the piping and equipment to prevent thermal shock. Warm-up steaming may take several hours. All atmospheric vents should be cracked open. As the cool metal condenses, steam condensate will collect and must be drained to prevent waterhammer that can damage lines or valves. Waterhammer is especially damaging to thermostatic steam traps. After the system has been warmed, the steam pressure is slowly raised and the distribution system blown down, bypassing all steam traps, which should be blocked out of the system. Blowdown of steam lines is important for removing particulate debris that may have entered the lines during repairs. Particulates, metal slag from welding, oyster shells, dirt, or any solids will cause steam traps to fail and their resultant failure will result in increased energy expenses. Blowdown continues until the steam distribution system is clean of particulate debris. Condensate drainage should continue until dry steam flows from each vent. Then all the steam traps are placed in service and their correct operation verified. Steam headers should then be placed on standby until steam is needed in the systems. If several steam headers are supplying steam at different pressures (15, 30, 120 pound steam, etc.) each header should be started-up separately, starting with the high-pressure header and working toward the lowest pressure header. Each header should be steam trapped separately, to ensure that condensate does not build up in the header while it is on standby. See Table 18-3 for a list of steam properties for typical plant steam systems.

Plant and Instrument Air Systems. Instrument and plant air systems, if not already started up, should be started at this time. Plant air operates at approximately 125 pounds per square inch gauge (psig) and instrument air at approximately 105 psig. The instrument air system should have drying facilities to remove condensate and particulates from the air. ***Plant air*** is general-purpose air and may be contaminated with light oils, dirt, and moisture. Plant air cannot substitute for instrument air because its contaminants will damage instruments or impair their function.

Instrument air is plant air that has been filtered and dried. Plant air enters the instrument air system, is dried by desiccants, and then passes through fine particle filters to keep foreign materials out of the instruments. These filters are usually in pairs, with one online and one on standby. They should be changed out or cleaned before system startup. If the unit has its own air compressors, operators will start the air compressors and bring plant air and instrument air headers up to standard pressure. If the site utility unit is supplying plant and instrument air, and if these systems are blocked out, the block valves at the battery limits should be opened and the systems brought up to pressure. Technicians should check the system for leaks using a soap solution. Once everything is operating properly, the air headers are placed on standby so they will be ready for startup.

Nitrogen System. The nitrogen header on the unit may receive nitrogen at pressures of 400 to 600 psig. In most cases nitrogen is supplied by pipeline from a contractor to the plant site and then distributed to the various units. Operators should check the pipeline station coming into the unit to ensure all valves, meters, and other instrumentation are working. They should then check that all shutdown blanks have been removed from the piping and the system is not leaking.

Natural Gas or Fuel Gas System. Units that have fired heaters require a source of fuel for the heaters. If the fuel is natural gas, it will be supplied to the unit header at pressures (250 psig) dictated by procedures. In most cases fuel gas will be supplied by pipeline from another source. Technicians should check the pipeline station coming into the unit to ensure all valves, meters, and other instrumentation are working. They then should check that all shutdown blanks have been removed from the piping and the system is not leaking.

Control Systems: Instruments, Alarms, and Circuits
Electrical circuits should be energized, instrument systems activated, and alarm systems

Table 18-3 Steam Properties for Typical Plant Steam Systems

Absolute Pressure psia	Temperature °F	Heat Content			Typical Use
		Sensible (hf) BTU/lb.	Latent (hfg) BTU/lb.	Total (hg) BTU/lb.	
29.7	249.8	218.4	946.0	1,164.4	15 psig Low pressure steam used in tracing, jackets, tank heaters, etc.
30.7	251.7	220.3	944.8	1,165.1	
31.7	253.6	222.2	943.5	1,165.7	
32.7	255.4	224.0	942.4	1,166.4	
33.7	257.2	225.8	941.2	1,167.0	
34.7	258.8	227.5	940.1	1,167.6	
36.7	262.3	230.9	937.8	1,168.7	
264.7	406.1	381.7	821.2	1,202.9	250 psig Intermediate pressure steam for process use.
274.7	409.3	385.3	817.9	1,203.2	
284.7	412.5	388.8	814.8	1,203.6	
294.7	415.8	392.3	811.6	1,203.9	
304.7	418.8	395.7	808.5	1,204.5	
314.7	421.7	398.9	805.5	1,204.4	
554.7	477.8	462.0	742.8	1,204.8	550 psig High pressure steam for process use.
574.7	481.6	466.4	738.1	1,204.5	
594.7	485.2	470.7	733.5	1,204.2	
614.7	488.8	474.8	729.1	1,203.9	

checked prior to startup. Turnaround activity may have accidentally blown fuses and tripped circuit breakers. Instrument airlines may not be correctly connected or not connected at all. Instrument technicians make these checks and operators should verify the checks have been made and the systems work.

During a turnaround many control valves are pulled for repair or replacement. Once the repair is complete and they have been installed on the unit, the inside technician should stroke each valve and verify they function properly. It should be verified that the valve moves in the proper direction in response to the flow controller signal, and that the valve is completely opened or closed when the controller indicates it is.

Alarm and interlock instrumentation must be verified that they are activated by the proper signals and respond at the proper set points. Never assume instruments and alarms have been installed by experts and they don't make mistakes. Everyone makes mistakes. Technicians should not make the mistake of failing to verify that the instruments and alarms in their area work properly. Failure to do so may result in a delay of several hours during startup. That delay could cost tens of thousands of dollars. Or even worse, the defective alarm or interlock can result in injuries or deaths and damaged equipment.

ELIMINATION OF AIR AND WATER IN VESSELS AND PIPING

Although we have covered this material earlier when we discussed process unit shutdown, it is so important that it does not hurt to lightly review it again. In Chapter 16 we purged to remove chemical residues and make the equipment safe for entry and hot work. When we opened equipment up for entry, we let air inside. In this chapter we purge to clear all equipment and piping of air before allowing hydrocarbons back into them. Failure to do so can result in an explosive mixture of hydrocarbon and air. Typically, air is removed from unit piping and vessels by steam purging, purging with an inert gas, or by water flooding. Equipment purging must continue until all process equipment contains 1 percent or less oxygen. This is verified by sampling and testing for oxygen using an air-in-steam analyzer when steam purging or a portable oxygen analyzer when purging with an inert gas.

Steam Purging

Steam is often used for purging because it is less expensive to use than inert gases like nitrogen, but purging may take longer. Since steam will condense in a cold system, the system must first be slowly brought up to steam temperature before purging can begin. Steam also presents several problems after purging is completed.

- As the system cools, steam will condense and create a vacuum that can damage equipment not designed to withstand even small vacuums.
- If the system pulls a vacuum (without damage) and if there are any small leaks, air will be drawn back into the system and purging must be repeated.
- Steam condensate, if not drained from the system, will remain as water. When the system is brought up it might flash to steam and over-pressure the system, causing damage.

Trained technicians conscientiously doing their jobs and following proper procedures can prevent these problems.

Steam should be let into the system at the feed line and flow through preheaters, furnace tubes, drums, and towers in the normal way. Coolers and condensers to be steamed should be drained free of water because the steam will heat the water and cause it to expand. If the waterside is blocked in, the expanding water may exert enough pressure to rupture the equipment. The vent and drain valves on equipment being steamed should be left open until steaming is completed and the system has cooled. If this is not done, water will be trapped in the equipment and flash to steam. Remember that a gallon of water will vaporize into steam that occupies a volume approximately 1,600 times greater. Towers and vessels should have steam introduced at the lowest point. When steam has been flowing from all vents on the system for about two hours, use an air-in-steam analyzer to check the volume of air in the venting steam. If the analyzer indicates 1 percent or less oxygen in the exiting vapor, the system is considered to be air free. During the steaming process, all low points should be constantly checked to verify condensate is draining. Do not permit steam to come in contact with instruments or pump seals because the high temperature of the steam may damage or destroy this equipment.

When steaming is complete, prevent pulling a vacuum on the system as steam is shutoff and the system begins to cool down by introducing nitrogen to replace the volume occupied by the condensed steam.

Water Flooding
Sometimes water flooding may be used if it has been determined that the facilities are designed to hold the required weight of the water. Remember that water is 20 to 30 percent heavier than hydrocarbons. Water should enter the feed line and follow the normal flow of the process, pushing equipment air to the highest vents. When towers, tanks, or other large vessels are being flooded, water flow should start at the bottom and push the air out ahead of it as the vessels and towers fill. Water should overflow all vents. After water flooding has been completed, the water must be replaced with an inert gas or a process gas before unit startup. A system can be considered water free when water cannot be found at any of the drain outlets on equipment.

Nitrogen or Inert Gas Purging
Sometimes an inert gas such as nitrogen or carbon dioxide may be used as a purge medium. When this method is used, a tightness test should be made during the first addition of purge gas. After a tightness test has been completed, a continuous purge or dilution purge method can be used to air-free the system. In the dilution purge method, the system is pressured up to three times and the gas vented after each pressure buildup. Normally, after the third venting, an analyzer test will indicate the system is air free. The continuous purge method allows the gas to flow through the system with analyses being made at the vents to determine when the system is air free.

Nitrogen can be hazardous. It has been said that nitrogen has killed more workers than any other gas in the petrochemical and refining industries. These fatalities have occurred when workers entered vessels that had been purged with nitrogen. Mankind walks around in an atmosphere of almost 80 percent nitrogen, so the gas is not poisonous. However, when nitrogen dilutes the oxygen content of a vessel to 17 to 18 percent, workers become drowsy.

If the oxygen content is about 15 percent, workers become drowsy very quickly and may sit down for a moment to rest. Then they fall asleep and die of hypoxia, a lack of oxygen.

INSPECTION OF PIPING AND EQUIPMENT SUPPORTS

Mobile supports and hold-downs of piping and equipment should be checked to prevent damage to sensitive equipment and piping failure. Mobile supports allow the thermal expansion of piping and equipment undergoing a temperature change. If these supports do not function as intended, vessels, heat exchangers, and piping may be damaged. The function of all supports should be understood and the supports should be inspected. Also, piping is held in place by various types of hold-down devices, such as brackets, clamps, gussets, snubbers, etc.

Thermal growth in horizontal vessels is guided and controlled by anchoring one end and letting the other slide by means of slotted bolt holes. (Remember Chapter 7 and the section on linear expansion?) Occasionally, a long horizontal vessel may require more than two supports, requiring it to be anchored in the middle and allowed to expand at both ends. Technicians should check that no foreign matter (scrap metal, tools, etc.) is trapped in the slots that will prevent sliding. Check that sliding plates are free to travel and that the sliding base plates are not bonded by rust or any other matter to its platform or support structure. Long straight runs of pipe subject to expansion should be checked for bowing or for support shoes that may have slipped off base supports when piping was brought up to temperature. Expansion joints serve to remove the excessive stress of linear expansion (Figure 18-2). Bowing overstresses piping and causes it to fail. The rule of thumb is that bowing is excessive if it can be seen. Bowing's probable cause is that the line was unintentionally anchored at both ends.

Piping is subject to severe vibration from fluid flow, directional changes of fluids, sympathetic vibration from reciprocating compressors, and other rotating equipment, etc. The nuts of U-bolts and clamps will vibrate loose or completely off after a period of

Figure 18-2   Expansion Joint

Figure 18-3   Piping Supports

time and allow excessive movement of the piping, which can result in piping failure. Technicians should inspect piping hold-downs and supports to verify they are in place and tight (see Figure 18-3).

BLIND REMOVAL

When the unit was shut down, blinds were installed in critical lines to prevent foreign materials from leaking into systems that were opened for maintenance or inspection. It is important that these blinds be removed before process materials are brought into the unit for startup. Any forgotten blind will prevent flow, abort the startup process, and require the system to be shut down and opened up again to remove the blind. The parts of the system that were opened will need to be purged again to remove air. The blind list should have been used to check off blinds as they were pulled. On that blind list should be an operator's initials verifying they personally inspected that the blind was removed. Only furnace fuel blinds should remain in place until it is time to light the furnace burners. They should be removed in the presence of a technician who will check-off their removal.

SAFE FLUID DYNAMIC TESTING OF EQUIPMENT

Some systems may be completed before the entire turnaround is complete. The equipment within that system can be run to simulate operating conditions. This is done with a safe fluid, such as air, water, or an inert gas. Dynamic testing of equipment using safe fluids may reveal defects that the most rigorous inspections during repair and construction will not reveal. Only the testing of system components can do this. Because equipment can be repaired more easily when it contains only inert gases, air, or water, omitting safe fluid dynamic testing increases risks and may result in costly delays later. Dynamic testing serves the following purposes:

- It proves the mechanical functioning of the equipment at design conditions. Problems are discovered early and can be fixed right away. The opposite condition is for operating personnel under pressure to fix a defective system rapidly as thousands of dollars of feedstock vents to flare.
- It helps determine instrument settings and responses.
- It removes much dirt and metal debris that might plug strainers, columns, downcomers, and lines.
- Boosts confidence in technicians as they realize how many systems have been run-in and are up to standard and dependable. They can be more focused on their duties and only have to worry about the areas of the plant that have yet to be run-in and proven.

Operations is responsible for checking pump pedestals, stuffing boxes, seal flush coolers, cooling-water connections to bearing jackets, process piping connections, installation of bleeds and drains, suctions strainers, availability of power, proper instrumentation, and lubrication of drivers and couplings. As it has been stated in other chapters: *Never assume maintenance or contract personnel did everything perfectly.*

All repaired pumps should be run in. The term **run in**, or **break in**, means to test the pump without using process materials. Normally, water is used as the fluid to be pumped. Motor loads should be checked with ammeters. With centrifugal pumps, the suction valve of the pump is opened wide and the discharge valve closed, or opened slightly if the pump is a high-capacity or high-speed pump. The discharge of a positive displacement pump must be opened enough to prevent excessive pressure buildup. Next, bleed the pump until it is liquid full and start it with the discharge valve throttled. If no pressure appears on the discharge gauge, stop the pump and determine why. If the discharge pressure is satisfactory, slowly open up the discharge valve. Check the pump and driver bearing for signs of excessive vibration or overheating and for mechanical seal leaks. Temporary screens in the pump suction lines should remain through the initial startup or for at least one month after startup. They prevent damage to the pumps by trapping foreign matter not flushed from the lines during the testing and preparation stage.

Many startups are aborted because of problems related to the pumps and compressors. Besides ensuring that all rotating equipment repairs are completed before startup, care must be exercised during the startup to ensure that equipment is lined up, warmed up, and started up properly to prevent damage to equipment. This includes early start up of lube oil or seal oil systems on larger machines, checking trips and alarms to verify that they will function, and close monitoring of pressures and temperatures as the equipment is brought online.

PRESSURE TIGHTNESS TEST

After the shutdown blinds have been removed, purge pressure should be re-established and increased to the specified process test pressure to check piping and equipment for leaks. Pressure should be held at this setting while operators check for leaks at all joints, flanges, manways, and plugs. On vacuum systems, a check should be made at full vacuum, checking the joints with a soap solution. A vacuum test is generally considered successful if a vacuum gauge does not drop more than one inch of mercury per hour.

TURNAROUND WASTE DISPOSAL

A large number of dumpsters will be full of bad piping, insulation, empty paint buckets, unsalvageable equipment, etc. These must all be removed from the area and hauled away. Any hazardous waste contained in waste drums, FRAC tanks or waste tanks will be picked up or pumped out by a contractor and hauled to a hazardous waste site. The Environmental Protection Agency's (EPA's) Cradle to Grave mandate ensures that these materials will be disposed according to regulatory compliance and not threaten human health or the environment.

INTRODUCING FEEDSTOCK TO THE UNIT

Up to this point the process unit has had large numbers of maintenance and contractor personnel working on it. By now, the majority of repair and maintenance jobs will have been completed and only a few non-unit personnel will still be working on last minute jobs. Now, when feedstock that might be flammable, toxic, or hazardous in some way is about to be introduced into the unit, the potential for an accident is greatly increased. Leaks, opening the wrong line or valve, and hot work (a source of ignition) greatly increase the hazard level. All non-unit personnel (contractors and maintenance) must come under strict scrutiny by the unit process technicians. The technicians must know what fluids have been introduced into the unit and into which systems, lines, or vessels. Failure of the technician to have this information greatly lessens the value of contractor scrutiny. What is the value of monitoring contract workers if you unwittingly allow them to open a line that has recently had process fluids introduced?

If the system starting up has a large gas compressor or blower, residual purge gas should be minimized or eliminated before startup. Most compressors are designed to operate in a stable condition, pumping certain materials with a relatively small range of molecular weights. If an inert gas does not fall into that molecular weight range and too much inert gas remains in the system, it will cause a compressor or blower to surge. Since compressors are expensive and are usually not spared, they will delay the startup if they are damaged.

Note: **Surging** is a sudden change in the flow and pressure, and if not eliminated quickly, can cause severe and extensive damage to the equipment, plus create a dangerous operating condition for the technician. Surging centrifugal compressors whose internals have disintegrated and torn through the compressor case have killed technicians.

INTRODUCING PROCESS MATERIALS

The introduction of process materials into the unit means the unit is ready to be come up and be lined out. This is usually a time of high stress. Fortunately, a sense of humor helps alleviate the stress (see Figure 18-4).

Before the initial process materials are introduced, the pressure on the purged system must be lower than the process material supply system's pressure. This prevents the purge material from backing into the supply system. If nitrogen was used as a purge, it should be vented to the atmosphere until the pressure in the system is about 2 to 3 psig above atmospheric pressure. This is enough pressure to prevent air from leaking into the system and also prevents having an excessive amount of nitrogen in the system during startup. The remaining nitrogen can be vented and purged to the flare using process materials for the purging. This must be done under controlled conditions because large quantities of nitrogen entering the flare could extinguish it.

Figure 18-4   Technician Anti-Stress Kit

After the systems have been purged and tested, the purge material should be replaced with the initial inventory of process materials. Petrochemical plants and refineries may use fuel gas, natural gas, or methane for this initial inventory. **Fuel gas** is a combination of hydrocarbon gases, such as methane, ethane, propane, and butane. The fuel gas sweeps the nitrogen from the vessels and piping into the flare header where the sweep gas is burned. This is called **backing in** fuel gas. Depending on the equipment's ability to handle the material, a solvent such as hexane may be used for this initial inventory or even actual system feed materials.

If the initial feed is a gas such as fuel gas or methane, it should be introduced at the highest point of vessels while draining water or condensate off of the bottom of the vessel. Watch the system pressure and prevent pulling a vacuum while draining the water or condensate. If a vacuum is accidentally pulled during this process, the affected parts of the system should be purged again because of the possibility that air was drawn into the system. While draining the system, stand up-wind in case the process material starts flowing from the drain line. *Do not leave the drain valve unattended when hydrocarbons are being introduced.* Hydrocarbon vapors could be released into the surrounding area and atmosphere if the valve is not closed once all the water has been drained.

The system can be made free of any trapped water by circulating liquid process material through the system in a normal flow pattern. The circulating material will carry any water left in the equipment to the low points of the system where it can be drained off. The system can be dried out further by slowly heating the circulating process material. If the water is heated slowly, it will not vaporize suddenly and create a pressure wave. Residual water in the unit will slowly vaporize and be carried along with the process material to the cooler parts of the unit. Water vapor carried into the reflux drums, product rundown lines, coolers, and condensers will fallout as condensed water and can be drained off at the knockout pots located at the low points of the equipment. A knockout pot is a low protrusion on the bottom of a piece of equipment. Water, being heavier than most hydrocarbons, will tend to collect in them.

All low points of lines and equipment should be checked for water during the heating and drying process. While the system is being heated and dried out, do not introduce materials into the system with temperatures above 200°F. A sudden vaporization or flashing of water to steam will occur with potential damage to equipment.

BRINGING THE UNIT ONLINE

The unit is brought online by following the unit startup procedure and gradually adjusting flows, temperatures, pressures, and levels to their normal operating values. Initial product being made during the early stages of startup will be off-specification and routed to a slop tank. Establishing normal operating conditions and producing quality product are the final steps in a successful unit startup. Once process materials are circulating through the unit, process technicians should be working to bring process variables to standard conditions. When these conditions are reached, they should start checking product quality and continue to adjust conditions as necessary to reach product specifications. Once product quality is acceptable, they should continue increasing unit flow rates until they achieve the normal production rate. When the unit is lined out and laboratory results indicate that the product meets specifications, product can be routed to the rundown tank or proper location.

POST-STARTUP REVIEW

After the unit is lined out, a post-startup review committee should meet to determine what they learned from the startup. They should analyze what went well and what did not. Contractors and suppliers are always an issue. Contractors whose work is sloppy or inferior and who ignore safety rules or are ignorant of them will be deleted from the contractor list for the next turnaround if they had not already been terminated while the turnaround and startup was underway. How well did maintenance and operations personnel interact? Were there communication problems? There is always room for improvement. Information from a post-startup review will be used to make the next startup even more successful.

SUMMARY

Startups must be as carefully planned as shutdowns. Like shutdowns, startups are not events that occur frequently so preparation for a startup is similar to that for a shutdown. Since a process unit may only be started up once every few years, the steps in a startup must be carefully planned so that nothing is overlooked. Startup may be divided into two phases, a preparatory phase and a startup phase.

Often, fires, equipment damage, and injuries occur during startup principally because unit procedures were not up-to-date, were not followed, or verification of turnaround repairs were not thorough.

Before startup, auxiliary equipment and services should be checked out or started-up before they are needed. Never assume that the auxiliary systems will start up trouble free. Auxiliary systems typically consist of the flare, heat transfer fluids, lubrication, cooling water, and refrigeration systems. Utility systems typically consist of water, steam, air, nitrogen, natural gas, and electrical systems.

Some systems may be complete before the entire turnaround is completed. The equipment within that system can be run to simulate operating conditions. This is done with a safe fluid, such as air, water, or an inert gas. Dynamic testing of equipment using safe fluids may reveal defects that even the most rigorous inspections during repair and construction will not reveal.

After the unit is lined out a post-startup review committee should meet to determine what they learned from the startup. They should analyze what went well and what did not. There is always room for improvement. The information will be used to make the next startup even more successful.

REVIEW QUESTIONS

1. List five activities of unit personnel when preparing the unit for startup.

2. Describe two activities of unit personnel when actually starting the unit up.

3. List several departments outside of the unit about to be started up that must be notified of the date of startup.

4. Explain why electric motors must be megged.

5. Describe several duties technicians must perform on their equipment before unit startup.

6. Why is it necessary to blind equipment that has been valved out of service for inspection or maintenance?

7. List the plant auxiliary systems.

8. List the plant utility systems.

9. State the hazard of condensate in steam lines.

10. List the various water systems on a process unit.

11. Explain why it is important that the flare system be operational before process materials are brought into the unit.

12. Explain the three problems steam presents after purging equipment.

13. Describe the inspection duties of technicians regarding piping and equipment support.

14. Explain the harm to unit startup if a blind is left in place.

15. How can you be sure all unit blinds are removed before startup?

16. Discuss the advantages of safe fluid dynamic testing.

17. What does the term *running in* mean?

18. Explain the advantages of running in equipment.

19. The most hazardous time on a unit about to start up is when _____ is introduced into the unit.

20. List two disadvantages of bringing a unit up too fast.

21. List two disadvantages of bringing a unit up too slowly.

22. Explain the importance of a post-startup review.

23. What two questions should be asked of newly constructed or modified equipment installed on the unit?

CHAPTER 19

Abnormal Situations

Learning Objectives

After completing this chapter, you should be able to

- *Define an* abnormal situation.

- *Explain how automated control systems weaken technician response to abnormal situations.*

- *State the cost per year to the U.S. economy due to abnormal situations.*

- *List several protection layers processing units have for dealing with abnormal situations.*

- *Name the OSHA standard that is concerned with abnormal situations.*

- *Explain how to deal with electrical power failure during operations.*

- *Explain how to deal with a cooling water failure during operations.*

- *Explain how to deal with an instrument air failure during operations.*

- *Describe how weather conditions can affect operations.*

INTRODUCTION

An *abnormal situation (or condition)* is any unit upset or event that requires immediate action to prevent serious consequences, such as harm to personnel, damage to equipment, or a major environmental release. Such situations are usually due to failure of a major piece of process equipment, failure of a process utility system, or a major upset in unit operating conditions.

The more time it takes to discover and correct abnormal process conditions, the greater the loss and disruption to business operations. Abnormal conditions range from those that cause lower quality or reduced production rates to those that cause catastrophic shutdowns. These conditions can result from equipment failure or degradation, variability in raw materials, process drift, and operator error.

Abnormal situations, also known as abnormal conditions, are typically caused by a combination of events that are not normally expected to occur at the same time. These are not adequately addressed by the safety interlocks and exception logic in a conventional control system and may be difficult for the operators to detect and resolve. One example is a chemical plant accident in Bhopal, India, in 1984 that led to thousands of casualties. In this accident, the amount of methyl isocyanate stored in a tank was beyond its permitted limit, the temperature alarm was turned off, the pressure gauge reading was ignored and believed to be inaccurate and, in addition, workers had little guidance to deal with the rapidly developing abnormal condition. There is no protection against this combination of events.

CONTROL SYSTEMS AND ABNORMAL SITUATIONS

Today, managing abnormal conditions is harder than ever. In the chemicals, oil, and gas industry, for example, the number of control loops that technicians must manage on large units has increased from 200 to 800 per operator over the last 20 years. Increasingly complex processes and sophisticated control strategies contribute to the problem. Technicians find it more difficult to pay sufficiently close attention to every aspect of an operation. Control systems notify the operator of process excursions outside of predetermined limits (a variable is too high or too low), but these limits are set wide to avoid false alarms. As a consequence of these wide limits, many problems are not discovered until a production disruption is imminent or is already taking place.

Control systems are also limited in their ability to diagnose and correct process problems. They give the operator little help in determining why a variable is too high or too low. Limit excursions are symptoms, not root cause problems. They trigger an alarm, which only indicates that a problem exists but does not point out the source of the problem or the best course of action to take. Part of the problem is the number of standing alarms that operators must manage and the alarm floods that result when problems occur. Too much time is spent trying to figure out what went wrong. Even after the root cause is identified, operators do not necessarily execute the ideal corrective response. Delays and wrong responses are costly. Each minute wasted trying to figure out what went wrong is a minute that the unit is making off-specification product, which ultimately wastes manpower and utilities.

Consequences of Abnormal Situations

The consequences of abnormal process conditions are significant. According to the U.S. National Institute of Standards and Technology (NIST), the inability of control systems and operating personnel to manage abnormal conditions costs the U.S. economy at least $20 billion a year. Most abnormal conditions do not result in catastrophic results such as explosions and fires; however, abnormal conditions frequently lead to costly impacts on operational performance and profitability. The impacts are displayed in Table 19-1.

Table 19-1   Impact of Abnormal Situations

Off-specification production
Wasted materials and manpower
Expensive unplanned shutdowns
Schedule delays
Equipment damage
Environmental harm
Safety problems
Damaged public image
Liability
Increased insurance premiums

Protection Layers

Processing plants typically have multiple layers of protection against the dangerous impacts of abnormal conditions. The first layer is the process control system, which is usually a distributed control system (DCS) or a programmable logic controller (PLC) based system. It provides safety interlocks and exception logic. The next layer of safety may be provided by a dedicated safety shutdown system. Its purpose is to eliminate unnecessary trips or transients due to single point failures, reduce or mitigate the undesirable effects of human performance errors, provide a safe and controlled response to unexpected conditions, and protect personnel safety, equipment, and the environment. Another layer of safety might be a fire protection and/or gas protection system.

TYPES OF ABNORMAL SITUATIONS

OSHA's Process Safety Management standard mandates procedures and operating guidelines for abnormal situations. There may be emergency procedures for a shutdown due to loss of electricity, loss of cooling water, serious fire, etc. Because emergency incidents can be unique in nature, there is no substitute for a technician's thorough understanding of the physics and chemistry of their process, knowledge of their process equipment, and good judgment. The latter, good judgment, is not possible without competency in the first two mentioned. Most of the time a process technician's job will be routine and the same day after day but when an abnormal situation occurs the technician's skills and knowledge will be challenged. During an emergency they must act to prevent harm to coworkers, damage to equipment, an environmental release, or to keep the operation running (shutdowns are expensive). Examples of abnormal situations include

- Intentional changes in operating conditions to produce experimental products or to evaluate experimental process conditions
- Loss of utilities (power failure, loss of cooling water, etc.)
- Weather-related problems

Operating personnel have control over intentional changes because they know when and what they are going to do. It is planned. The last two situations come out of nowhere like an act of God, often with no prior warning.

Operating Outside of Standard Operating Conditions

There are several reasons why it may be necessary to operate a process unit outside the normal range of standard operating conditions (SOCs):

- Marketing wants to modify SOCs to create new product.
- The technical department wants to alter SOCs in an attempt to make the process run more economically or produce more throughput.
- Equipment problems may restrict normal operations and SOCs must be altered until the problem is fixed.

When a research, marketing, or technical group proposes an operating change they will provide some operational guidance for the unit. They will submit an official request for the change and include a set of modified operating procedures and/or conditions. These documents will state the reason for the modified operations, the advantages and costs of such changes, and safety implications associated with the change. Operating personnel will review the suggested modified procedure for safety and operational competency.

When non-routine operations are caused by problems with a piece of major equipment (or system), a decision must be made to shut down or continue operating at altered conditions. The unit manager, along with operating personnel, will review the feasibility of continuing to operate by asking critical questions, such as those that follow. Can the unit still make on-specification product? Will the safety of personnel be at an unacceptable level of risk? Will the safety of equipment be an unacceptable level of risk? What guidelines are needed to operate successfully under this non-routine situation? Process technicians should be involved in these decisions because they have the best feel for how the unit will operate under the new conditions, and they will have the responsibility of operating the unit. The new hazards created by running outside of standard conditions must be listed, discussed, and included in the temporary operating procedures.

KEEP RUNNING OR SHUT DOWN?

When a serious unit emergency occurs, the inside technician (board person) has the greatest control over the unit. They will have to make some quick decisions. This is when their knowledge of the unit physics, chemistry, and equipment is critical. The inside technician will have to use their troubleshooting and analytical skills to determine if the crew can keep the unit running and recover from the upset or if the situation is serious enough to initiate an emergency shutdown. Sometimes one section of the unit, the section with the problem, can be brought down and the rest of the unit remain up and re-circulating. Repairs can be made on the down section and it can be brought up and integrated into the rest of the unit, which stayed up. This is much less expensive than bringing the whole unit down. If the situation is very hazardous and can result in injury to personnel or harm to equipment, prudence will demand the inside technician initiate the emergency shutdown. Most often, time permitting, the decision to bring the unit down will come from upper management. However, there may be instances when the inside technician will not have time to wait for unit engineers or management to come to the unit and review their data and decision. Situations that may require an emergency shutdown of an entire processing unit are displayed in Table 19-2.

Failure of major equipment or utility systems other than those listed in Table 19-2 will cause unit upsets; however, this should not require a complete shutdown in most cases. If the unit

Table 19-2   Situations That Can Initiate a Process Unit Shutdown

Total and sustained failure of one of the following: electrical power failure, cooling water failure, or instrument air failure
A major loss of steam pressure
Serious and uncontrollable leakage of a toxic or flammable material
Serious and uncontrollable fire or explosion
Unexpected combination of events
Runaway exothermic reaction
Failure of a key piece of equipment that is not spared

involved makes intermediate products used downstream by several other units, every action will be taken to keep the unit up and brought back to normal conditions so as not to interrupt the production of downstream units. If the unit is a batch operation, or a small-scale continuous process, which is easier and less expensive to shut down and restart, decisions are easier to make about shutting the unit down. In all cases, the safety of personnel and process equipment are the major factors in the decision to run or shut down.

Some situations that should not require shutting down a critical unit include

- Failure of equipment when an installed spare unit is available
- Situations that can be mitigated or controlled by rate reductions
- Failure of individual instruments where a bypass or manual mode of operation is possible
- Momentary failure of any utility whose effects can be turned around or abated as soon as the utility is restored

Causes of Preventable Plant Shutdowns

Many process unit shutdowns are avoidable. They are not due to acts of God, such as lightning, hurricanes, and earthquakes. Approximately 63 percent of unplanned unit shutdowns are due to human error. Failure to follow procedures (forgetting a step or skipping over a step) accounts for 40 percent of unplanned shutdowns. Mechanical or instrumentation problems account for 31 percent. Inadequate procedures (poorly written or not updated), account for 23 percent and internal process event problems account for 6 percent.

Clearly written and updated procedures are critical to the safe and profitable operation of a process unit. Also critical is a well-trained workforce. This is why the Process Safety Management standard (PSM) was created. Sections of it mandate critical technician training periodically and the management of change (MOC) of operating procedures and technical information.

RESPONDING TO EMERGENCIES

Today many process units are highly automated because it reduces the size of the workforce required to run a process unit, and because it gives better control of the unit. However, there are serious downsides to this automation. One of the downsides is that technicians spend less time interacting with the process and become less familiar with the process. Automation takes away the day-to-day critical thinking required by a technician about the steps in their process because the unit is literally running itself. A second downside is that technicians adopt the philosophy that since the control software is so smart and has so many built-in safeguards, there is no need to learn the physics and chemistry of the unit. Another problem with effectively responding to emergencies is that because many units come up and run for three to five years before coming down for a turnaround, technicians cannot remember things they do not do on a frequent basis. When an emergency occurs, unless it is a small unit or a very simple process, technicians will have to take the time to consult emergency shutdown procedures. Those procedures should be readily available in the control room as both hard copies in the procedure manual and electronically on the computer.

In general, efficiently recovering from abnormal situations will involve

- Keeping process flows, levels, pressures, and temperatures within safe bounds
- Controlling heat and energy sources
- Controlling reaction systems (if reactors are involved)
- Restoring or maintaining the integrity of utility systems

To prevent or mitigate abnormal situations, many process units have automated safeguards, such as safety interlocks (SIs) or emergency shutdown systems (ESDs), engineered into the operating system design.

RESPONDING TO SPECIFIC ABNORMAL SITUATIONS
Electric Power Failure

A major electrical power failure is not a common occurrence in the processing industry. Process operations depend on a reliable electrical supply and many safeguards are built into the supply system to prevent sudden failures. If a failure does occur, it is usually related to a problem in the electrical substation supplying power to the plant or weather-related damage to the power transmission lines supplying the plant (see Figure 19-1).

A partial or total loss of electrical power is a major upset to a processing unit because all pumps, compressors, agitators, cooling fans, and other process equipment driven by electric motors will stop functioning. All flows will stop. If the failure occurs at night, emergency lights will come on but technicians will be working in a reduced lighting situation. To prevent this costly and hazardous situation from occurring, most processing units will have electrical emergency systems in place that reduce the effects of an electrical power failure. Such systems could include

- Emergency lighting in critical locations that automatically comes on when electrical power is lost. The emergency battery-powered lights in hotel and office building stairwells are an example.
- Diesel-driven electric generators that provide power to run critical pieces of equipment necessary to maintain unit control and bring the system down safely. These may be programmed to start automatically when a power failure is sensed.
- An **uninterrupted power source (UPS)** consisting of a bank of wet cell batteries that supply power for computer and other electrically powered control instrumentation (see Figure 19-2).

Figure 19-1   Electric Power Substation

Figure 19-2   UPS System

- Turbine-driven pumps and other rotating equipment powered by steam that continues supplying critical services. These are frequently provided as spares for essential electrically driven units, such as cooling water pumps.

Most electrical power failures are short-term events of five to ten minutes and then power is restored. In such an event, instrumentation with battery back-up systems will continue to run but electrically driven equipment will shut down and flows will stop. Technicians

313

responding to a brief loss of electric power should direct their efforts toward maintaining control of systems and equipment that are still operating, then work to restart equipment shut down by the power failure. Flow should be reestablished in critical systems while keeping pressures, temperatures, and levels within safe ranges. Competent operators can regain control of the unit and prevent a total shutdown.

In the event of a total and sustained power failure (approximately 30 minutes or longer), the unit must be shut down according to emergency procedure as quickly as possible. The unit will be notified of a sustained power failure by management or security. Some of the steps to this type of emergency shutdown are

1. Switch control systems to battery power (usually done automatically).
2. If the unit is DCS controlled, align system screens to provide maximum possible overview of the process to monitor changes to the entire unit.
3. Start emergency generators and turbine-driven process equipment (usually done automatically). Using this back-up power source, start up the equipment necessary for an orderly shutdown. Each unit will have a list of such critical equipment.
4. Monitor process variables for dangerous or damaging situations.
5. Vent to flare, depressurize to downstream equipment, and drain to a collection sump as necessary to gain and maintain control of unit process variables.
6. If power has been lost to the cooling tower fans, reduce the cooling tower load as much as possible by cutting back and/or shutting down heat-generating equipment.
7. Stop boil-up to distillation towers that have lost reflux flow.
8. Shut down and purge or vent any vessel or system that does not respond to control efforts.
9. Reduce temperatures and flows until the unit can be safely shut down.

All of these points will be addressed in a process unit's emergency shutdown procedure(s).

Cooling Water Failure

Complete cooling water failure is rare. Most plants will have a central cooling water unit with variable pump drivers (turbine driven and electric motor driven) so that if the plant lost electric power, the steam-driven pumps would be operable and if steam was lost, electrically driven pumps would be available. The plants will also have supplementary supply headers and redundant instrumentation for the cooling water system. Complete cooling water failure can have a catastrophic potential for units that rely on cooling water to maintain control over exothermic reactions or to prevent excessive pressure increase in the heated vessels.

The following actions are normally taken if a total sustained cooling water failure occurs and time allows for an orderly shutdown. Again, the unit's emergency shutdown procedure covers these actions.

● Cut back fuel to fired furnaces and other heat generating sources while avoiding thermal shock to equipment. Once burners and other heat sources are shut down, stop process flow to the heaters.

314

- Shut down, vent, and block in (in some cases, flush) reactors that use cooling water for temperature control.
- Continually monitor process variables.
- Vent to the flare or downstream systems as necessary to prevent excessive pressure increases.
- Keep liquid levels in towers, columns, and drums as close to normal as possible to make unit startup easier.

Instrument Air Failure

Instrument air activates the pneumatic control valves that control the flow, pressure, level, and temperature on the process unit. A sudden failure of the instrument air system could leave the inside technician without much control of their unit, a terrifying situation when the unit is filled with fluids that are toxic or highly flammable. True, interlocks and ESDs should prevent a catastrophic accident but a serious accident (technician injury and equipment damage) might still occur. Because the instrument air system is so critical, a redundant source of instrument air is usually designed into the plant when it is built. Redundant instrument air can be supplied by

- Spared instrument air compressors
- Multiple header supply sources
- Nitrogen systems
- Jumper systems from other units

As a result of the aforementioned possibilities, a major unit instrument air failure is highly unlikely.

The failure mode of the control valves provides another layer of safety when the instrument air system fails. Control valves can be designed to fail open or fail closed when instrument air is lost. During design of the instrumentation system, each control valve is evaluated to determine how it should respond (fail) if instrument air is lost. Its failure mode assists in the control of unit problems. For instance, assume a reactor had four control valves, two feed valves, one pressure relief valve, and one product valve. Now, assume instrument air is lost to the reactor. Control system designers would have planned for the feed valves to fail closed to stop the flow of feed into the reactor and the build up of heat and pressure caused by the reactants chemical reaction. They would have planned for the product valve to fail open so the reactor would dump its contents, reducing the heat and pressure within the reactor. And, as an extra measure of safety, they would have the pressure relief valve fail open also.

If the impossible happened and a total, sustained instrument air system failure occurs, the inside technician must coordinate with the outside technicians to maintain control over the unit while bringing it down safely. On their console screen they will note the changing process variables, those threatening to exceed limits, and radio outside technicians to manually operate valves and equipment to control the variables. The outside technicians will be scrambling around like people on a damaged warship in the heat of battle. Though the control valves have gone to their pre-selected failure mode, if flow is needed through the valve for the unit shutdown where a control valve has failed closed, the outside technician can open a bypass line around the valve. Or, the control valve may have a hand-wheel mounted on top for manual operation of the valve.

Additional actions that need to be taken for an orderly shutdown include

- Cut back on fired furnaces and other heat generating sources while avoiding thermal shock to equipment. Once burners and other heat sources are shut down, stop process flow to the heaters.
- Stop the flow of catalysts and reactants to reactors and, if possible, flush the reactors.
- Monitor process variables.
- Vent to flare or downstream systems as necessary to prevent excessive pressure increases. Shut down, vent, and block in any system with high, uncontrollable temperature or pressure.
- Keep liquid levels in towers, columns, and drums as close to normal as possible to enable quick start up of the unit.
- Reduce flows and temperatures to the point where the unit can be shut down.

Steam System Failure

Steam systems have several safeguards that make a complete system failure rare. Large processing sites will have two or more boilers, thus if a boiler failure occurs, the plant can still run at reduced rates on the remaining boilers. The boilers will also be well instrumented so that warning of a problem will occur before a complete shutdown occurs. Operating problems that could affect multiple boilers include loss of fuel gas supply and loss of boiler feed water. A major problem with either could put most or all of the plant's boilers down at one time. Also, a major problem with the high-pressure steam distribution header(s) can cause a steam system failure.

The steam system will use different distribution headers to supply a high-pressure steam (500+ pounds per square inch gauge [psig]), a medium-pressure steam (100 to 200 psig), and a low-pressure steam (<100 psig). The medium- and low-pressure headers are usually supplied steam from the high-pressure steam header. An automatic letdown valve or a steam-driven turbine can be used to reduce the pressure to the desired lower level. A failure of either the medium- or low-pressure system will not affect the high-pressure system. Steam will still be available.

If a process unit experienced a serious steam distribution problem, technicians should determine the extent of the problem. They should determine (1) which areas of their unit are affected, (2) which equipment needs attention first, and (3) which distribution systems are at fault. Is it all three, high-, medium-, and low-pressure steam? Is it just the low and medium-pressure headers? Also, technicians should know if the unit has emergency plans that dictate rerouting steam from an unaffected header to the problem area.

Complete steam system failure may have severe consequences for plants that rely on steam to maintain boil-up in multiple distillation columns, maintain reactions, prevent vapor systems from condensing, prevent liquid systems from freezing, or prevent polymer or heavy crude oil systems from becoming too viscous to pump. The following actions should be taken if a total, sustained loss of steam occurs:

1. Cut back on steam requirements by reducing flow rates through equipment and/or lowering operating temperatures.

2. Shut down, vent, and block in reactors that use steam for temperature control. This prevents an excess of unreacted materials from accumulating downstream.
3. Shut down, flush, and block in any system with materials that could solidify if its material becomes too cold.
4. Keep liquid levels in towers, columns, and drums as close to normal as possible to make unit startup quicker.

Steam turbines require special attention during a steam system failure because problems with such systems can snowball if demand exceeds steam supply. This occurs because turbine governors sense a slowdown in rotor speed and open up to admit more steam, which in turn lowers steam pressure further, which cause the governors to ask for even more steam, and so on. Under these conditions it is necessary to reduce turbine load or speed as much as possible.

Fires, Leaks, and Other Abnormal Situations

No leaks and no fires—that is the goal for a processing unit using toxic or flammable materials. However, leaks still occur and, if the material leaked is flammable, a fire can also occur. Often such situations can be brought under control without incident if they are quickly detected. Most processing units have leak detectors and alarms in critical areas. Every alarm requires immediate investigation. The outside technician for the area involved should be radioed and told to investigate the alarm. To do this safely, they should check the wind direction, note the location of safety equipment, and the location of any potential ignition sources before entering the area. Using the appropriate personal protective equipment (PPE), they should approach the area from upwind with caution while maintaining radio contact. When the technician identifies the problem, they should communicate it to the control room and request additional help, if needed. If the leak is significant, the technician should describe the incident, back off to a safe location, and wait for the emergency response team.

The response to the leak or fire will depend on the size of the leak or fire and the toxic nature of the leaked or burning materials. Response may also be dependent upon the location of the leak with respect to the equipment and occupied building, the proximity to the known ignition sources, and the wind direction and velocity. Hazwoper procedures for the process unit or plant will dictate who should respond and the extent of the response.

Runaway Exothermic Reactions

Most reactions are temperature dependent. Increasing temperature increases the reaction rate. If it is an **exothermic reaction** (a reaction that gives off heat), increasing the reaction rate increases the temperature, which further increases reaction rate. Without proper engineering controls, this scenario can result in major equipment damage and injuries to personnel when process or equipment limitations are reached and equipment ruptures. Runaway exothermic reactions and the rupture of reactors are not a common occurrence. Operators should be familiar with the chemistry of their unit and know which reactions are exothermic and have the potential for a runaway. They should also know the reactor temperature and pressure SOCs. Reactors capable of serious runaway conditions will have numerous engineered and automatic safeguards to shut down the system before loss of control. Safeguards include temperature and pressure interlocks, pressure safety valves (PSVs), and ESDs.

Evidence of an impending runaway reaction includes (1) a sudden increase in reactor temperatures, (2) little or no response to reactor temperature adjustments, and (3) unusual consumption of reactant materials. Technicians noting such situations should consider (1) shutting down reactor feeds and (2) cooling the reactor with water spray.

WEATHER-RELATED ABNORMAL SITUATIONS

Weather conditions—lightning and high winds, freezes, flooding—can affect the operation of processing plants. Most plants have building codes that take weather situations into account during construction. Plants are designed and constructed with strengths far greater than those normally required, allowing for severe weather conditions. As an example, a plant may be built in an area subject to hurricanes, but the most severe hurricane experienced in the area may have been one with winds of 135 mph. To be on the safe side, the plant may be built to withstand winds up to 160 mph. Plants will also have emergency procedures to deal with such situations when they occur. Many sites subscribe to a local weather forecasting system and 24-hour weather radio weather station that provides updated weather reports.

Electrical Storms

Electrical storms dissipate tremendous amounts of energy in lightning strikes. A typical lightning strike may release power on the order of 10^{12} horsepower in a few millionths of a second, which unleashes a tremendous destructive power. Most plants rely on protective systems and changes in plant operations when anticipating electrical storms. Protective systems reduce the hazards of lightning strikes and include lightning rods and masts and grounding and bonding of equipment, structures, and buildings. During electrical storms administrative controls mandate that technicians should not operate crane booms or climb towers or tanks.

Hurricanes and Tornadoes

Hurricanes and tornadoes generate damaging high winds. Hurricanes also can cause flooding and storm surge damage. The best protection against such storms is advanced preparation. Plants build in hurricane or tornado-prone areas that are designed to resist high wind and flooding. Emergency procedures describe how to prepare for hurricanes. Hurricanes can be tracked for days before they hit, so there is adequate time to prepare for their arrival. Unfortunately, tornadoes do not give much warning before they hit, so minimum time is available for preparation.

Freezing Weather

Sites operating in areas with severe winters will have permanent facilities for heating piping and instrumentation, plus additional insulation on critical piping and equipment. All plants should have emergency procedures in place for dealing with freezing temperatures. Water, condensate, and process lines that contain materials that can freeze should be circulated to keep materials from becoming stagnant and freezing. Electrical or steam tracing along pipes keeps these systems above the freezing point. Some areas that may require special attention are discussed in the following paragraphs.

Fire Protection Systems

The major fire water headers are often installed underground below the frost line. Aboveground fire hydrants and fire monitors should be insulated or otherwise protected to prevent freeze-up. Sprinkler and deluge systems should be dry-pipe designed or filled with an antifreeze solution.

Safety Showers and Eye Wash Stations. Safety showers and eye wash stations may have their piping heat traced to keep the water from freezing. Care should be taken to ensure the water is warm but not too hot for its intended use.

Steam Traps and Condensate Lines. Stagnant steam traps and condensate lines can freeze and rupture. It may be necessary to drain the steam traps of condensate and open bypass lines around them to prevent their freezing.

Process Instrumentation. If condensate builds up in the instrument impulse lines or instrument housings, the instruments can freeze and will no longer provide accurate information. If the frozen instrument loop is part of a control system, alarm, or interlock, these functions are no longer working either. Instrument housings should be insulated to protect the instruments and impulse lines should be heat traced to prevent freezing.

Condensate Drains. Storage tanks and distillation reflux tanks may have condensate collection traps (knockout pots) that remove residual water from process materials in these systems. They are frequently drained via a level-controlled automatic valve. If this control system freezes, water can accumulate in the process equipment and cause serious problems. During freezing weather these systems should be checked to ensure that they are working properly. Those that are not automatically drained during freezing weather must be manually drained frequently.

SUMMARY

An abnormal situation (condition) is any unit upset or event that requires immediate action to prevent serious consequences, such as harm to personnel, damage to equipment, or a major environmental release. Such situations are usually due to failure of a major piece of process equipment, failure of a process utility system, or a major upset in unit operating conditions. Managing abnormal conditions is harder than ever. In the chemicals, oil, and gas industry, for example, the number of control loops that operators must manage has increased dramatically along with increasingly complex processes and sophisticated control strategies. It is increasingly difficult for operators to pay sufficiently close attention to every aspect of an operation.

The types of abnormal situations are (1) intentional changes in operating conditions to produce experimental products or to evaluate experimental process conditions, (2) loss of utilities (power failure, loss of cooling water, etc.), and (3) weather-related problems. When an abnormal situation occurs, a decision must be made to keep running and try to maintain control of the unit or to shut down. In all cases, the safety of personnel and process equipment must be the deciding factor.

Situations that can initiate a process unit shutdown are loss of a utility (steam, instrument air, electricity, etc.) or an auxiliary (flare, cooling water, etc.), serious fires and leaks, and weather conditions.

REVIEW QUESTIONS

1. Define *abnormal situation*.

2. Abnormal situations range from _____ to _____.

3. Explain why managing abnormal situations are harder than ever.

4. The cost per year to the U.S. economy due to abnormal situations is _____.

5. List five impacts of abnormal situations.

6. List several protection layers processing units have for dealing with abnormal situations.

7. Name the OSHA standard concerned with abnormal situations.

8. List three reasons why a processing unit may deliberately operate outside of standard operating conditions.

9. Describe the two major factors that determine whether a unit experiencing problems will attempt to keep running or shut down.

10. The percent of unplanned unit shutdowns due to human error is _____.

11. Explain how automated control systems weaken technician response to abnormal situations.

12. Explain why the loss of electric power is a serious concern for processing units.

13. List three sources of emergency electrical systems.

14. Name the two drivers for cooling water pumps.

15. List three backup sources of instrument air.

16. Name the two failure modes of control valves.

17. If instrument air is lost and a control valve failed closed, how would technicians route flow through or past the control valve if they had to?

18. Discuss why steam system failures are not common.

19. Low-pressure steam is suddenly unavailable to the unit. Does this mean that the unit will also experience loss of medium- and high-pressure steam? Explain your answer.

20. List four systems for types of equipment affected by freezing weather.

CHAPTER 20

Process Troubleshooting

Learning Objectives

After completing this chapter, you should be able to

- *Explain why interpersonal skills are a necessary part of troubleshooting.*

- *Discuss why troubleshooting skills are important to the process industry.*

- *List five basic tools used in troubleshooting.*

- *List the five phases of troubleshooting.*

- *Define abnormal situation.*

- *Discuss why each of the five phases of troubleshooting is important.*

- *Explain why it is important not to go "outside the boundaries" of your operating envelope.*

INTRODUCTION

As stated in Chapter 19, disruptions in industrial processes cause $20 billion or more in damages and economic losses each year. Most of that $20 billion can be recouped through better management of emergencies—by catching them in the earliest alarm stage or preventing them entirely. A workforce well trained in analytical abilities and troubleshooting skills is critical to the saving of billions of dollars.

Today, plants in the process industry are the safest they have ever been, thanks to the investment in sophisticated process control equipment and improvements in operating reliability. Technicians running these processes are much better trained than their predecessors a generation ago. In the past the only hiring requirements for a process technician was good health and a high school diploma. Many technicians today have a two-year associate degree in process technology. Technicians regularly encounter technical problems on their units, and in most cases, they resolve the problems and the operation keeps running. Occasionally the problem is significant enough to draw management's attention and these problems, too, are normally resolved. However, there have been times when operations personnel failed to solve a serious problem and the consequences were painful. The consequences may be the destruction of part of a processing unit (very expensive), an emergency shutdown of the unit (very expensive), or the local community forced to evacuate their homes or sheltering in place (very bad for public relations). Numerous small problems not effectively handled can have a significant impact on the bottom line of the process unit. Poorly handled operating problems can lead to expensive problems listed in Table 20–1.

As we just mentioned, the consequences of production problems and their cost can be huge. That's why the role of process technicians in solving problems—*troubleshooting*—is so important. Troubleshooting requires tools and skills, which we will discuss in detail in this chapter. Most technicians when troubleshooting a problem and asked to describe their methodology and why they do it that way can't explain their methodology. They just do it because that is the way they have learned through experience. However, experience is both a harsh (injuries and deaths) and expensive (damaged equipment, down time) teacher: Companies cannot afford to destroy $70,000 of equipment each time a new problem occurs and operators learn by experience.

Many of the principles of physics and mechanics that apply to process applications—distillation hydraulics, phase separation, heat transfer, etc.—have been well known for decades. There is no mystery to the physics. The challenge in troubleshooting consists of untangling the influence of several factors when they combine factors like human error, mechanical failure, and corrosion. One fact all technicians should keep in mind when they are about to make process changes is that most process problems are initiated by human error—a never-ending source of ingenuity and surprise.

Table 20-1   Operating Problems that Can Become Expensive

1. Wasted utilities
2. Increased down time
3. Increased raw material usage
4. Decreased production efficiency
5. Emergency staffing
6. Increased maintenance work
7. Releases to the environment (fines)
8. Environmental remediation
9. Wasted manpower

TROUBLESHOOTING AND INTERPERSONAL SKILLS

"Are you an idiot? There's no way changing that could have caused this problem!" Definitely, the preceding statement (1) is no way to build confidence in the team member who offered a suggestion, (2) does not encourage other team members to contribute to solving the problem, and (3) does not in any way display the interpersonal skills required for team work and troubleshooting. The capability to solve complex problems will depend on the interpersonal skills of the team members. Technicians should be able to put forth their own suggestions without fear of being ridiculed or sidelined. Problem solving by deductive reasoning functions through objective viewpoints offered by knowledgeable people. Team members should also be an active participant in the troubleshooting process rather than a passive on-looker. In today's complex plants it is rare that a lone individual is a troubleshooter. Troubleshooting is a team effort that pools the mixture of expertise, experience, knowledge, and skills of the team.

TROUBLESHOOTING IN THE PROCESS INDUSTRY

Red Adair, a famous oil well firefighter, told all his employees "The first thing I like to do is get it right the first time." Makes good sense when you're stepping into a roaring inferno. Getting it right the first time is important to any business endeavor. Who wants to lose an important sales contract because they didn't get it right the first time, or a patient undergoing open-heart surgery? Getting it right the first time is not a matter of luck, it is a matter of methodology, knowledge, and analytical skills.

The definition of the term **troubleshooting** is *the successful repair or correction of a problem when the cause of problem is not immediately known.*

Three key terms are in the definition. They are: (1) successful repair or correction, (2) cause of the problem, and (3) not immediately known.

1. The term *successful repair or correction* means that when the troubleshooting process is finished the unit process or equipment has been restored to normal condition. The operation may have been shut down and equipment repaired or replaced but the goals and requirements of the business—safety, production, reliability, cost—have been met.
2. The term *cause of the problem* means fixing a problem requires finding the root cause. Repairing a problem with a known cause does not require troubleshooting.
3. The term *not immediately known* means that troubleshooting involves finding the cause of the problem. Finding the cause involves a combination of technical know-how and calculating probabilities.

Troubleshooting versus Trial and Error

The process industry cannot afford to troubleshoot by trial and error, which is very unsystematic and inefficient. That is an attitude of "Let's try this," and one hour later, "Uh-oh, that didn't work. What are we going to try next?" Can you see how expensive trial and error can become? Troubleshooting is a methodical process that employs logical techniques to solve a problem. Troubleshooting is a process that can be broken down into a series of actions that involves a complex set of decisions. The process has distinct and sequential phases and steps and skipping any one of them impedes the goal of finding and fixing the problem. The more knowledge about the technical subject (the process unit) a technician has, the better they will be at troubleshooting.

People are rarely trained in any formal method of troubleshooting. Most well-known troubleshooters are self-taught, intuitive, and rely principally on their own experience. For troubleshooters who have learned by experience, it is often a case of "been there, done that." They know what works because they have been on the unit for 25 years and experienced a lot of "been there, done that" situations. As a result, some people considered expert troubleshooters normally don't make good teachers of troubleshooting methods because they don't have a method. Another form of troubleshooting is by following a troubleshooting procedure and doing exactly what it says. This is a great method if the procedure exactly matches the situation at hand. Airline pilots get this kind of training on flight simulators. However, the procedural method does not often work for a complex processing unit because of the almost infinite number of ways things could interact to create problems. Both ways to learn troubleshooting—through experience or by procedure—have their disadvantages.

Successfully solving operating problems requires technical knowledge of the process unit, experience, and a systematic method that can be applied to a wide variety of situations. With troubleshooting, the clock is always ticking because the process exists to make money, hence there is pressure to find the problem and fix it fast. The Sandia National Laboratory did a study of the relationship between time and problems in a control room environment. Their data confirmed what anyone who works on a processing unit already knows. If the technician takes only 10 minutes to solve a problem, the odds of getting the right answer are 50 percent. Give the technician an hour and a half and the odds of getting the right answer improve to 99.9 percent! The only problem is that time costs money. Saving some of the billions of dollars lost to emergency situations lies in getting better at getting to the right solution as quickly as possible.

COMPUTERIZED PROCESS CONTROL
As plant designs evolve, more and more layers of advanced controls, interlocks, and alarms are being added in an attempt to minimize the effects of plant upsets. As a result, the primary role of many technicians has shifted from the day-to-day operations of the plant to monitoring the process for developing problems and responding to abnormal operations. Today, the first line of defense against abnormal situations often is not the technician but the process control system. When the control system cannot cope with the situation, then the technician is expected to take control. Developments over the past 30 years have made a technician's job more difficult because processes have become more complex. Advanced control has pushed plants to operating on the edge for maximum productivity. This is great for bottom line numbers, but operating on the edge makes it more difficult to control systems when upsets occur. With modern distributed control systems (DCS), plants are running with smaller operating staffs, which increases an individual technician's area of responsibility. DCS systems have the capability to

- Control the entire process
- Monitor process variables
- Provide alarms
- Log changes
- Trend variables

Notice, no place did it say that DCS systems troubleshoot and solve problems. The DCS systems sound alarms and contain a lot of trend data, but a human (technician) still has to determine what the problem is and fix it.

The Computerized Control Room

The computerization of instrumentation and the replacement of board-mounted instruments with a DCS system revolutionized process plant control. It has its advantages and disadvantages.

Advantages. Inside technicians (board persons) can control the unit from the console without getting up and moving from controller to controller (see Figure 20-1). The control system can quickly calculate results rather than reveal only raw data, which greatly enhances plant safety. It stores immense amounts of data, such as trends, alarms, logs, etc. It also frees engineers and supervisors from time-consuming routine tasks, allowing them more time to optimize operations and investigate problems. It can control a whole series of instruments, rather than one or two. Programs are developed that will put a distillation tower on line, optimize the reflux, minimize steam consumption, and systematically shut down the column in both emergency and normal modes. Such systems can be used for plant accounting purposes (material and energy balances). Also, flows can be integrated and shipments from tanks can be measured and recorded from a control room.

Disadvantages. Operators tend to trust computerized data more than they once did instruments, yet the accuracy of sensing elements has not been improved. A fouled thermocouple will give a false reading whether it is coupled to a gauge, an analog recorder, or a computer. The numbers represented are averages, spikes are not shown, and records are therefore less precise. A skilled operator can scan as many as 20 or more board-mounted

Figure 20-1 Inside Technician and DCS Screens

instruments in one glance, giving him a rapid assessment of the performance of that section of the plant. Getting the same information from one console requires more effort and time. When operations are unsteady, it is difficult to watch all the required variables at the same time. Without additional screens, extra operators are useless.

BASIC TROUBLESHOOTING TOOLS

There are a few troubleshooting tools based on physical science and common sense. Technicians may know these tools but aren't aware of them as tools (see Figure 20-2):

- Material balance—What goes in must come out or an unstable condition exists. A common and sometimes quick method of verifying process indicators is to compare process values using the material balance tool. Levels or pressure or both are affected when a process is not in material balance.
- Energy balance—Most processes require constant temperature for proper operation. Heat energy input must be consumed, transformed, or removed or the process temperature will not remain stable. Temperature will either increase or decrease.
- Interrelationship of variables—It is unusual for a single variable indication to change in a process without affecting another variable.
- Expectations—Common sense tells us what to expect in a sequence of events. This logic applies in most processes and should be used to verify data. When the output of a flow controller is increased, an increase in flow should result.
- Process knowledge—The more you know about the chemistry, physics, and equipment of your processing unit, the more valuable your troubleshooting skills will become. A technician's best source of information about his unit's normal operating conditions and equipment specifications are the unit's technical and procedure manuals.
- Process data—There is a lot of process data starting from the present and going back months. The data is contained in logs, rounds lists, the DCS

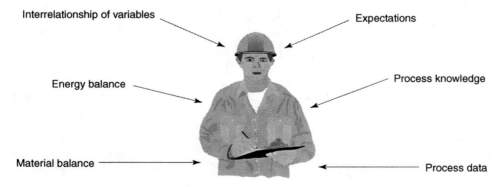

Figure 20-2   Troubleshooting Tools

system, and the laboratory information management system (LIMS). It can yield valuable clues and trends and help answer *who*, *what*, and *when*.

THE FIVE PHASES OF TROUBLESHOOTING

Often operators are taught troubleshooting by cause-and-effect training, also called the procedural method. They are taught how to respond to specific process problems. When individuals are taught by the procedural method we negate one of their most valuable tools: their analytical skills. We have asked them not to think and reason, just to respond to a scenario. Two important keys to process troubleshooting are process knowledge and a systematic approach to solving problems.

A SYSTEMATIC APPROACH TO PROBLEM SOLVING

The rest of this chapter describes a systematic approach to problem solving. Basically, troubleshooting can consist of five or more phases. Depending on which troubleshooting method you use, the phases may be stated differently but essentially they mean the same thing. As an example, phase 1 may be stated as "Recognize the Abnormal Situation" or "Identify the Problem." Phase 2 may be stated as "List and Evaluate Potential Causes" or "Identify Potential Causes." For our methodology, troubleshooting will consist of the following five phases:

1. Recognize the abnormal situation.
2. List and evaluate potential causes.
3. Choose and verify the cause.
4. Correct the problem.
5. Track the effect of the action.

Phase 1: Recognize the Abnormal Situation

Not every problem needs troubleshooting of the sort that you will learn in this chapter. Some problems simply require correction because the cause is known. If you run out of gas on the highway and your gas gauge indicates empty, you don't troubleshoot your car, you buy some gas. A key role of process technicians is to determine which situations require troubleshooting and which ones don't. Problems that have the potential for serious consequences if left uncorrected are called *abnormal situations*.

Process industry experts have studied process operations and provided a description of abnormal situations. They describe abnormal situations as "events that are outside of the desired operating conditions" and are generally characterized by the following symptoms: (1) alarm flooding, (2) the activation of protective system programs (ESDs, interlocks, etc.), and (3) accidents, such as leaks, fires, and explosions. The U.S. National Institute of Standards and Technology (NIST) estimates that abnormal situations cost the U.S. more than $20 billion annually, of which $10 billion was from the petrochemical industries.

In Chapter 19 we said an abnormal situation was "any unit upset or event that requires immediate action to prevent serious consequences, such as harm to personnel, damage to equipment, or a major environmental release." The only thing we might want to add to that is a clause that states "and the cause must be understood to correct the problem."

The troubleshooting process begins when you recognize an abnormal situation. From this definition of an abnormal situation, there are three conditions required for the problem to be considered an abnormal situation that requires troubleshooting:

1. There are significant potential consequences, such as harm to people, equipment, environment, and lost production.
2. The problem requires correction.
3. The cause must be understood to correct the problem.

Troubleshooting the abnormal situation starts with the recognition of an abnormal situation. This is the essence of troubleshooting. To help determine what is the abnormal situation you ask yourself the three Ws: *what*, *when*, *who* (Figure 20-3).

- Answering the *what* question determines (1) what specifically is the problem and (2) what indicates that a problem exists?
- Answering the *when* question determines (1) when did the problem actually begin and (2) where is the problem located?
- Answering the *who* question determines (1) who reported the situation and (2) who else has information about the problem?

Keep in mind that troubleshooting abnormal situations in process operations is serious business. Changing operating conditions in response to an abnormal situation—feeds, flows, temperatures, pressures—requires an understanding of cause before a new set of variables is introduced into the operation and further complicates the situation. As we have stated several times, there is no substitute for knowledge about the operation of a process and the process equipment.

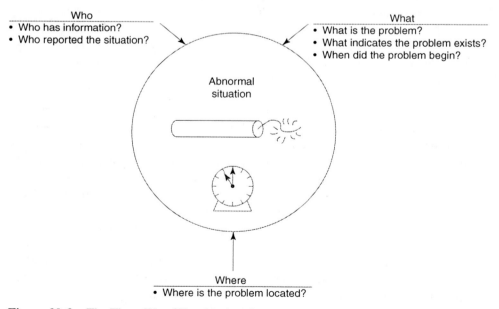

Figure 20-3   The Three Ws of Troubleshooting

Phase 2: List and Evaluate Potential Causes

Once the abnormal situation has been recognized, the next phase is to determine the potential causes of the situation. How should a skilled and knowledgeable process technician find the cause of an abnormal situation? The more someone knows about the subject, the more explanations—potential causes—they are capable of identifying. Spending time thinking about what causes might explain the abnormal situation is a phase in the troubleshooting process many people are tempted to shortcut. Experienced and well-trained technicians are capable of diagnosing the majority of all the possible causes of an operating problem. Many of the biggest failures in troubleshooting occurred because knowledgeable staff didn't consider all possibilities. Encouraged by groupthink, they jumped to a conclusion.

Some of the reasons for the failure to consider all potential causes are

- Taking action based without fully understanding the situation
- Choosing the obvious solution without recognizing all the factors in play
- Going with the most obvious cause because of a desire to solve the problem quickly

Wasting time trying an ill-considered solution because of panic or impetuousness may create bigger problems, such as losing production, increasing the cost and seriousness of the problem, adding new variables that complicate problem solving, or allowing the problem to escalate.

What Is a Potential Cause?   We've been referring to the list of causes as *potential* causes. A **potential cause** is anything that might explain the abnormal situation. Where do we look to find potential causes? One source is your technical knowledge of the unit or area where the problem is located. Another source is your operating experience. What are the odds that your potential cause is the real problem? Probabilities play a very important role in troubleshooting because the probability of a particular cause being the problem can vary significantly. It is important to classify the potential causes by their probabilities, which are high, medium, and low.

How to Evaluate a Potential Cause.   Evaluation of potential causes is a simple process of selecting based on evidence. Comparing possible causes against what is known about the abnormal situation shortens the list of potential causes. That's one of the advantages of taking the time to learn what is known about the abnormal situation: *what*, *when*, and *who*. Selecting isn't difficult: it is simply a matter of finding inconsistencies based on what is known about the situation. Do the data, information, and instrument readings support or reject a potential cause? Knowledge about the process again plays a critical role in selecting causes. People ignorant of their process unit can never contribute to the troubleshooting process. The danger to the troubleshooting process is to stop to making a list of potential causes too soon.

One tool used to rank the feasibility and effort of solutions is a probability/effort grid (see Figure 20-4). Assume your group came up with five potential causes for the present upset state of the unit. Now, you need to agree on which one or ones are the best to tackle first.

Using the probability/effort grid, your group can rank the problems you have identified by the amount of effort or time required to correct them. Also, the grid ranks the probability of that potential cause. To get started, the group lists and numbers each potential problem. Now, you draw a table with two columns and five rows to the right of the list. Rate each probability as high (H), medium (M) or low (L) and each effort as little (1), moderate (2) or great (3). Plot each solution's values on the 9-block grid. The upper left corner represents the highest probability and lowest effort. This is where you start.

Success in solving problems depends upon the quality of the information available to work with. It is critical once an abnormal situation develops that information be written down because events will build up and move faster. Facts and figures not written down will become vague and recalled incorrectly. All data should be dated and its source identified. In today's computerized environment, especially with DCS and LIMS systems, there is a wealth of data that can go back from the present time of the problem to hours or days ago.

Phase 3: Choose a Potential Cause and Verify It

From the list of potential causes, choose the one you want to test first. If the potential causes number more than one, judgment is required in selecting which cause to test first. Remember that troubleshooting is defined as *successful repair or correction.* If troubleshooting

Potential Causes	Probability	Effort
1. Bad temperature controller	M	1
2. Plugged line	L	3
3. Steam leak	M	2
4. Exchanger leak	H	1
5. High level of air in steam	H	2

Probability \ Effort	1 Little	2 Moderate	3 Great
High	4	5	
Medium		3	
Low	1		2

Figure 20-4 Probability/Effort Grid

takes more resources and time than the problem deserves, it can't be judged successful. Pick the most appropriate cause based on all the facts you have collected about the abnormal situation. Every potential cause has a cost measured in time and effort to test it. Not every cause will require the same amount of time and effort to test. Start with the causes that have the highest probabilities and are easiest to test. If you can rule out a fast and easy cause, one that takes five or ten minutes and minimum effort, it's often worth starting there first.

Methods to Isolate the Cause. In many cases, the leading candidates for your choice will be obvious by the facts and your knowledge about the process. But what if there isn't a cause with the kind of probability that justifies its choice? There are four methods that are effective and efficient ways to begin the isolation of the cause:

1. Half splitting—Sometimes there are a large number of potential causes with equal probabilities. The half-splitting method is the most efficient way to find the correct answer when there isn't a high probability cause at the top of the list. You divide the number of potential causes in half and check one of the halves. If the problem wasn't there, divide the remaining number of potential causes in half and check one of the halves. Keep dividing the potential causes in half and test the causes in each half portion until you've run out of causes.
2. Bracketing—The bracketing method breaks a process into components and then identifies which sub-process might contain the cause. Following this method for a car engine that won't start, the sub-processes that can initially be bracketed follow the "fuel-fire-power" breakdown. If the starter is getting power, the problem isn't with the electrical power system. If you have fuel going into the fuel injector, then the problem isn't lack of fuel.
3. Input/Output—This method works well in finding the source of product quality problems and some intermediate stream problems. Each major step in an integrated process (such as a reactor or distillation tower) is considered as a process. Assume we have a problem with the finished product leaving a reactor. The inputs to that process, both feed material and equipment and controls are tested to see if all is correct and functioning properly. Then the output and its equipment and controls are measured in the same way. This method will isolate the problem to a particular piece of equipment or system.
4. Series—This technique follows the entire system from beginning to end, searching for the problem one step at a time. It isn't very efficient and takes time, but it is sometimes the only way to find the source of some problems.

Verify the Cause. At this point in the troubleshooting process you have recognized the existence of an abnormal situation, collected the data, considered the potential causes, and selected what the team thinks is the most likely cause. You're now ready to take action but before you do, take the time to verify. Common sense says to take five minutes and double check your information about what is known about the situation before deciding to act on the selected cause. Verify that your theory explains why all of the data—the *what, when,* and *who*—supports your choice of potential cause. Do any of the facts or data contradict your conclusion about the cause?

Troubleshooting is a methodology, not an exact science. Though most of the facts and data support your conclusion about the cause, some may not. Your odds of being right are good but not great. Have the discipline to persist and seek more data. There will be pressure on you, and seeking more information will add a few minutes to the troubleshooting process, but what is the cost of being wrong about the cause because you didn't commit another five minutes to seeking more information? How much is another hour of production on your unit worth?

Phase 4: Correct the Problem

Training and experience often dictate what the most appropriate action is. Two points to keep in mind that even experienced technicians sometimes lose sight of in the anxiety of system problems are

1. There are often different solutions to solve the same abnormal situation. You want to choose the solution that best meets the critical goals of the company: safety, cost, and production.
2. The solution should never take you outside the boundaries of your approved operating authority and responsibility.

Read the last point again. Now consider the implications of your action. Will the solution you choose set in motion a new round of actions that may affect other processing units and greatly increase the original problem? If so, that will mean you will need to watch the process carefully and deal with the consequences of your corrective move. Corrective action is supposed to be a solution *but it can be the cause of an even bigger problem*. Make sure you do not step outside of the boundaries of your approved operating authority and responsibility. When in doubt, consult with someone of higher authority.

Taking Action and Correcting the Problem.   Finally, you get to take action. This is the only step in the troubleshooting process that involves real operating activity; every other step involves what sets man apart from the other animals: thinking, data analysis, and data collection. That speaks volumes about the nature of the troubleshooting process and the skills now needed by process technicians. Troubleshooting is a *thinking process that requires good analytical skills*. Always record the action you take to correct the problem. Remember, you can't remember everything and things will become blurry and move fast. Written documentation can be a blessing if your first move isn't your last move, which means you (the team) guessed wrong. You still have the original problem plus you have to reset conditions back to where they were before you guessed wrong. If you don't do that, you've made the problem more complicated.

Phase 5: Track the Effect of the Action

It is entirely possible that the first attempt at a solution won't be the correct one. Remember, troubleshooting is a methodology, not a science. If the first attempt doesn't succeed, one potential cause can be ruled out. Now we proceed to the next ranking suspect cause. But it isn't as easy as "try this, try that" when trying to correct an operational problem. Even where troubleshooting has identified the correct cause, it takes time for the results of a process change to appear in the process. It may be that the first move to address the abnormal situation will not be sufficient to bring the process back to normal operating conditions.

In many situations, successful corrective action will require a series of steps, each of which bring the process closer to normal. It is critical to monitor the results of the steps and look for the expected results. Recording the actions taken and knowing the amount of time required for the change to work its way through the system is essential. You should know the answers to these questions before you take corrective action:

- How long will it take for the effect of your correction to show up in the process?
- What parameters will show you the result of the correction?
- If the repair or correction works, how will you know?
- If it does not work, how will you know?

If the situation is not corrected we go back to the list of potential causes and choose the next appropriate choice. At this point, you will know what cause you can now rule out, and you'll have additional information about what may have changed in the situation from the initial condition. Henry Ford said it best: "Failure is the opportunity to begin again, this time more intelligently."

When Everything Fails

Some times there just isn't enough time. You're testing your third potential cause and the risk to people and equipment has become too high. Do you call it quits and shut the process down? Heroic efforts to save the unit may be the stuff of legend, but they are also the stuff of major accidents. It is always better to shut the process down than put people and equipment at risk. You've lost production but you haven't lost lives or equipment.

SUMMARY

Troubleshooting is the process of detecting problems, determining their causes, and then correcting the problems. An important part of determining the cause of a problem is knowing how the process operates under normal conditions. Troubleshooting consists of five phases:

1. Recognize the abnormal situation.
2. List and evaluate potential causes.
3. Choose and verify the cause.
4. Correct the problem.
5. Track the effect of the action.

One tool used to rank the feasibility and effort of solutions is a probability/effort grid. Using the probability/effort grid, your group can rank the problems you have identified by the amount of effort or time required to correct them. Also, the grid ranks the probability of that potential cause.

Input/output testing involves comparing the condition of a process entering a subsystem or vessel to the condition of the process leaving the subsystem or vessel. When bracketing is used to establish the boundaries of a problem, anything upstream or downstream of the boundaries is eliminated as a possible cause of the problem. Once a problem is bracketed to within a certain part of a process, using the serial method or the half-splitting method can narrow the source of the problem. The serial method of troubleshooting consists of checking components in a process one at a time, in sequence. Each time that the half-splitting method is used to check a part of a process, the remaining number of possible causes of a problem is reduced by half.

REVIEW QUESTIONS

1. Write the definition of *troubleshooting*.

2. List the five phases of troubleshooting.

3. Explain why troubleshooting is important to the processing industry.

4. List four operating expenses that increase when operating problems are handled poorly.

5. Define *abnormal situation*.

6. What launches a troubleshooting situation?

7. Define *potential cause*.

8. List the four methods used to isolate a cause.

9. Why is it important to verify the cause?

10. Why is it important for a technician not to step outside of the boundaries of their authority and responsibility?

Glossary

A

abnormal situation (or condition)–any unit upset or event that requires immediate action to prevent serious consequences, such as harm to personnel, damage to equipment, or a major environmental release.

absolute viscosity–the viscosity of a liquid that takes into account the weight, distance of flow, and speed of flow of a liquid.

accident–a happening that is not expected, foreseen or intended, frequently resulting from negligence and ending in injury, loss, or considerable damage.

affected employee–the person whose job requires them to use locked out or tagged out equipment on which servicing and maintenance is being performed, or whose jobs requires them to work in areas where such work is being performed.

air mover–an electrically driven fan or blower used to force air into a vessel.

alignment–the positioning of two or more pieces of equipment to form a line. Example: Motor and pump.

analyzer–an analytical device used to measure physical or chemical characteristics of a material.

API gravity–an arbitrary scale for measuring the density of oils adopted by the American Petroleum Institute. It runs from 0.0 (equivalent to a specific gravity of 1.076) to 100 (equivalent to a specific gravity of 0.6112). The greater the API gravity the lighter the oil. Water has an API gravity of 10.

appraisal costs–the cost of inspecting produced materials or services to ensure that it is error-free.

area rounds–an activity of the outside technician. Area rounds require the outside technician to spend time in the field walking through his area of responsibility to keep up-to-date on the status of operations and equipment.

assembly point–a meeting place to which employees report during plant-emergencies for the purpose of personnel accounting.

authorized employee–the person who locks or tags out machines and equipment.

autoignition temperature–the temperature of a flammable material that is so high that the material itself can serve as the ignition source for a fire. Each material will have its own unique autoignition temperature.

auxiliary system–a plant system that supports process operations, such as cooling water and flares.

B

backing in–the use of fuel gas or some unit material to sweep purge gas from vessels and piping into the flare header where the sweep gas is burned.

bag-house–a pollution control device that removes particles from a stream of air or gas.

batch operations–chemical processes or operations that proceed one step at a time making discrete batches of a material or product. Such an operation will complete one batch, then start the whole process all over again.

battery limits–the boundaries of area of responsibility for a process unit.

belt conveyer–a mechanical conveyer that uses a flat or troughed belt to transport bulk solids.

blanketing–the process of putting nitrogen into the vapor space above the liquid in a tank to prevent air leakage into the tank. Also referred to as a "nitrogen blanket."

bleed–the process of withdrawing a small portion of a contained material from a line or vessel by slightly opening a valve in the line or vessel.

bleeder valve–the valve used to bleed a line or vessel of fluid.

blind (blank)–a solid metal disc that is put into a process line, usually between two flanges, to insure that no material can flow through the line. Blinds are used when equipment is isolated from the process for maintenance or inspection work.

blind list–a list of all blinds that were installed in process lines during a shutdown, turnaround or similar operation. The blind list is used to account for all blinds and to prevent a start-up with some blinds still in place.

block in–to isolate a piece of equipment by closing the valves in all the lines to and from the piece of equipment.

block valve (isolation valve)–any valve that is intended to positively block (stop) flow.

blowdown–taking material out of a vessel to reduce the level or impurity concentration.

board man–see *Inside Technician*.

body language–postures and stances that communicate feelings.

boiler feed water (BFW)–water of high purity that has been chemically and/or physically purified of most suspended and dissolved solids.

boil up–the vapor evolved when heat is applied to a liquid.

bomb (sample)–a metal (or glass) cylinder with a valve on each end used to take gas or liquid samples. Bombs can be designed for high pressures (1500 psia or greater) or low pressures (10 to 25 psia).

bonding–a technique to dissipate static charges to prevent them from acting as ignition sources. Bonding involves electrically tying two pieces of equipment or piping together with a wire or other conductor to prevent dissimilar charges from forming on them.

bottom sediment and water (BS&W)–analytical term for emulsions of oil, water, and mud that come out of crude oil during storage.

breakthrough–the appearance of substances that should have been retained by an absorption, filtration, or adsorption system.

brine–an aqueous solution of calcium or sodium chloride used in a refrigeration system.

British thermal unit (BTU)–the amount of heat required to raise the temperature of 1 pound of water 1°F.

bucket elevator–a mechanical conveyer consisting of a series of buckets mounted on a chain or belt inside a casing; used for transporting materials vertically or along an inclined path.

bulk density–refers to the weight of a quantity of solid material per unit of volume, such as pounds of plastic pellets per cubic foot in a container.

burner–a device used to introduce, distribute, mix, and burn a fuel.

C

carry-over–unwanted liquid or solid material that is carried out of a fractionating column, absorber, or vessel in the overhead.

CARSEAL–a wire loop used on the hand wheels of certain critical valves. The wire loop is not to be broken except within safety guidelines.

cascade–refers to a control system involving two or more controllers.

cascade control–ties the operation of one controller to the operation of a second controller. The output of one controller is used as the setpoint for the other controller.

catalyst–any substance that affects the rate of a chemical reaction but is not part of the reaction.

catwalk–name given to walking areas on top of structures or tanks.

caustic–sodium hydroxide or potassium hydroxide. A basic solution.

cavitation–a condition inside a pump wherein the liquid being pumped partly vaporizes due to temperature, pressure drop, and so on.

certificate of analysis (CofA)–a signed certificate listing and testifying to the analytical values of products undergoing custody transfer.

certification sample–a sample collected from a batch or lot of finished product that is analyzed to provide data that guarantee or certify that the particular material meets all quality requirements.

channeling–an open flow path through the bed allows the feed to avoid intimate contact with the adsorbent.

check sample–a sample submitted to investigate or confirm a suspicious or abnormal condition.

chief technician–see *Lead Technician.*

chemical test–an analytical test that destroys or changes the chemical structure of the substance being sampled.

clear–preparing process piping or equipment for maintenance, inspection, or other work. Clearing involves removing all traces of process material by draining and purging with steam or an inert gas like nitrogen.

cloud point–the temperature at which paraffin waxes or other solid substances begin to crystallize out or separate from a solution when oil is chilled under specific conditions.

coalesce–to cause small droplets to combine to form larger drops that settle by gravity.

color–(the color of a chemical.) Generally, the clarity and hue of yellow represents a poor quality. A high quality is water-white or clear.

communication filter–perceptions receivers of a communication use to interpret the message. They include knowledge level, biases, special points of view, prejudices, and so on. Filters can alter the message, or even block communication altogether.

composite sample–a sample composed of aliquots taken at different times during an operation, or from different locations in the material sampled. The blended aliquots form one sample that represents the entire time period or quantity of material involved.

composition–a process variable determined by testing. A quantitative test that determines the amounts of the various chemical compounds in the sample.

condensate–liquid that is collected by cooling a vapor.

condense–to transform from the gaseous state to the liquid state.

condenser–a heat exchanger used to condense a vapor to a liquid.

conductivity–the measurement of a substance's ability to conduct heat or electricity.

cone roof tank–a type of atmospheric storage tank that has a fixed roof and a cone-shaped roof with one or more internal support columns.

confined space–safety term indicating an area with limited opening for entry and exit that makes egress difficult in an emergency.

console–a panel (usually associated with a computer) containing gauges and controls for controlling a process.

constant level oiler–an oil bowl mounted on the equipment by a connecting pipe. Oil flow is driven by gravity and initiated by oil level in the equipment sump.

continuous operations–constantly running equipment or processes to produce material or product. The equipment is operated at a standard set of conditions while raw materials are continuously added to the equipment and the finished product is continuously removed. No unique batches of material are produced.

continuous purging–a common purging technique in which the gas or vapor to be purged is diluted or displaced with nitrogen, causing the gas to be vented from the vessel. Also referred to as sweep purging.

control charts–graphical representation of process statistical data that can be used to identify process variations and trends.

control loop–a collection of instruments that work together to automatically control a process.

control room–a room from which a process unit is controlled.

control room technician–the process operator is responsible for running a unit or plant from the control room and for all phases of the operation for the entire unit (or plant).

control sample–a sample of known quality that is used as a reference when testing other samples.

control valve–an automated valve used to regulate and throttle flow.

controller–an instrument designed to maintain a process variable at setpoint.

copper strip corrosion test–a physical test that will indicate the presence of substances that will corrode copper and other metals.

corrosion–the gradual destruction of a metal by chemical processes.

cost of quality (COQ)–the costs specifically associated with the achievement or non-achievement of product or service quality.

critical path–a work plan for a job, or operation that arranges the sequence of jobs to best utilize the time and resources available.

cryogenics–extremely cold gases and liquids which are at a temperature of $-148°F$ or colder.

custody transfer–the transfer of ownership of product.

cut–an individual fraction obtained during the distillation of petroleum products.

cut point–the temperature in a distillation tower that is used to obtain a desired product.

D

daily operating instructions (DOI)–instructions written by the unit superintendent or operating engineer for the tasks to be done during the day shift. Also called daily order book (DOB).

daily order book (DOB)–see *daily operating instructions.*

daily operation records–routine checklists completed during the shift, data sheets filled in as the technician completes area rounds, and shift reports maintained as historical records.

dead spot (dead leg)–a place in process piping or equipment where flow can stop and material can stagnate.

dehydrogenation–the removal of hydrogen.

deluge system–a water spray system used in controlling fires or heat.

demister–equipment used to remove mist from vapors.

dense phase transport–pneumatically conveying solid powders or granules through a pipe or tube by pushing slugs of solid materials down the pipe with bursts of air.

density–refers to the amount of space occupied by a particle of a certain weight. The smaller the amount of space occupied for a given weight, the greater the density of the particle.

derailer–a device that fits over the rails on a railroad track and prevents oncoming tank cars from running into the spotted tank car.

dew point–the temperature at which a vapor or gas begins to condense to a liquid.

differential pressure–the pressure difference (ΔP) between the inlet and outlet pressure at two different points.

dike (bund, fire wall)–a wall (earthen, shell, or concrete) built around a piece of equipment to contain any liquids if the equipment ruptured or leaked.

dilute phase transport–pneumatically conveying solid powders or granules through a pipe or tube by suspending the solid particles in a continuously moving air stream.

dilution purging–a technique to remove residual process materials or air from a process system which involves repeatedly pressuring and then venting the system using an inert gas.

dip tube–a tune that extends down into a vessel usually for level or interface measurement or for sampling a particular location in a vessel.

distillate–the product of distillation formed by condensing vapors.

distillation–the separation of a mixture based on boiling point using the process of evaporation and condensation.

distributed control system (DCS)–a process control system that uses a computer(s) for the control of a process.

diverter–a valve designed to redirect the flow of material within a system.

drilling–a technique for cleaning plugged process lines and equipment, by mechanically drilling out the plug. This technique is frequently used to clean heat exchange tubes.

drumming–the filling and labeling, either manually or automatically, of drums (5, 42, 55 gallon) with product for shipment.

E

emergency response–dealing with a situation that poses a threat to either the safety of the process and personnel or to continued operation of the process. Such a threat may be a fire, explosion, toxic release, and so on.

emergency situation–a situation that poses a threat to either the safety of the process and personnel or to continued operation of the process. Such a threat may be a fire, explosion, toxic release, and so on.

end point (EP)–an analytical term, the highest temperature indicated when a hydrocarbon is subjected to a laboratory distillation analysis. Also called final boiling point (FBP).

endothermic reaction–a chemical reaction that takes in heat.

entrainment–liquid droplets or solid particles carried along in a vapor stream, generally due to the velocity or turbulence of the vapor.

erosion–the wearing away of material by fluid moving across its surface.

exothermic reaction–a chemical reaction that generates heat. Such reactions can get out of control if the temperature of the system is not controlled.

expansion joint–a type of joint used in piping that contains a telescoping section or bellows to absorb the strain caused by the expansion or contraction due to changes of temperature in a metal.

extruder–a device that forces ductile or semi-soft solids through die openings to produce a continuous film, strip, or tubing.

F

fail closed–a term used to designate the failure mode of a process control valve if the operating source, usually instrument air, is lost. A fail-closed valve will close if an instrument air failure occurs.

fail open–a term used to designate the failure mode of a process control valve if the operating source, usually instrument air, is lost. A fail-open valve will open fully if an instrument air failure occurs.

fail-safe mode–a term used to designate the safe failure mode of a process control valve if the operating source, usually Instrument air, is lost. The fall safe mode may be open or closed depending on the process.

failure costs–the cost of doing something wrong. These are costs involved in damaged equipment, off-specification product, injured employees, public relations, etc.

feed tray–the tray located immediately below the feed line in a distillation tower.

feedback–the two-way flow of information during communication that provides each party with some assurance that the other party has understood the message.

feeder–a mechanical device that stimulates the movement of solid particles and meters bulk material to a process.

field technician (outside technician)–an operator with responsibilities for an outside area of a unit. They spend much of their working time in the field monitoring process conditions and operating equipment.

filter receiver–a piece of equipment used for removing dust from air. Typically, it will have a cyclone bottom section and an upper section equipped with a multiplicity of cloth or synthetic filter bags to retain fine dust particles (similar to a vacuum cleaner bag).

final boiling point (FBP)–the maximum temperature required to boil the heaviest component of a complex mixture; also called the end point.

fire alarm–a signal that provides warning of a fire or potential fire in some specific area.

fire eye–a device used to detect the interruption in a fire or flame, normally associated with furnaces.

fire watch–a technician who continually or periodically checks an area where hot work is occurring in a hazardous area.

first break–the initial opening of a process line or process equipment after it has been in service. The first break is done assuming that process material is still present in the system and appropriate precautions are taken.

fixed orifice trap–a steam trap that contains a set orifice in the trap body and continually discharges condensate. They are self-regulating.

fixed shift schedule–a working schedule that assigns each employee to a permanent shift. If working a fixed shift schedule, the employee will not rotate shifts.

flame arrestor–an in-line device installed under equipment vents to prevent flame propagation through the line.

flame impingement–a flame striking the surface of equipment, such as furnace tubes.

flame retardant clothing (FRC)–clothing that has been made of material resistant to flash fire. Such clothing can provide some protection to the technician in the event of a sudden flash fire.

flammability range–the concentration range of a mixture of a flammable material in air (or oxygen) that will burn if an ignition source is present. Each material has a unique flammability range.

flammable mixture–a mixture of a flammable material in air (or oxygen) that will burn if an ignition source is present.

flare–equipment designed to safely burn excess hydrocarbons.

flash point–the minimum temperature at which a liquid will give off enough vapor to form a flammable mixture with the surrounding air.

flash steam–low energy steam that flashes from hot condensate when condensate pressure is reduced.

float and tank gauge–an automatic tank gauging system designed for atmospheric tanks and low-pressure tanks that continuously measure level in bulk storage tanks. It uses a float attached to a tape to indicate the level in a tank.

float trap–consist of a ball float and a thermostatic bellows element. As condensate flows through the body, the float rises or falls, opening the valve and releasing condensate according to the flow rate.

floating roof tank (FRT)–a storage tank that uses a roof that floats upon the surface of the stored liquid and is used to decrease the vapor space and reduce the potential for evaporation.

flowability–a measure of a material's ability to flow steadily and consistently as separate particles.

fluidized–mixing fine solids with a gas, such as air or nitrogen, for the purpose of making the solid act like a fluid and flow. The fine solid (dust-like solid) will flow just like a gas or liquid.

fouling–a buildup of foreign material on the surface of the containing vessel or tube.

Frac tank–wheeled trailers that serve as temporary storage tanks and that can provide up to a 20,000-gallon capacity storage per tank.

fractional distillation–the separation of a liquid by distillation into fractions (desired groups).

fractionating tower (distillation tower)–a vessel that separates the component of a mixture by a distillation process.

free water–water that is not dissolved in or suspended in a sample. Water that has separated from the sample.

friable–materials that break apart easily and must be handled by equipment designed to minimize damage to the fragile particles.

function deteriorating breakdown–performance deterioration due to wear and tear, loss of fit between parts, low-voltage, poor insulation, or leakage.

function reduction–equipment that does not operate correctly but can continue to operate, but may result in defective product, reduced output, frequent stoppages, noisy operation, reduced speed, or unsafe conditions.

fuel gas–a combination of gases, such as methane, ethane, propane, and butane, which are used in furnace burners.

G

gas chromatograph (GC)–an instrument which uses the difference in molecular weight or attraction of different materials to analyze the composition of process streams. GCs only analyze materials in the vapor state.

GC analysis–measurement of the chemical composition of materials using a gas chromatograph for the measurement.

gesture–a form of communication that connotes feelings involving the hands, eyebrows, eye contact, and so on.

grab sample–a sample of a material or process stream collected at an instant in time. The grab sample represents the material sampled at some unique time and/or place.

gravimetric feeding–a method of weighing and metering bulk solids in which a gravimetric feeder maintains a set feed rate by weight per unit of time.

gravity lubrication–a low-pressure lubricating system that uses the forces of gravity to provide lubricant flow.

grounding–establishing an electrically neutral system by attaching an electrical conductor to a part of the system and also to an object at ground or zero potential. Grounding serves to reduce or eliminate static charge.

H

hand wheel–a device attached to the stem of a valve that permits the opening and closing of the valve.

hazard–something that poses a risk or danger.

head pressure–the pressure exerted by a column of fluid.

heat of vaporization–the heat absorbed or given up when a substance changes state from a liquid to a vapor.

heavies–see *heavy ends.*

heavy ends–a distillation term referring to the material boiling at the highest temperature.

heel–liquid material left in a cargo trailer or tank car that is supposed to be empty.

hidden defects–causes of equipment breakdown that evade notice.

high-pressure gas cylinders–contain gases (hydrogen, argon, oxygen, and so on) under high pressure (2000–2400 psig). They are commonly made of carbon steel. Many of the gases are used for analyzer service.

hopper–a container, often funnel shaped, from which the contents can be emptied slowly and evenly.

hot work–any work (grinding, sanding, welding, and so on) that can generate a source of ignition to flammable vapors.

housekeeping–the act of keeping a work area and equipment in a safe, clean, and usable condition.

hydraulic conveying–a method of moving bulk solids in which solid particles are suspended in a flowing liquid to form a slurry, which is pumped to the desired location.

hydrogenation–the chemical addition of hydrogen to an unsaturated compound.

hydrometer–an analytical instrument for determining the various gravities of liquids.

hypoxia–a lack of oxygen.

I

idlers–pieces of equipment used on conveyer belts that guide, support, or carry the belt.

ignition source–any item or action that can generate sufficient energy to ignite a flammable mixture. Sparks, friction, high temperatures, electrical arcs, and so on can act as ignition sources.

incident–a safety term indicating any event that leads to or could have led to an accident.

inert blanketing–a technique used to prevent air intrusion into systems containing flammable or reactive materials. Blanketing uses a gas like nitrogen to form a low-pressure pocket of inert gas over the liquid in a container or piece of equipment.

inert gas–a gas which is not chemically active, such as helium, argon, and nitrogen.

initial boiling point (IBP)–an analytical term related to distillation. The temperature at which the first drop of material distilling fall into the distillation receiver.

innage–a measure of a depth of liquid in a tank or other container. Normally, tank gauges are innage gauges.

inside technician (board man)–the technician responsible for the operation and control of the entire process unit from the board screen (console) in the control room and for making adjustments as needed.

instrument air (IA)–plant air that has been filtered and/or dried for instrument use.

interlock–devices that automatically break a circuit when an unsafe situation is detected. Often used to shutdown systems.

intermodal container–a container, used for shipping chemical materials, that can be transferred as a package from truck, to rail car, to barge, or ship without ever unloading the contents of the container. Intermodal containers are available for solids, liquids, and gases. Also referred to as an ISO container.

ISO container–see *intermodal container.*

isolation–in reference to energy sources, is the use of lockout and tagout devices to isolate a piece of equipment from all sources of energy.

inverted bucket trap–a trap that has a "bucket" that rises or falls as steam and/or condensate enters the trap body. When steam is in the body, the bucket rises and closes a valve. As condensate enters, the bucket sinks down and opens a valve that allows the condensate to drain.

J

job aid–a memory jogger or short list of steps taped to equipment with a signature of an approving authority and dated.

K

kinematic viscosity–the ratio of the absolute viscosity of a liquid to its specific gravity at the temperature that the viscosity is measured.

knockout drum–a drum located between the flare header and the flare designed to separate liquid hydrocarbons from vapors being sent to the flare.

knockout pot–a small vessel or lowest section of a vessel used to remove (knockout) liquid droplets or impurities.

L

latent heat–the heat energy that produces a phase change in a material without causing a temperature change.

lead technician (chief technician)–a technician that accepts responsibility for a crew and performs both management and technician work. Management work may consist of recording time sheets, calling out overtime workers, and so on.

leak test–a test for locating leaks in piping and equipment by filling the piping and equipment with a gas (nitrogen) and applying liquid soap to the exterior area to be tested.

leg–a vertical line on the bottom of larger lines used to drain the larger line.

light ends–(1) The lighter fraction of hydrocarbons in a mixture. (2) The first four hydrocarbons, methane, ethane, propane, and butane. Also called lights.

lights–see *light ends.*

line up–to open the necessary valves from point A to point B.

linear expansion–the increase in length of a metal as it is heated and begins to expand.

liquefied petroleum gas (LPG)–a mixture of hydrocarbon gases that has been liquefied by increasing system pressure.

live steam–saturated or superheated steam above atmospheric pressure. Live steam is useful as an energy source in contrast to flash steam.

lockout–a technique that uses a mechanical lock, to isolate process equipment from their source of energy, or inactivate the equipment so that it can be worked on safely.

loss-in-weight feeder–a feed mechanism that works by keeping track of and controlling the amount of weight lost from the feeder container. A typical loss-in-weight feeder includes a scale and a controller.

lower explosive limit (LEL)–the lowest concentration of flammable material in oxygen or air that will support combustion. Any mixture with less flammable material will not burn.

lubricant–any substance that can provide a protective film between moving mechanical parts. A lubricating film will cushion and reduce friction, providing protection against wear and shock.

lump breaker–a mechanical device with teeth that aids in the flow of solids. The rotating cutter teeth inside the lump breaker break up the lumps and enable the material to be moved without clogging up the conveying systems.

M

M–abbreviation symbol for one thousand.

MM–abbreviation symbol for one million.

makeup–the feed needed to replace that which is lost by leakage or normal use in a closed circuit, recycle operation.

making rounds–a scheduled systematic surveillance of an area of responsibility that may include collecting samples, recording data, monitoring equipment, and minor preventative maintenance.

man-hours–the number of hours required to do a particular job.

manifold–a pipe with one inlet and several outlets, or the exact opposite, a pipe with one outlet and several inlets.

manual lubrication–lubricant supplied by squirt-type or plunger-pump oilcans and grease supplied by manual grease guns.

material handling–the loading, unloading, or transfer of bulk materials from or between cargo tanks, various vessels storage tanks, or process units.

material safety data sheet (MSDS)–a document that provides information concerning the hazards of a chemical material and provides information on using the material safely.

mean-time-between-failure (MTBF)–the average time a piece of equipment runs before it fails (breaks down).

mechanical conveying–the use of mechanical equipment, such as conveyer belts and bucket belts to move bulk materials.

megging–measuring the electrical resistance between the rotating and stationary coils of an electric motor to determine if sufficient moisture is present to cause a short circuit on start-up.

motor control center (MCC)–a room or building that contains electrical switchgear, circuit breakers, transfer switches, and so on.

motor starter–a device used to control current to an electric motor. Also called switchgear.

N

net positive suction head (NPSH)–the equivalent head in feet of liquid necessary to push the required fluid into the impeller.

non-routine operation–operating a chemical process outside the normal standard operating range.

O

off-Specification (off-spec)–a product that does not meet purity specifications.

oil mist lubrication–lubrication designed to provide microscopic air-borne oil particles that lubricate rotating equipment and other equipment.

on-test–product that is within specification.

operating procedures manual–a document that provides information on correctly running a piece of equipment, system, or entire unit.

operator logbook–a notebook or other written document used by the process technician to record the events of their working shift.

operator maintenance–maintenance performed by operating technicians on the equipment in their area of responsibility.

oscillating conveyor–see *vibrating conveyor.*

outage–the difference from the top of a container to the surface of the container material. Also referred to as ullage.

outfall–the place of discharge of a drain or pipe. Usually associated with environmental samples.

outside technician–see *field technician.*

overhead–the product or stream coming off the top of a distillation tower.

P

packed tower–a tower filled with packing material instead of trays.

packing–(1) Special material used in packed towers to aid in distillation. (2) Material used to prevent leaks in valves and pumps.

packing gland–a sealing device that prevents leakage around the stem of a mechanical valve.

pad–a concrete slab.

palletizing–the loading of small cargo containers on a pallet and banding or shrink-wrapping them to the pallet.

partial pressure–the amount of pressure attributable to a particular gas present in a confined gas mixture.

pelletizer–a device with rotating knives that cut the strands of a polymer as they leave the die holes at the exit point of an extruder, creating pellets of the polymer.

permissive–a special type of interlock that controls a set of conditions that must be satisfied before a piece of equipment can be (is permitted) started.

personal protective equipment (PPE)–safety equipment used by an individual for protection against injury. PPE include gloves, safety glasses, respirators, safety shoes, and so on.

petrochemicals–chemicals derived from petroleum, which can be used to produce consumer products rather than being used for fuel.

pH–a symbol for the measured strength of an acid or base.

physical change–a change that does not affect the chemical structure of a substance.

physical test–an analytical test that does not affect the chemical structure of a substance.

plant air (service air)–the supplied air system used for pneumatic tools and other air operated equipment. Normally, it is not filtered or dried.

pneumatic conveying–transport of granular or powered solid materials using moving air or other gas.

polymers–large molecules built by the repetitious combining of small simple chemical units.

portable liquid containers–these are usually drums of varying size and may be made of high density plastic or metal. These may contain special solvents or other materials not needed in great volume but in a volume normally bigger than a drum.

potential cause–anything that might explain what is causing an abnormal situation.

pour point–analytical term for the lowest temperature at which a liquid examined under prescribed conditions will flow under the influence of its own weight.

power failure–a process emergency that is caused when electrical power to the process is suddenly lost.

348

predictive maintenance–the use of special instrumentation to monitor important operating parameters on rotating equipment to detect incipient failure and predict when failure is imminent.

pressure–force exerted per surface area.

pressure lubrication–lubricant circulated by a pump to high-speed equipment.

pressure purging–see *dilution purging.*

pressure regulator–a mechanical instrument that is used to control the discharge pressure of gas from a high pressure cylinder or other pressurized container.

pressure storage tank–a tank used to store volatile liquids that have a Reid vapor pressure greater than 18 psig. The three types of pressure storage vessels are drums, spheres, and spheroids, and are mainly used for the storage of liquefied petroleum gases (LPG), such as butane and propane.

prevention costs–the cost of preparing for an activity and performing it error-free.

prime–to displace air or vapor from a pump with liquid.

process stream–process materials making their way through the unit before they become a finished product or consumed in the process.

process safety management (PSM)–an OSHA standard, 29CFR1910.119, concerning the safe management of highly hazardous materials.

process unit–an integrated group of process equipment used to perform a task or produce a product.

product specification sheet–see *specification sheet.*

programmable logic controllers (PLC)–microprocessors used in industrial process control.

puking–a condition that occurs when the vapor pressure of a liquid is so great that it forces the liquid up a distillation tower or out the overhead line.

pumparound–a means of removing heat from the middle or lower part of a fractionating tower by pumping a drawn off component through heat exchangers and back into the tower.

punch list–a maintenance list of work that needs to be done on equipment. Also called a work list.

purge–a procedure in which an inert gas is used to remove undesired gases from a vessel so as to reduce the undesired gas to a safe or acceptable level.

Q

quality–conforming to all aspects of customer requirements.

R

rattling–the use of a hammer or some device to cause vibrations in piping or valves that may be plugged. The vibrations may loosen the plug or fragment it.

reactor–a vessel used to mix two or more substances together and initiate and sustain a controlled reaction to create a new substance(s).

reflux–the portion of condensed vapor that is returned to the top tray of a distillation column.

reflux drum (accumulator)–a drum that receives condensed vapors that are to be used as reflux.

Reid vapor pressure test–an analytical method used for the determination of the vapor pressure of fuels for spark-ignition engines.

relaxation–the allowance of time for a fluid transferred to a storage tank to slowly dissipate any charge of static electricity built up during the transfer process.

ring oiler–a lubrication system that uses a ring of brass or steel, with a diameter about one and one-half times that of the rotating shaft. The ring is placed on the shaft so that the two will rotate together and the ring will splash oil from the oil sump onto internal moving parts.

rodding–a technique used to mechanically unplug a process system by poking and probing with a metal rod.

root cause–the cause, which when identified, eliminates the problem forever.

rotary valve–a type of feeder that is basically a rotor with vanes that turns inside a housing and is used to assist and control material flow.

rotating shift schedule–a working schedule that rotates the employee from day shift to night shift and then back again on some periodic frequency, with time off usually scheduled between shift changes.

rounds list–a list, either digital in a data gatherer or hardcopy, that identifies what equipment technicians are to look at, what they are to record, and what tasks they are to perform.

runaway reaction–a exothermic reaction that has become unstable and propagates at an increasing rate because of the heat generated by the reaction.

rundown–the finished product being pumped from a process unit to storage.

rundown tank (day tank)–tanks that hold the current unit production until product purity is determined and the tank material can be transferred to field storage tanks.

run in (break in)–running new or repaired equipment under no load or a light load to verify operability.

running blind–this is a blind with the function of blocking certain lines, vents, or drains at all times when the unit is running. They are called running blinds because they blind equipment while the unit is running.

rupture disc–a pressure relief device that will burst or tear if the pressure reaches its bursting limit. Usually, a thin piece of metal or composition material designed to rupture at a certain pressure.

S

sack container–a container for collecting and transporting granular or powdered solids, which is made from woven material. In appearance, a sack container may look like a large fabric shopping bag capable of holding several hundred or a thousand pounds of material.

safety–eliminating or avoiding hazards in your work area or procedures.

safety manual–a document that serves as a reference for plant safety procedures and policies.

sample–a representative portion of a material collected for analysis.

sample cylinder (bomb)–a cylindrically shaped container used to collect hazardous process samples which may be at high pressures or temperatures. The container has valves at both top and bottom to help control the sampling operation when such a cylinder is used.

sample schedule–a document that provides information on what materials to sample, when to sample, where to collect the sample, how often to sample, and what testing to do on the sample.

sample thief–a device used to draw sample from a storage tank.

sampling–the process of collecting a representative quantity of material for analytical or evaluation purposes.

saturated steam–steam that has been reduced in temperature to a point slightly above condensing.

Saybolt color test–a test for determining the color of gasoline and burning oils; values range from +30 for gasoline to –16 for furnace oils. Water white is +30.

Saybolt viscosity–the time in seconds required for 60 cc of oil at a given temperature to flow through a calibrated orifice.

scale–a surface coating on equipment that grows by oxidation (rust) or chemical deposition or crystallization, interfering with flow and heat transfer.

screw conveyer–a conveyer consisting of a helical screw that rotates upon a single shaft with a stationary trough or casing, used to move bulk material.

screw feeder–a mechanism for handling bulk solids in which a rotating helicoid screws moves the material forward, toward, and into the process unit.

scrubber–a piece of equipment designed to remove a particular component from a gas or vapor stream, usually by countercurrent contact with a liquid.

scrubbing column–a vessel designed to act as a scrubber.

seal flush–used in conjunction with a mechanical seal to minimize or dilute possible product leakage. Also acts as a cooling medium.

seal oil–oil used as the sealing medium to prevent the passage of fluid or gas from one chamber to another.

sensible heat–the amount of heat required to raise or lower the temperature of a substance without causing a change of state.

setpoint–the point at which the controller is set to regulate the process.

shelter in place–a release of gases or vapors into the community that causes homeowners, seeking to avoid harm from the release, to stay indoors, close all windows and doors, and shut off air conditioning.

sidestream–the product removed from the side of a distillation tower.

skills–actions which demonstrate the ability to perform tasks required in a job situation.

slinger rings–a ring, usually found in rotating equipment, that rotates on a shaft, taking oil from a sump and splashing it onto gears and bearings.

slop–material that should be an intermediate stream or final product but is not. Off-spec material.

slow roll–operating a piece of equipment at a speed below its normal speed.

slurry–a mixture of solid particles and a liquid.

SOCs–see *standard operating conditions.*

SOPs–see *standard operating procedures.*

specific gravity–is a measure of the ratio of mass of a given volume of material at 23°C to the same volume of deionized water.

specific heat–the quantity of heat required to increase 1 pound of a substance by 1°F.

specification sheets–documents that provide quality information about a product. Specification sheets will usually provide the product name or code, the various product properties that are analyzed, The test methods used for analysis, the goal, and the maximum and minimum range for acceptable product.

sphere–a pressure storage vessel shaped like a ball and supported above grade on tubular columns mainly used for the storage of liquefied petroleum gases.

splash lubrication–lubrication technique in which mechanical parts move in a reservoir of oil. As the parts move they splash the oil onto other surfaces and parts that need to be lubricated.

spotting–the placing of a railcar, tank truck, or barge in the correct position at a loading rack or dock, and inspecting it before the transfer has begun.

Staging–the critical placement of temporary buildings, power, air, tool cribs, field parts, materials warehouses, and runoff containment barriers, in preparation for a turnaround.

standard operating conditions (SOCs)–the operating ranges for the variables of a process. SOCs will include flow rates, pressures, temperatures, and levels of all important process parameters. Information will usually include a goal and a maximum and a minimum value for all important process parameters.

standard operating procedures (SOPs)–a set of directions or instructions that defines the particular steps to take when a certain situation or condition occurs.

standby–jargon for a technician who observes a potentially dangerous task and is prepared to render aid if something goes wrong.

start-up–the process of starting the operation of a unit or an individual piece of equipment which has not been operating for some period of time.

static tank–a tank that is neither receiving product nor releasing product. There is no flow in or out of the tank.

statistical process control (SPC)–a statistical tool used to continuously monitor a process's performance with charts and graphs.

steam condensate–water of almost distilled water quality that contains a substantial amount of heat, both latent and sensible.

steam tables–a document that provides information about the physical properties of steam.

steam tracing–tubing installed adjacent to piping and covered with insulation that supplies heat (from steam) to the piping.

steam trap–a device used to remove condensate from the steam system.

strapping of a tank–measuring the amount of liquid in a tank.

stroke a valve–testing a control valve to determine if it can be accurately controlled.

stuffing box–the area around a moving shaft or stem that contains packing material designed to prevent the escape of process fluids.

suction–normally, refers to the inlet side of a pump, compressor, fan, or jet.

suction pressure–the pressure in the line leading into the pump.

sump–a pit or tank that receives and temporarily stores drainage at the lowest point of a circulating or drainage system.

superheated steam–steam that has been heated to a higher temperature than its boiling point. Superheated steam can release both sensible and latent heat.

SuperSack™–a trade marked name for a sack container.

surging–a sudden change in the flow, level, temperature, or pressure.

sweep purging–see *continuous purging*.

swing line–a line or section of pipe that may be raised or lowered for pumping into or from a tank.

switch personnel–workers who operate locomotives or railcar movers and move railcars about within the plant. They move the empty tank car to the track scale, tare it, and then move it to the loading rack.

switch rack–a panel containing electrical switches and breakers for motors and pumps.

switchgear–a device used to control current to an electric motor. Also called a motor starter.

T

tagout–a technique which uses an information tag to isolate process equipment from their source of energy or inactivate the equipment so that it can be worked on safely. System tag out is used when there is no place to use a lock for a system lockout.

tails–the process flow taken off the bottom of a distillation tower, sometimes called the "bottoms."

tank car–a railroad car designed to carry liquid materials.

tank farm–an area containing storage tanks.

tank gauging–the generic name for the static quantity determination of liquid products in bulk storage tanks.

tank truck–a truck trailer designed to carry liquid materials.

TAR–see *turnaround*.

technical manual–a document that provides technical information about the design and construction of a chemical plant or process unit and its equipment.

thermal shock–the sudden expansion or contraction of a material when subjected to a large temperature change.

thermodynamic trap–a trap that has a disc that rises and falls depending on the variations in pressure between steam and condensate. Steam will tend to keep the disc down or closed. As condensate builds up it reduces the pressure in the upper chamber and allows the disc to move up for condensate discharge.

thermostatic trap–a trap that has a metallic corrugated bellows that is filled with an alcohol mixture that has a boiling point lower than that of water. The bellows will contract when in contact with condensate and expand when steam is present.

throttle–the control of flow through a line with a valve.

thrust–the force exerted endwise to a shaft to give forward motion.

thrust bearing–a bearing that prevents the axial movement of a shaft.

tightness testing–testing a process system for leaks by applying pressure (or vacuum) and inspecting the joints. Tightness testing may also evaluate a maximum operating pressure for the system.

tote bin–square portable containers of varying volume for the transportation of small quantities of liquids or solids. They have lifting handles, and depending on liquid or solid service, valves (liquid) or hatches and silo bottom configuration (solids).

trap–a device installed in a line to remove condensate and conserve steam.

trip–a term used when a piece of electrical or mechanical equipment is stopped either manually or automatically.

troubleshooting–to find the cause of a problem and correct it.

turnaround (TAR)–the shutdown period for a process unit, usually for mechanical reconditioning and upgrades.

U

ullage–see *outage*.

uninterrupted power source (UPS)–an emergency electric power supply consisting of a bank of wet cell batteries that supply power for computer and other electrically powered control instrumentation.

unit technical manual–a manual containing detailed information about the unit process and equipment.

upper explosive limit (UEL)–the highest concentration of flammable material in oxygen or air that will support combustion. Any mixture with more flammable material present will not burn.

utilities–one of the non-processing (support) facilities of a processing plant, usually producing steam, electricity, cooling water, refrigeration, utility and instrument air, de-ionized water, and effluent treatment.

V

vacuum breaking valve–a valve that opens when exterior pressure on a vessel is greater than the pressure inside the vessel.

vacuum purging–is used on vessels that can withstand a vacuum. Once the vessel to be purged has been pumped out, a vacuum pump is used to evacuate the vessel. Injecting nitrogen then breaks the vacuum, time is allowed for mixing, and the vessel is vented. The process is repeated until the vessel is considered to have a safe level of residual vapors.

vent–to release or depressure to atmosphere or vent header.

vibrating conveyor–a mechanical conveyer in which a vibrating or oscillating action causes material to move. Also called an oscillating conveyor.

viscosity–the resistance to flow displayed by a fluid.

viscosity test–a test that measures a fluid's viscosity.

volume expansion–the tendency of a liquid to increase in volume as it is heated.

volumetric feeding–a method of metering bulk solids that feeds material into a process at a set volume per unit of time.

W

water flooding–a technique used to purge a system free or residual process materials (or air during a start-up) which involves filling the system and vessels with water and allowing the water to overflow and flush the system.

water hammer–shock forces generated in a steam system when slugs of condensate remain in the system.

wear ring–rings within a pump that allow the impeller and casing suction head to seal tightly together without wearing each other out.

weigh-belt feeder–a feeder that weighs solids on a moving belt. Material is withdrawn from a hopper, transported onto the belt, weighed on the belt, and then discharged to the process unit.

work request–a written document that is used to request specific work. Work requests are usually used when maintenance work is needed on equipment.

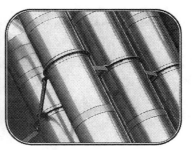

Bibliography

Ambs, Richard. "Choosing a Dump Bag Station for Effective Dust Control and Operator Safety." *Powder and Bulk Engineering* (May 2003): pages 45-49.

Bloch, Heinz P. "Quantifying the Effects of Moisture-contaminated Lube Oil on machine Life." *Hydrocarbon Processing* (June 2003): page 8.

Bloch, Heinz P., and Fred K. Geitner. *Major Process Equipment, Maintenance and Repair*. Houston: Gulf Publishing Company, 1997.

Clark, David, and Paul Hayes. "Turnaround Performance Optimisation." *Petroleum Technology Quarterly* (Autumn 2003): pages 133-139.

Dobyns, Lloyd, and Clare Crawford-Mason. *Quality or Else*. Boston: Houghton Mifflin Company, 1991.

Driedger, Walter. "Controlling Vessels and Tanks." *Hydrocarbon Processing* (March 2000), www.hydrocarbonprocessing.com.

Feig, James. *Process Technology: Operations*, self-published, 2000.

Fitzgerald, F., Gans, M., and Korpes, S. "Plant Startup—Step by Step." *Chemical Engineering* (October 3, 1983), www.che.com.

Gallagher, T.A. "Floating Roof Technology Advances." *Hydrocarbon Processing* (September 2003): pages 63-68.

Goetsch, David L. *Occupational Safety and Health*. 3rd ed. Upper Saddle River, NJ: Prentice Hall, 1999.

Hudson, Sam, and James Spigener. "Behavior-based Technology is the Next Step to Operations Excellence." *Hydrocarbon Processing* (January 2002), www.hydrocarbonprocessing.com.

Kukuk, Michael. *Troubleshooting Resources*, self-published, 1999.

Laan, Thomas J., and Paul R. Smith. *Piping and Pipe Support Systems*. NY: McGraw-Hill, 1987.

Lieberman, Elizabeth, and Norman Lieberman. *Working Guide to Process Equipment.* 2nd ed. NY: McGraw-Hill, 2003.

Lieberman, Norman. *Troubleshooting Process Operations*, 2nd ed. NY: McGraw-Hill, 2003.

McCabe, Warren L., Julian C. Smith, and Peter Harriott. *Unit Operations of Chemical Engineering.* NY: McGraw-Hill, 1985.

Motylenski, Robert J. "Proven Turnaround Practices." *Hydrocarbon Processing* (April 2003): pages 37-42.

Siecke, Michael J., and R. J. Weber. "Is Regulatory Compliance a Burden or an Opportunity?" *Hydrocarbon Processing* (June 2002), www.hydrocarbonprocessing.com.

Stoess, H. A. *Pneumatic Conveying.* NY: John Wiley and Sons, 1983.

Weirauch, Wendy. "The Turnaround Season." *Hydrocarbon Processing* (April 2002), www.hydrocarbonprocessing.com.

Wilson, Dean. "Boiler Basics." *MRO Today* (February/March 1999), www.mrotoday.com.

Woll, Dave. "Time to Rethink Process Plant Roles and Responsibilities." *Hydrocarbon Processing* (September 2004): page 15.

Ural, Erdem. "NFPA 68: How the Revised Explosion Venting Guidelines Affect Your Plant." *Powder and Bulk Engineering* (June 2003): pages 13-22.

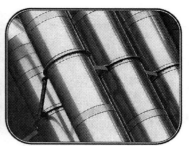

Index

Note: Italicized page numbers indicate illustrations.